LIPID OXIDATION

Edwin N. Frankel

University of California
Davis, California, U. S. A.

THE OILY PRESS
DUNDEE

Copyright © 1998 THE OILY PRESS LTD,
6 Dunnottar Place, West Ferry, Dundee, Scotland DD5 1PJ

All rights reserved. No part of this publication may be reproduced, stored in a retrieval system or transmitted in any form or by any means: electronic, electrostatic, magnetic tape, mechanical, optical, photocopying, recording or otherwise, without permission in writing from the publisher.

ISBN 0 9514171 9 3

British Library Cataloguing-in-Publication Data. A catalogue record for this book is available from the British Library.

This is - Volume 10 in the Oily Press Lipid Library

(**Volume 1** - "Gas Chromatography and Lipids" by W.W. Christie; **Volume 2** - Advances in Lipid Methodology - One" edited by W.W. Christie; **Volume 3** - "A Lipid Glossary" by F.D. Gunstone and B.G. Herslöf; **Volume 4** - "Advances in Lipid Methodology - Two" edited by W.W. Christie; **Volume 5**, "Lipids: Molecular Organization, Physical Functions and Technical Applications" by K.Larsson; **Volume 6** - "Waxes: Chemistry, Molecular Biology And Functions"edited by R.J. Hamilton; **Volume 7** - "Advances in Lipid Methodology - Three" edited by W.W. Christie; **Volume 8** - "Advances in Lipid Methodology - Four" edited by W.W. Christie); **Volume 9** – "*Trans* Fatty Acids in Human Nutrition" edited by J.L. Sébédio and W.W. Christie).

Printed in Great Britain by Bell & Bain Ltd., Glasgow

PREFACE

The oxidation of unsaturated fatty acids is one of the most fundamental reactions in lipid chemistry. When unsaturated lipids are exposed to air, the complex, volatile oxidation compounds formed cause rancidity, decreasing the quality of foods containing lipids as well as foods in which oils are used as ingredients. Another important process in the oxidation of unsaturated fats proceeds through activated species of oxygen. Singlet oxygen produced by photooxidation, in the presence of a sensitizer such as chlorophyll, is an important reactant that has attracted much interest to organic and biological chemists. Products of lipid oxidation have been implicated in many vital biological reactions. The revival of the field of lipid oxidation in the last two decades may be attributed in large part to the accumulating evidence that free radicals and reactive oxygen species participate in tissue injuries and in degenerative diseases.

Since the early 1960's, our understanding of the autoxidation of unsaturated lipids has advanced considerably as a result of the application of powerful new analytical tools. The characterization of cyclic peroxides from autoxidized linolenate, structurally related to the physiologically active prostaglandins, has also generated widespread interest in the field. Much of the work reported on the oxidative decomposition products of unsaturated fats emphasizes the role they play in causing rancidity in foods and cellular damage in the body. Decomposition of fat hydroperoxides creates a wide range of carbonyl compounds, hydrocarbons, furans, and other materials that contribute to flavor deterioration of foods. However, materials from secondary oxidation are also implicated in biological oxidation. Although fat hydroperoxides are the recognized precursors of volatile secondary products, the mechanism of their decomposition is not clearly understood.

This book aims at integrating a large body of interdisciplinary information on the oxidation of unsaturated lipids in order to develop the basic principles involved in the methodology and mechanisms of free radical oxidation. Understanding these principles is necessary for gaining insights into how unsaturated lipids in foods are subject to oxidative deterioration, and for developing appropriate control measures. The question of how biological systems undergo oxidative damage is very complex and the literature accumulates at a geometric rate.

This is an advanced textbook for graduate students, academic and industrial scientists concerned with the many phases of the complex series of lipid oxidation reactions. Starting from the basics of free radical and hydroperoxide chemistry, and methodology, the book progresses into topics of increasing complexity including control methods, antioxidants, and oxidation in multiphase systems, foods, frying fats and biological systems. Investigators working with polyunsaturated fatty acids and lipids must be seriously concerned with their oxidation, since these products have caused numerous problems in analyses, particularly in stability tests. Unfortunately, in spite of the great strides made in the field of lipid oxidation, many contemporary studies employ nonspecific and unreliable methods to follow lipid oxidation. Lipid oxidation is often studied under drastic conditions and with simple model systems that do not simulate complex multi-phase foods and biological systems. Much of the classical concepts of lipid oxidation come as a result of elegant kinetic studies, in which a wholly theoretical picture is developed in support of mechanistic hypotheses. The conditions and systems often employed in these studies may be simplistic, however, and do not represent those of real complex food systems undergoing oxidation.

This book discusses these and other pitfalls inherent in basic studies that use artificial radical initiators with unrealistic models and homogeneous systems. The application of simple research models may lead to numerous problems in the ultimate interpretation of results, because lipid oxidation proceeds by a complex sequence of reactions influenced by many factors, all of which become extremely difficult to unravel in real food and biological systems. These systems are multi-phased and controlled by complex colloidal phenomena affecting different sites of oxidation and antioxidation. In interpreting the effects of prooxidant and antioxidant compounds their "effective" concentrations in different phases must be considered. A dimension of lipid oxidation that is important to better understand control methods deals with the relative partition of oxidants and antioxidants in multiphase systems. This topic has not received sufficient attention.

One of the main objectives of this book is to develop the background necessary for a better understanding of what factors should be considered, and what methods and lipid systems should be employed, to achieve suitable evaluation and control of lipid oxidation in complex foods and biological systems. Now, and throughout its 50-year history, the field is still developing and more progress can be expected in improving our understanding of the complex phenomena of lipid oxidation.

CONTENTS

	Pages
Preface	
Introduction	1
Chapter 1. Free Radical Oxidation	13
Chapter 2. Hydroperoxide Formation	23
Chapter 3. Photooxidation of Unsaturated Fats	43
Chapter 4. Hydroperoxide Decomposition	55
Chapter 5. Methods to Determine Extent of Oxidation	79
Chapter 6. Stability Methods	99
Chapter 7. Control of Oxidation	115
Chapter 8. Antioxidants	129
Chapter 9. Oxidation in Multiphase Systems	167
Chapter 10. Foods	187
Chapter 11. Frying Fats	227

Chapter 12. 249
Biological Systems

List of abbreviations 293

Index 295

INTRODUCTION

Lipids are important structural and functional components of foods. The specific definition used in this book for the term *lipids* refers to fatty acids or closely related derivatives or metabolites. In foods, lipids have a significant effect on quality even when they constitute a minor component. The main lipid components of foods are triacylglycerols and are referred to as *fats* and *oils*. They not only contribute to flavor, odor, color and texture, but also confer a feeling of satiety and palatability to foods. Nutritionally, lipids provide a concentrated source of calories (9 kcal per gram), that is about two-fold higher than either carbohydrates or proteins. They also furnish essential nutrients, including linoleic acid, linolenic acid, and fat-soluble vitamins (A, D, E and K). The subject of *lipid oxidation* will be introduced briefly with the classification of lipid components in foods and biological systems, with minor non-glyceride components, and with a few basic concepts of free radical organic chemistry.

A. CLASSIFICATION OF LIPIDS

The principal classes of lipids with which we will be concerned include free fatty acids, triacylglycerols consisting of fatty acids esterified to glycerol, and phosphoglycerides consisting of fatty acids esterified to glycerol containing phosphoric acid and organic bases.

1. Fatty Acids

These long chain aliphatic carboxylic acids are essential parts of most lipids in foods including triacylglycerols (also referred to as triglycerides) and phosphoglycerides (also referred to as phospholipids). Fatty acids can be either saturated or unsaturated (Figure 1). The fatty acids of major vegetable oils and animal fats include saturated, monounsaturated and polyunsaturated fatty acids (Table 1).

The common range of saturated fatty acids has a straight chain, even number of carbons and chain length of 12 to 22 carbons. *Myristic acid* with 14 carbons is designated as 14:0 (or 14 carbons with no double bond), *palmitic acid* with 16 carbons as 16:0, and *stearic acid* with 18 carbons, as 18:0.

The monounsaturated *oleic acid* with one double bond in the cis (or Z) configuration in the 9-carbon position (relative to the carboxyl group) is designated as 18:1 n-9. The "n" symbol indicates the number of carbons from the terminal end of the 18-carbon chain, a nomenclature favored by

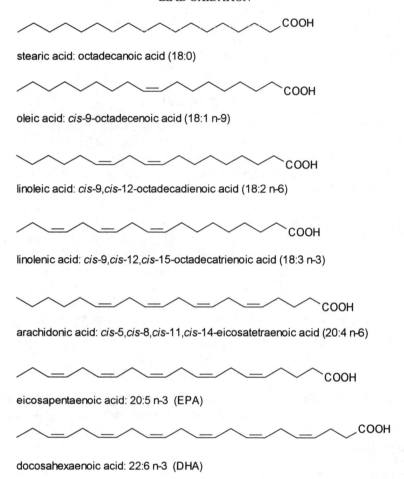

Fig. 1. Line drawing representations of the structures of saturated, monounsaturated and polyunsaturated fatty acids.

biochemists. *Elaidic acid* has the same structure as oleic acid except that the double bond is in the *trans* (or *E*) configuration. This acid is present in only small amounts in milk fat. It is also produced by partial hydrogenation of polyunsaturated fats for the manufacture of margarines and shortenings. These hydrogenated products contain other *cis* and *trans* isomers of monounsaturated fatty acids in which the double bond has migrated between the carbon-8 and carbon-12 positions.

The polyunsaturated fatty acids are characterized by having two or more *cis* double bonds separated by a single methylene group, or a 1,4-diene structure. *Linoleic acid* with two *cis* double bonds in the 9,12-positions is designated as 18:2 n-6 (9,12). *Linolenic acid* is 18:3 n-3 (9,12,15); *arachidonic acid* is 20:4 n-6 (5,8,11,14); *eicosapentaenoic acid* is 20:5 n-3 (5,8,11,14,17) and abbreviated as EPA; and *docosahexaenoic acid* is 22:6 n-3 (4,7,10,13,16,19)

TABLE 1
Average fatty acid composition of vegetable oils and animal fats (% by weight).[a]

Fats	14:0	16:0	18:0	18:1	18:2	18:3	Others
Corn	—	13	3	31	52	1	—
Cottonseed	—	27	2	18	51	trace	2
Peanut[b]	—	13	3	38	41	trace	C_{20-24} 5
Olive	—	10	2	78	7	1	2
Palm	—	44	4	40	10	trace	2
Rapeseed[c]	—	4	2	56	26	10	20:1 3
Safflower	—	7	3	14	75	--	1
Soybean	—	11	4	22	53	8	2
Sunflower	—	6	5	20	69	trace	—
Lard	2	27	11	44	11	—	5
Beef tallow	3	27	7	48	2	—	13

[a] Gunstone et al. (1994).
[b] Also termed groundnut oil in Europe.
[c] Low-erucic acid, also known as canola oil in Canada and USA

and abbreviated as DHA. Linoleic acid and linolenic acid are essential fatty acids because they are not synthesised in the body. EPA and DHA are also considered important in modulating the biosynthesis of oxygenated derivatives of arachidonic acid called *eicosanoids*, which have hormone-like functions and play an important role in inflammation processes (see Chapter 12).

Fatty acids occur mainly as esters of glycerol called glycerolipids, including triacylglycerols, or triglycerides, and phosphoglycerides or phospholipids. The alcohol can also be a long-chain compound or a sterol.

2. Triacylglycerols

Triacylglycerols are important storage lipids and the main constituents (~ 99 %) of vegetable oils and food lipids. Natural triacylglycerols contain generally a mixture of different fatty acids and their composition reflects the relative fatty acid concentrations. However, when the proportion of one fatty acid is higher than 33%, the oil contains a large amount of triacylglycerols containing one fatty acids, such as trioleylglycerol (or triolein, designated as OOO) in olive oil, and tripalmitoylglycerol (or tripalmitin, designated as PPP) in palm oil (Table 1). The number of possible triacylglycerols increases markedly with the number of different fatty acids. With two different fatty acids, the number of possible triacylglycerols is eight. For example, with palmitate (P) and oleate (O), we can have the following triacylglycerols: PPP, POP, OPO, OOO, OPP, PPO, POO, and OOP. The pairs OPP/PPO and POO/OOP are unsymmetrical enantiomeric pairs (or mirror image forms). With three different fatty acids, the number of possible triacylglycerols is 27, or equal to n^3 for n = the number of different fatty acids. The physical, texture and rheological properties of triacylglycerols is greatly dependent not only on their fatty acid composition but also on their distribution among the 1-, 2- or

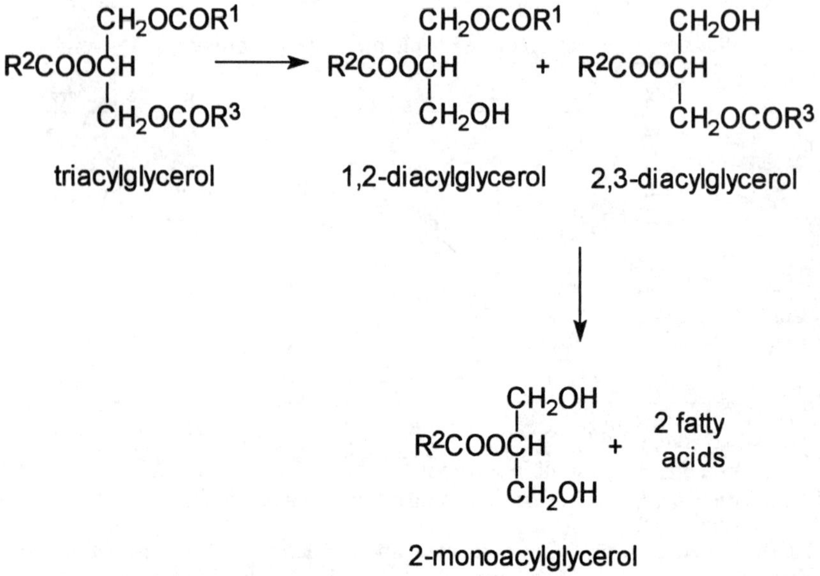

$$\begin{array}{c}\text{CH}_2\text{OCOR}^1\\|\\\text{R}^2\text{COOCH}\\|\\\text{CH}_2\text{OCOR}^3\end{array} \longrightarrow \begin{array}{c}\text{CH}_2\text{OCOR}^1\\|\\\text{R}^2\text{COOCH}\\|\\\text{CH}_2\text{OH}\end{array} + \begin{array}{c}\text{CH}_2\text{OH}\\|\\\text{R}^2\text{COOCH}\\|\\\text{CH}_2\text{OCOR}^3\end{array}$$

triacylglycerol   1,2-diacylglycerol   2,3-diacylglycerol

$$\downarrow$$

$$\begin{array}{c}\text{CH}_2\text{OH}\\|\\\text{R}^2\text{COOCH}\\|\\\text{CH}_2\text{OH}\end{array} + \text{2 fatty acids}$$

2-monoacylglycerol

Fig. 2. Hydrolysis of a triacylglycerol by the enzyme pancreatic lipase.

3-position (also designated as *sn*-positions) in the glycerol molecule.

Triacylglycerols can be hydrolysed either chemically or enzymatically to produce a mixture of diacylglycerols (or diglycerides), monoacylglycerols (or monoglycerides), free fatty acids and glycerol. Chemical hydrolysis with aqueous alkali or acid occurs randomly to produce two fatty acids from the *sn*-1 and *sn*-3 positions (or α-positions) and one fatty acids from the *sn*-2 (or β-position). Enzymatic hydrolysis by lipase occurs selectively at the α-positions to yield 1,2- and 2,3-diacylglycerols, followed by 2-monoacylglycerol and two fatty acids. Dietary lipids are assimilated in the body following the hydrolytic action of pancreatic lipase which takes place in the small intestine (Figure 2).

3. Phosphoglycerides

These are important structural lipids in foods and cell membranes, which are also known as phospholipids or phosphatides. They are based on phosphatidic acids which are 1,2-diacyl esters of 3-glycerolphosphoric acid, linked to organic bases or other moieties (Figure 3). Phospholipids are minor components of the crude vegetable oils obtained either by solvent extraction or by pressing oilseeds. A large portion of these materials is removed by a hydration process known as degumming (see Figure 7-1), where they are separated as a mixture with triacylglycerols known as *lecithin*. The *sn*-1 position in phosphatidylcholine (PC) of animal origin is largely esterified by

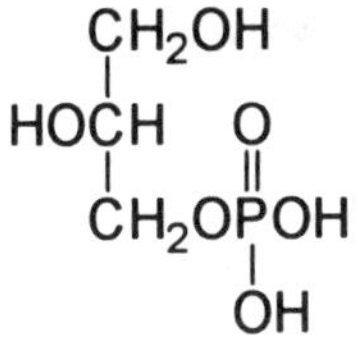

INTRODUCTION

$$\begin{array}{c} CH_2OH \\ | \\ HOCH \quad O \\ | \quad \quad \| \\ CH_2OPOH \\ | \\ OH \end{array}$$

3-glycerophosphoric acid

$$\begin{array}{c} CH_2OCOR^1 \\ | \\ R^2COOCH \quad O \\ | \quad \quad \| \\ CH_2OPO{-}X \\ | \\ OH \end{array}$$

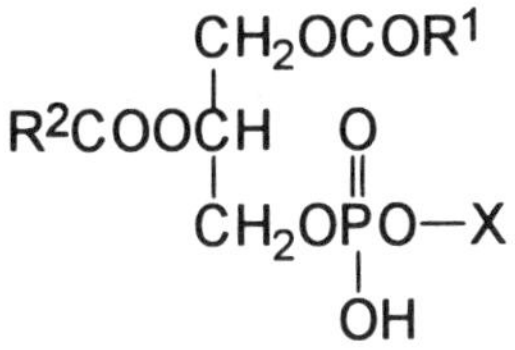

phosphatidic acid (X = H) (PA)

phosphatidylcholine (X = $CH_2CH_2N^+Me_3$ or choline) (PC)

phosphatidylethanolamine (X = $CH_2CH_2N^+H_3$ or ethanolamine) (PE)

phosphatidylserine (X = $CH_2CH(NH_3)^+COO^-$ or serine) (PS)

phosphatidylinositol (X = $C_6H_{11}O_5$ or inositol) (PI)

phosphatidylglycerol (X = $CH_2CH(OH)CH_2OH$ or glycerol) (PG)

Fig. 3. The structures of the common range of phospholipids.

saturated fatty acids while the *sn*-2 position contains polyunsaturated fatty acids. The lecithin obtained from soybeans is a mixture of PC (20-23%), phosphatidylethanolamine (PE) (16-21%) and phosphatidylinositol (PI) (12-18%), and is used as an important emulsifier in foods. These phospholipids play an important role in stabilizing vegetable oils during processing (Chapter 7), and as synergists by reinforcing the activity of antioxidants (Chapter 8). When dispersed in water, phospholipids form multi-layers consisting of *bilayers*, which are important components of cell membranes, meat and fish products (see Chapters 9 and 10).

Phosphatidylcholine can be completely hydrolysed with aqueous acid to produce fatty acids, glycerol, phosphoric acid and choline. With alkali, the fatty esters are preferentially hydrolysed, leaving the glycerophosphorylcholines, which can in turn hydrolyse slowly into glycerophosphoric acid and choline. Enzymatic hydrolysis occurs selectively at different ester sites by several phospholipases to produce a range of products. Phospholipases A1 and A2 are found in snake venom and hydrolyse the fatty acid at the *sn*-1 and *sn*-2 positions respectively, to produce lysophosphatidylcholine. Phospholipase B is a mixture of phospholipase A1 and A2 that hydrolyses fatty acids from both sn-1 and sn-2 positions. Phospholipase C yields 1,2-diacylglycerols and phosphorylcholine. Phospholipase D is found in plant tissues and hydrolyses the phosphate ester to produce choline and phosphatidic acid (1).

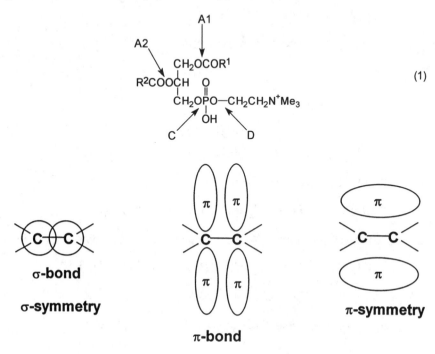

Fig. 4. Electronic representation of an unsaturated bond.

B. MINOR COMPONENTS

Fats and oils contain varying amounts (1-2%) of non-glyceride components including hydrocarbons such as squalene and carotenoids, sterols, tocopherols and tocotrienols, chlorophyll, vitamin A (retinol), and vitamin D (cholecalciferol). Only the tocopherols, carotenoids and chlorophyll have a significant effect on the oxidative stability of fats and oils. The *tocopherols* (Figure 8-1 and Table 8-4) are the most important free radical chain-breaking antioxidants (Chapter 8). Because the tocopherols are thermally stable, a relatively large proportion (60-70%) remains in the oils after processing. The antioxidant activity of tocotrienols, present mainly in palm oil, is not fully established. The *carotenoids* are effective inhibitors of photosensitized oxidation by quenching singlet oxygen, a very active form of oxygen responsible for this type of photooxidation (Chapter 2). For most edible oils the carotenoids are largely removed and destroyed during conventional processing by deodorization at elevated temperatures. For some oils such as virgin olive oil and corn oil, the yellow color imparted by carotenoids is desirable and maintained by partial or full omission of the bleaching step, or by deodorizing at lower temperatures. The *chlorophylls* are very active promoters of photosensitized oxidation (Chapter 2). These greenish pigments must be removed as completely as possible by bleaching to prevent the damaging effects of light oxidation. *Squalene* and *sterols* have little effects on lipid oxidation, but some types of sterols are claimed to have protective effects by inhibiting polymerization of oils during frying (Chapter 11).

C. CHEMICAL REACTIONS OF UNSATURATED FATTY ACIDS

The electronic picture of a double bond consists of two components (Figure 4):

i. a sigma (σ) bond that has an electronic symmetry that overlaps the two olefinic carbons endwise, or has σ-symmetry; this bond is as strong as an ordinary single bond.

ii. a pi (π) bond that consists of a pair of electrons which occupy a position in which one is above and one below the plane of the paper, and overlap sidewise, or has π-symmetry. The π-bond is weaker and of lower energy because the π electrons are less firmly held between the two nuclei and more exposed. The π electrons are consequently responsible for the greater reactivity of unsaturated compounds. Because the π bond is more readily polarizable than the σ bond, it imparts *nucleophilic* character to double bonds which react with *electrophilic* reagents.

Organic reactions can be divided mechanistically by the formation of either heterolytic ionic or homolytic radical intermediates.

1. Heterolytic Ionic Reactions

In these reactions a covalent bond is formed by one species donating and the other accepting an electron pair. *Heterolytic cleavage* of the resulting covalent bond produces two species, one negatively charged ion accompanied by the electron pair, and one positively charged ion (2).

$$X{:} + Y \longrightarrow X{:}Y \longrightarrow {:}Y^- + X^+ \quad (2)$$

The negatively charged ion Y^- is an electron-donating or *nucleophilic* species, and the positively charged ion X^+ is *electrophilic*, or accepts an electron pair. These charged ions follow laws of electrostatic attraction and repulsion.

2. Homolytic Radical Reactions

In these reactions a bond is formed by two *free radical* species with odd electrons, both donating a single electron. *Homolytic cleavage* of the resulting covalent bond produces free radicals by splitting the pair of electrons.

$$X^{\cdot} + {^{\cdot}}Y \longrightarrow X{:}Y \longrightarrow X^{\cdot} + {^{\cdot}}Y \quad (3)$$

LIPID OXIDATION

These free radicals containing odd electrons are highly reactive but have no formal charge. The free radicals and charged ionic intermediates have distinctly different characteristic behaviors that can be summarized as follows:

Free radical reactions:	Ionic reactions:
i catalysed by light, heat, by other free radicals and peroxides. ii inhibited by antioxidants (radical acceptors reacting with other free radicals). iii proceed in vapor phase or non-polar solvents. iv autocatalytic and have an induction period.	i unaffected by light, free radicals and peroxides; often catalysed by acids or bases. ii unaffected by radical acceptors or antioxidants. iii rarely occur in the vapor phase; strongly affected by solvents. iv follow first or second order kinetics.
Examples of free radical transformations include, oxidation, most polymerization and photochemical reactions and reactions using peroxides as catalysts.	Examples of reactions involving ionic intermediates include alkali isomerization, dehydration, hydrolysis and saponification.

The autoxidation of unsaturated fatty acids is a chain process occurring autocatalytically through free radical intermediates. Although autoxidation implies that this reaction occurs spontaneously under mild conditions, it is generally initiated by trace metals and peroxides or hydroperoxides present as ubiquitous impurities in food and biological lipid systems (see Chapter 1).

Several radical transformations can be considered in the context of lipid oxidation. These reactions are facilitated by the presence of an odd electron in the free radical intermediates that weakens the surrounding bonds.

a. Hydrogen abstractions. Radicals can readily abstract a hydrogen from unsaturated fatty acids containing an allylic center, —CH=CH-CH$_2$ — . The formation of allylic intermediates is an important characteristic of free radical reactions that determines the nature of lipid oxidation products (Figure 5). The two allylic intermediates formed differ only in the distribution of electrons and can be represented as one resonance hybrid structure A. The odd electron in this allylic conjugated radical is delocalized and distributed through the three carbon structure. This phenomenon contributes to its strong resonance stabilization. Radicals conjugated with polyunsaturated fatty acid systems are similarly stabilized by resonance.

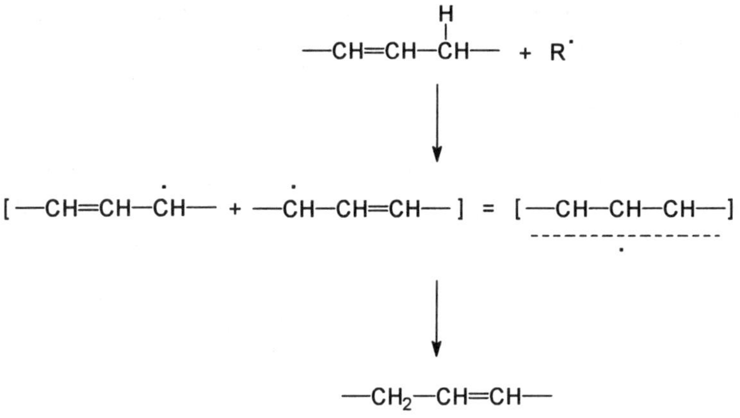

$$\text{—CH=CH—}\overset{H}{\underset{|}{\text{CH}}}\text{—} + R^\bullet$$

$$\downarrow$$

$$[\text{—CH=CH—}\overset{\bullet}{\text{CH}}\text{—} + \text{—}\overset{\bullet}{\text{CH}}\text{—CH=CH—}] = [\text{—CH}\cdots\text{CH}\cdots\text{CH—}] \quad (A)$$

$$\downarrow$$

$$\text{—CH}_2\text{—CH=CH—}$$

Fig. 5. Double bond migration by hydrogen abstraction in a free radical mechanism.

b. Addition reactions. One radical can attack the π electrons of a double bond to be transformed into another radical (4).

$$R\text{—CH—CH}^\bullet + \text{—CH=CH—} \longrightarrow R\text{—CH—CH—CH—CH}^\bullet \quad (4)$$

The new radical thus formed can undergo further reactions, such as cyclization and polymerizations (Chapter 4). Allylic peroxyl radicals can thus add to an unsaturated fatty acid to produce a peroxide-linked dimer (5).

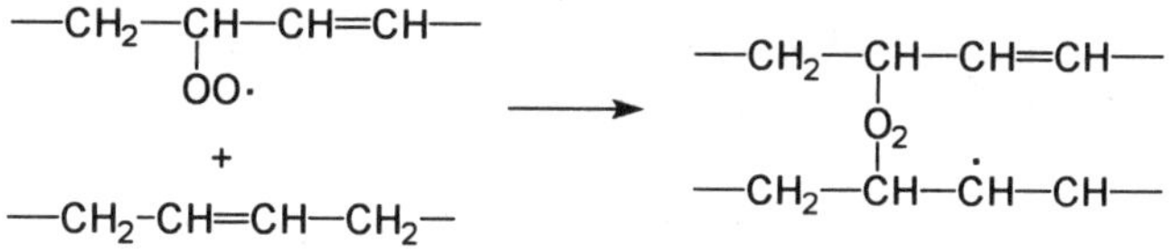

$$\begin{array}{c}\text{—CH}_2\text{—CH—CH=CH—}\\|\\\text{OO}\cdot\\+\\\text{—CH}_2\text{-CH=CH—CH}_2\text{—}\end{array} \longrightarrow \begin{array}{c}\text{—CH}_2\text{—CH—CH=CH—}\\|\\\text{O}_2\\|\\\text{—CH}_2\text{—CH—CH—CH—}\end{array} \quad (5)$$

c. Fragmentation reactions. Secondary alkoxy radicals may undergo decomposition by *β-scission* or elimination to form aldehydes and a simpler radical (6).

$$R'_2\text{—CH—O}^\bullet \longrightarrow R'\text{—CHO} + R'' {}^\bullet \quad (6)$$

These decomposition reactions become important at elevated temperatures or in the presence of metal catalysts (see Chapter 4).

d. Rearrangement reactions. A hydroperoxide can rearrange by reversibly losing a hydrogen radical and eliminating oxygen from a peroxyl radical intermediate (7).

$$\text{ROOH} \underset{+H^\bullet}{\overset{-H^\bullet}{\rightleftharpoons}} \text{ROO}^\bullet \rightleftharpoons R^\bullet + O_2 \quad (7)$$

Lipid hydroperoxides undergo rearrangement of the OOH group readily to produce mixtures of positional and geometric isomeric products (see Chapter 2).

Allylic peroxides can also undergo 1,3-peroxyl transfer either by a concerted (one step) pathway (8):

$$R_1-\underset{OO^\bullet}{CH}-CH=CH-R_2 \longrightarrow R_1-\underset{O-O}{CH---CH---CH}-R_2 \longrightarrow R_1-CH=CH-\underset{OO^\bullet}{CH}-R_2 \quad (8)$$

or by a dissociative pathway involving the removal and re-addition of oxygen (9):

$$R_1-\underset{OO^\bullet}{CH}-CH=CH-R_2 \longrightarrow R_1-\underset{+O_2}{CH---CH---CH}-R_2 \longrightarrow R_1-CH=CH-\underset{OO^\bullet}{CH}-R_2 \quad (9)$$

e. *Disproportionation reactions.* A hydroperoxide can be converted to a ketone by disproportionation involving the loss of a hydrogen radical from an alkoxyl radical intermediate (10).

$$R_2-CH-OOH \longrightarrow {}^\bullet OH + R_2-CH-O^\bullet \longrightarrow R_2-C=O + -H^\bullet \quad (10)$$

Similarly, an allylic hydroperoxide can undergo disproportionation into a mixture of ketone and hydroxy products (11).

$$R_1-\underset{OOH}{CH}-CH=CH-R_2 \xrightarrow{-{}^\bullet OH} 2\,R_1-\underset{O\bullet}{CH}-CH=CH-R_2 \longrightarrow$$

$$R_1-\underset{O}{\overset{\|}{C}}-CH=CH-R_2$$
$$+$$
$$R_1-\underset{OH}{CH}-CH=CH-R_2 \quad (11)$$

3. Radical Stability

The stability of a radical depends on the rate of its disappearance or formation according to the reaction conditions. The *thermodynamic stability* of a radical is related to the ease of dissociation of the R–H bond to give R$^\bullet$ and H$^\bullet$ or the R–H dissociation energy. The thermodynamic stability of a radical depends on double bond conjugation and electron delocalization stabilized by resonance in allylic systems produced by hydrogen abstraction in polyunsaturated lipids (see Chapter 2).

The *kinetic stability* of a radical is largely controlled by steric factors. When the radical center is crowded, the radical becomes less reactive and persists longer under normal conditions (it has a longer life-time). Aromatic compounds that can form allylic radicals show similar benzylic stabilization. If the radical center is sterically crowded by bulky tertiary butyl substituents, the allylic radical intermediates formed by hydrogen transfer have *kinetic stability* that imparts important antioxidant properties (see Chapter 8). Thus, when phenolic compounds contain three bulky tertiary butyl substituents, they form persistent radicals after hydrogen donation and inhibit lipid oxidation by interrupting the propagation of free radicals (see Chapter 8).

D. LIPID OXIDATION AND HEALTH

Rancidity of edible oils is a serious problem in many sectors of the food industry because of increasing emphasis on the use of polyunsaturated vegetable and fish oils, discontinuing the use of synthetic antioxidants, and fortification of cereal foods with iron. Oxidation of lipids not only produces rancid flavors in foods but can decrease their nutritional quality and safety by the formation of secondary products after cooking and processing (Chapter 11). Effective control methods against lipid oxidation include the use of metal inactivators, minimizing the loss of natural tocopherols and exposure to air and light during processing, hydrogenation and the use of antioxidants (Chapter 7). Hydrogenation of polyunsaturated oils is becoming less and less attractive because of recent evidence that *trans* isomers may have adverse nutritional effects. An alternate approach to hydrogenation is to mix different proportions of high-oleic oils with polyunsaturated vegetable and fish oils to prepare more stable edible oils with a wide range of desired fatty acid compositions.

Much research has been conducted to better understand the basic processes of lipid oxidation and singlet oxygenation of polyunsaturated lipids, antioxidant action, and the effects of decomposition products of lipid oxidation. Special attention has been given recently to the use of natural antioxidants (Chapter 8). Flavonoids found in many fruits and vegetables are potent antioxidants that may be especially important in protecting against human disease.

In the developed countries it appears that increased consumption of linoleate from vegetable oils in our diet may have reached too high a level in promoting a variety of chronic diseases, including coronary artery and inflammatory diseases. Excessive linoleate in our diet causes an undue release into blood of eicosanoids, oxygenated metabolites of arachidonic acid that have profound physiological effects. These eicosanoids produce abnormal cell functions in the body that result in predisposition to heart attacks (Chapter 12). Because n-3 polyunsaturated fatty acids are known to antagonize the metabolism of linoleic acid, many nutritionists have advocated that a portion of the vegetable oils in the diet be replaced with fish and fish oils rich in these n-3 fatty acids. However, this approach overlooks the possible risk factor from the high susceptibility to autoxidation of the polyunsaturated fatty acids in fish oils, especially 20:5 n-3 and 22:6 n-3. The oxidation of these highly unsaturated fatty acids causes rancidity and oxidative deterioration in fish oil products which cannot be effectively controlled by supplementation with vitamin E and other antioxidants.

Modifications of the polyunsaturated lipid intake in the diet require that we examine carefully the possible nutritional consequences of oxidative damage to biological systems. Oxidation of low-density lipoproteins (LDL) in circulating blood has been implicated in the etiology of coronary heart diseases (Chapter 12). Some of the nutritional problems due to lipid oxidation could be alleviated either by consuming more antioxidants, or by increasing the amounts of oxidatively stable oleic acid in our diet. There is evidence that diets rich in oleic acid and antioxidants increase the oxidative stability of LDL and may reduce coronary disease. Although in healthy subjects an important major antioxidant defense is to prevent metals from catalysing the generation of reactive oxygen species, excessive iron supplementation in the diet may overload this defense system. We need to improve our diet by reducing or minimizing the risk factors associated with oxidative deterioration of polyunsaturated dietary lipids that may also be aggravated by excessive iron supplementation. Flavonoid antioxidants represent a positive potential in our diet that require further research to improve our understanding of their mechanism of action (Chapter 12).

BIBLIOGRAPHY

Alexander,E.R. *Principles of Ionic Organic Reactions*. (1950) John Wiley & Sons, New York.
Chan,H.W.S. (editor) *Autoxidation of Unsaturated Lipids*. (1987) Academic Press, London.
Fossey,J., Lefort,D. and Sorba,J. *Free Radicals in Organic Chemistry*. (1995) John Wiley & Sons, Masson, Chichester and Paris.
Gunstone,F.D. *Fatty Acid and Lipid Chemistry*. (1996) Blackie Academic & Professional, London.
Gunstone,F.D., Harwood,J.L. and Padley,F.B. *The Lipid Handbook*. (1994) Chapman and Hall, London.
Lowry,T.H. and Richardson,K.S. *Mechanism and Theory in Organic Chemistry*. (1981) Harper & Row, New York.
Pryor,W.A. *Free Radicals*. (1966) McGraw-Hill Book Co., New York.
Walling,C. *Free Radicals in Solution*. (1957) John Wiley & Sons, London.

CHAPTER 1

FREE RADICAL OXIDATION

A. MECHANISM

Autoxidation is the direct reaction of molecular oxygen with organic compounds under mild conditions. Oxygen has a special nature in behaving as a biradical by having two unpaired electron ($^{\cdot}$O-O$^{\cdot}$) in the ground state and is said to be in a triplet state. The oxidation of lipids proceeds like that of many other organic compounds by a free radical chain mechanism, which can be described in terms of initiation, propagation, and termination processes. These processes often consist of a complex series of reactions.

1. Initiation

In the presence of initiators (I), unsaturated lipids (LH) lose a hydrogen radical (H$^{\cdot}$) to form lipid free radicals ($L^{\cdot}$) (1).

$$LH \xrightarrow[R_i]{I} IH + L^{\cdot} \qquad (1)$$

The direct oxidation of unsaturated lipids by triplet oxygen ($^3O_2$) is spin forbidden because the lipid ground state of single multiplicity has an opposite spin direction from that of oxygen of triplet multiplicity. This spin barrier between lipids and oxygen can be readily overcome in the presence of initiators that can produce radicals by different mechanisms.

(a) thermal dissociation of hydroperoxides (LOOH) present as impurities (2):

$$LOOH \longrightarrow LO^{\cdot} + {^{\cdot}OH} \qquad (2)$$

(b) decomposion of hydroperoxides catalysed by redox metals (M) of variable valency (3,4):

$$LOOH + M^{2+} \longrightarrow LO^{\cdot} + OH^{-} + M^{3+} \qquad (3)$$

$$LOOH + M^{3+} \longrightarrow LOO^{\cdot} + H^{+} + M^{2+} \qquad (4)$$

(c) exposure to light in the presence of a sensitizer such as a ketone (RCOR) (5):

$$RCOR + h\nu \longrightarrow RCO^{\bullet} + R^{\bullet} \qquad (5)$$

Because the reaction of unsaturated lipids with singlet oxygen ($^1O_2$) (Chapter 3) occurs at a significantly greater rate than with normal triplet oxygen ($^3O_2$), the hydroperoxides that are rapidly formed under photosensitized conditions can serve as important initiators of free radicals by reactions (2), (3) and (4).

The mechanism of initiation of lipid oxidation has been debated for many years. The most likely initiation process is the metal-catalysed decomposition of preformed hydroperoxides. The thermal oxidation of unsaturated lipids is generally autocatalytic and involves initiation by decomposition of hydroperoxides, which is generally considered metal-catalysed because it is very difficult or nearly impossible to eliminate trace metals that act as potent catalysts for reactions (3) and (4).

2. Propagation

The alkyl radical of unsaturated lipids (L$^{\bullet}$) containing a labile hydrogen reacts very rapidly with molecular oxygen (k_o) to form peroxyl radicals (6). This step is always much faster than the following hydrogen transfer reaction (k_p) with unsaturated lipids to form hydroperoxides (7).
This is the most widely occurring oxidation and describes the first stages of

$$L^{\bullet} + O_2 \xrightarrow{k_o} LOO^{\bullet} \qquad (6)$$

$$LOO^{\bullet} + LH \xrightarrow{k_p} LOOH + L^{\bullet} \qquad (7)$$

the oxidation of unsaturated lipids producing hydroperoxides as the fundamental primary products. Because this step is slow and rate-determining, hydrogen abstraction from unsaturated lipids becomes selective for the most weakly bound hydrogen.

The susceptibility of lipids to autoxidation thus depends on the availability of allylic hydrogens and their relative ease to react with peroxyl radicals to form hydroperoxides. A hybrid radical (A) is formed, with a partial free radical ($\delta^{\bullet}$) at each end of the allylic system (8). Oxygen attack at each end of the allylic system produces a mixture of allylic 1- and 3-hydroperoxides (10).

$$LOO^{\bullet} + R-CH_2-CH=CH-R' \longrightarrow$$

$$\underset{\delta^{\bullet} \text{----------} \delta^{\bullet}}{R-\overset{1}{CH}-\overset{2}{CH}-\overset{3}{CH}-R'} \text{ (A) } + LOOH \qquad (8)$$

$$\xrightarrow{O_2} \begin{array}{c} R-CH-CH=CH-R' \\ | \\ OO^{\cdot} \end{array} +$$ (9)

$$R-CH=CH-\underset{\underset{OO^{\cdot}}{|}}{CH}-R'$$

$$\xrightarrow{LH} \text{allylic 1- and 3-hydroperoxides} \quad (10)$$

We shall see in Chapter 2 that the hybrid radical (A) can assume different stereochemical conformations which determine the geometric configurations (*cis* and *trans*) of the isomeric hydroperoxides from various unsaturated lipids oxidized under different conditions.

3. Termination

At the last stages of oxidation after reaching a maximum the rate decreases, the peroxyl radicals react with each other and self-destruct to form non-radical products by the termination reaction (11).

At atmospheric pressure, termination (k_t) occurs first by the combination of

$$LOO^{\cdot} + LOO^{\cdot} \xrightarrow{k_t} \text{non-radical products} \quad (11)$$

peroxyl radicals to an unstable tetroxide intermediate followed rapidly by its decomposition by the Russell mechanism which yields non-radical products (12).

$$LOO^{\cdot} + LOO^{\cdot} \xrightarrow{k_t} [LOOOOL] \longrightarrow$$

$$\text{non-radical products} + O_2 \quad (12)$$

Many complexities can be considered when hydroperoxides undergo thermal or metal-catalysed homolysis generating peroxyl and alkoxyl radicals leading either to additional radicals that continue the chain or to non-radical end-products. Alkoxyl radicals can thus either react with the unsaturated lipid to form stable and innocuous alcohols (13), or undergo fragmentation into unsaturated aldehydes and other unstable products (14) causing rancidity in polyunsaturated lipids. Both of these reactions produce new radicals that continue the chain and further complicate the process of lipid autoxidation.

$$LO^{\cdot} + LH \longrightarrow LOH + L^{\cdot} \quad (13)$$

$$LO^\bullet \longrightarrow RCHO + L^\bullet \quad (14)$$

Other termination reactions can proceed by condensation of peroxyl, alkoxyl or alkyl radicals. At low temperatures, peroxyl radicals can combine to produce peroxyl-linked dimers (LOOL) with the formation of oxygen (15).

At low oxygen pressures and elevated temperatures, alkoxyl and alkyl

$$2\ LOO^\bullet \longrightarrow LOOL + O_2 \quad (15)$$

radicals can combine to produce ether-containing dimers (16), and carbon-carbon linked dimers (17).

$$LO^\bullet + L^\bullet \longrightarrow LOL \quad (16)$$

$$2\ L^\bullet \longrightarrow L{-}L \quad (17)$$

The decomposition of hydroperoxides and dimers will be considered in detail in Chapter 4.

One of the most important termination reactions (18) involves an antioxidant (AH).

$$LOO^\bullet + AH \rightleftharpoons LOOH + A^\bullet \quad (18)$$

The application of antioxidants will be considered in Chapter 8 as one of the important practical measures to control lipid oxidation.

B. KINETICS

Many of the classical mechanistic concepts of lipid oxidation were formulated on the basis of kinetic studies. Later developments in support of the general free radical mechanism of oxidation were based on structural studies of the primary hydroperoxide products. To simplify the kinetics of linoleate oxidation the reactions were studied at early stages of oxidation, at low levels of conversion, at lower temperatures and in the presence of an appropriate initiator. Under these conditions, the propagation reactions producing hydroperoxides in high yields are emphasized, and the decomposition of hydroperoxides is minimized or considered insignificant. However, the kinetics become much more complex when autoxidation is carried out to high conversions, or at elevated temperatures, or in the presence of metals. Hydroperoxides decomposition becomes significant at advanced levels of oxidation of linoleate and at earlier stages with linolenate and other polyunsaturated fatty acids containing more than three double bonds. A multitude of low-molecular weight and high-molecular weight decomposition products is formed by complicated pathways (13-17), which greatly complicate the kinetics of lipid oxidation.

Hydroperoxide Formation. At oxygen pressures greater than 100 mm Hg, the reaction of alkyl radicals $L^{\cdot}$ with oxygen is so rapid that the concentration of the resulting peroxyl radicals $LOO^{\cdot}$ is much greater than $L^{\cdot}$. Applying stationary state conditions, by assuming the concentrations of $L^{\cdot}$ and $LOO^{\cdot}$ do not change, the rate of change is zero.

$$d[L^{\cdot}]/dt = 0 \text{ and } d[LOO^{\cdot}]/dt = 0$$

Addition of these two equations leads to

$$R_i - 2k_t [LOO^{\cdot}]^2 = 0 \text{ and } [LOO^{\cdot}] = (R_i/2k_t)^{1/2}$$

Since $d[LOOH]/dt = k_p [LOO^{\cdot}][LH]$, the following expression can be derived for the rate of hydroperoxide formation:

$$d[LOOH]/dt = k_p (R_i/2k_t)^{1/2} [LH]$$

and the rate of oxidation is independent of oxygen pressure. The ratio of the rate constants $k_p/(2k_t)^{1/2}$ is referred to as "oxidizability" and used as a measure of the reactivity of unsaturated lipids to undergo autoxidation (Section C).

When oxygen is limited and the partial oxygen pressure (Po_2) is below 100 mm Hg, assuming a constant rate of initiation and amount of lipid substrate, the rate equation can be simplified as follows:

$$d[LOOH]/dt = A[Po_2/(Po_2 + B)]$$

where A and B are constant and the rate becomes oxygen dependent.

To study quantitatively the kinetics of lipid oxidation and antioxidation a standard way of controlling and measuring the rate of free radical initiation is to use thermally labile azo compounds. These artificial initiators generate radicals at a reproducible, well-established and constant rate. In the presence of initiators such as α,α-azobisisobutyronitrile (AIBN) or benzoyl peroxide, the overriding initiation can be directly related to the rate of production of the initiator radical. Also, by using either water-soluble or lipid-soluble azo dyes, these compounds can initiate radicals at known specific micro-environments. For example, the water-soluble 2,2'-azobis-(2-amidino-propane) dihydrochloride (AAPH), generates free radicals in the aqueous phase, and the oil-soluble, 2,2'-azobis (2,4-dimethylvaleronitrile) (DMVN), generates free radicals within the lipid phase. These azo compounds decompose thermally at a known and constant rate (K_d), to produce two radicals ($A^{\cdot}$) and nitrogen.

The azo radicals ($A^{\cdot}$) react rapidly with oxygen to give peroxyl radicals

$$A\text{—}N\text{=}N\text{—}A \xrightarrow{K_d} [A^{\cdot}N_2 A^{\cdot}] \longrightarrow 2 A^{\cdot} + N_2 \quad (19)$$
$$\text{cage}$$

($AO_2^{\cdot}$), which attack lipids by abstraction of hydrogen atoms to form lipid radicals leading to lipid hydroperoxides.

The initiator radicals initially formed in solution are held together briefly in

$$A^{\cdot} + O_2 \longrightarrow AO_2^{\cdot} \xrightarrow{LH} AOOH + L^{\cdot} \quad (20)$$

a cage of solvent molecules. This cage effect causes radical molecules to recombine and slows down their diffusion through the solvent. Therefore, the rate of initiation (R_i) depends on the rate of decomposition of the initiator (k_d) and on the fraction e of the initiator radicals that escape the solvent cage affecting the efficiency of the chain initiation reaction (20).

$$e = R_i / 2k_d [\text{Initiator}]$$

The rate of initiation is generally measured by an induction period method using an antioxidant (AH) that has a known stoichiometric factor n, defined as the number of radicals trapped by each molecule of antioxidant. Because α-tocopherol is known to have an n value of 2, a known concentration is used to determine the induction period, τ, during which the oxidation is inhibited.

$$R_i = n[\text{AH}] / \tau$$

This kinetic approach is generally used to study free radical oxidation in simple model systems. However, the azo initiators used in these studies are artificial systems that are not found in either foods or biological systems. The efficiency of these initiators is greatly affected by the lipid systems used for oxidation and by the solvent viscosity; the more viscous the solvent the lower their efficiency. The efficiency of initiators in escaping the solvent cage varies; DMVN is about 75% efficient whereas AIBN is 65% efficient. The diazo hydroperoxides (AOOH) formed by reaction (20) are known to interfere with the HPLC analyses of lipid hydroperoxides. Diazo initiators produce a big flux of peroxyl radicals that do not have time to branch and proceed to other reactions observed in real lipid systems. The quantitative kinetic data obtained with diazo initiators may thus be oversimplified and not relevant to either foods or biological systems.

C. RELATIVE RATES OF AUTOXIDATION OF UNSATURATED FATTY ACIDS.

The relative rates of autoxidation of different unsaturated fatty acid esters were compared on the basis of oxygen absorption measurements (Table 1-1). In neat systems without added initiator, linoleate was 40 times more reactive than oleate, linolenate was 2.4 times more reactive than linoleate, and

TABLE 1-1.
Rates of autoxidation of unsaturated fatty esters. [a,b]

Fatty esters	Number of -CH_2-	mole O_2 per 100 hr [a]	Relative rates [a]	Oxidizability $M^{-½} sec^{-½}$ [b]	Relative rates [b]
18:1	0	0.04	1		
18:2	1	1.63	41	0.020	1
18:3	2	3.90	98	0.041	2.1
20:4	3	7.78	195	0.058	2.9
22:6	5			0.102	5.1
Trilinolein	1 x 3	1.99	50	0.080	4.0

[a]Holman and Elmer (1947): neat methyl or ethyl esters were autoxidized at 37°C and rates were measured by oxygen absorption with a Warburg respirometer.

[b]Cosgrove *et al.* (1987): chlorobenzene solution were oxidized at 37°C in the presence of 2,2-azobis(2-methylpropionitrile) and rates were measured by oxygen absorption with a pressure transducer.

arachidonate was 2 times more reactive than linolenate. The oxidizability of polyunsaturated fatty acid (PUFA) esters was also compared on the basis of oxygen uptake measured kinetically by the induction period method described above, in solution in the presence of azo initiators. The oxidizability of 18:2, 18:3, 20:4, and 22:6 was linearly related to the number of bis-allylic positions present in the fatty esters. From this relationship, the oxidizability of each PUFA was increased approximately two fold for each active bis-allylic methylene group. Thus, the oxidizability of 22:6 was 5 times greater than that of 18:2.

The autoxidation of trilinolein was more complicated that that of the simple fatty esters and did not follow the same rate equation. The order of the reaction was about 0.84 compared to 1.0 for the simple fatty esters. The efficiency of the initiators used, DMVN, was increased to 100% in the trilinolein compared to about 75% in the simple fatty esters. The difference in kinetic behavior between trilinolein and methyl linoleate was attributed to the tendency of the triacylglycerol to form aggregates. The kinetic induction period approach used to measure oxidizability may be subject to errors because of the changes in efficiency of some of the artificial initiators used according to the system and the lipid substrate. Phenolic antioxidants such as α-tocopherol, are also affected by the colloidal properties of the lipids used in the oxidation test system employed (Chapter 8).

Further complications arise in comparing the relative rates of oxidation of PUFA which change with the medium and with the methods used to analyse oxidation products. The more highly unsaturated fatty acids produce hydroperoxides in lower yields because they are readily decomposed especially in the presence of contaminating metal catalysts. Thus, methods based on yields of hydroperoxides show decreasing rates with increasing number of double bonds.

D. METAL CATALYSIS

We have seen that the above kinetic formulation requires initiators that can generate radicals at a known and constant rate and are generally based on homogeneous systems in solution. However, the kinetics of lipid oxidation become complicated in the presence of metals. As the oxidation proceeds, the rate of lipid radical formation is not constant and increases by the accumulation of lipid hydroperoxides which decompose in the presence of metals. Further complications arise in heterogeneous systems where the metal and hydroperoxide initiators are distributed in different phases (Chapter 9).

Trace amounts of heavy metals, such as iron and copper, have a marked accelerating effect on the rates of lipid oxidation. The homolytic decomposition of lipid hydroperoxides is generally considered to be metal-catalysed because it is very difficult or nearly impossible to eliminate traces of metals that can act as potent catalysts. Metals may produce radicals by two pathways:

(a) by electron transfer:

$$M^{(n+1)+} + LH \longrightarrow M^{n+} + L^{\cdot} + H^{+} \qquad (21)$$

(b) by formation of metal-oxygen transition complexes ($M-O_2$) or metal-hydroperoxide complexes ($M-HOOR$), which catalyse autoxidation and decomposition by one-electron redox reactions:

$$M^{n} + ROOH \longrightarrow M^{n+1} + RO^{\cdot} + OH^{-} \qquad (22)$$

$$M^{n+1} + ROOH \longrightarrow M^{n} + ROO^{\cdot} + H^{+} \qquad (23)$$

Reactions (22) and (23) require metal ions which can exist in two oxidation states with a suitable redox potential. Reaction (23) is usually much more rapid than reaction (22) and the metal is converted mostly in its most oxidized state, so that the rate of chain initiation depends on reaction (23). In the presence of metals, these reactions thus initiate additional autoxidation chains which accelerate the rates of lipid oxidation.

Kinetically, if the reaction of metal and hydroperoxides is assumed to be the overriding initiation reaction:

$$LOOH + M \xrightarrow{k_i} X^{\cdot} \xrightarrow{LH} L^{\cdot} \qquad (24)$$

The following rate expression can be derived for the rate of metal initiation:

$$R_i = k_i [LOOH][M]$$

Substituting this equation into the rate expression formulated above for hydroperoxide formation, we obtain the following rate of the metal-catalysed oxidation:

$$d[LOOH]/dt = k_p[LH] \, k_i \, [LOOH][M]^{1/2} / 2k_t$$

The rate of autoxidation is thus proportional to the square root of the metal ion concentration. However, this rate does not hold when the concentration of metal initiator is too small to measure. In many food and biological systems, metal catalysts may be coordinated with ligands as complexes or may exist as dimers or higher molecular weight compounds. Other chelating materials may form strong complexes with metals and inactivate their catalytic effects in promoting hydroperoxide decomposition (Chapter 4).

In actual lipid oxidation, one cannot overlook the critical role of trace metals, which complicate the kinetic sequences of initiation, and decomposition of lipid hydroperoxides. These metals catalyse both initiation of free radicals and decomposition of hydroperoxides, which become particularly significant with polyunsaturated lipids containing more than two double bonds. With these polyunsaturated lipids, although the yields of hydroperoxides are reduced in the presence of metals, they produce volatile decomposition products that have a serious impact on flavor deterioration. In foods and biological systems, the mixture of trace metals and hydroperoxides is the most important initiator that plays a key part in the development of free radical oxidation and rancidity. The use of artificial azo compounds as initiators to study free radical oxidation are therefore not relevant.

Bibliography

Al-Malaika,S. Autoxidation, in *Atmospheric Oxidation and Antioxidants*, Vol. I, pp. 45-82 (1993) (edited by G. Scott), Elsevier, Amsterdam, The Netherlands.
Bateman,L. Olefin oxidation. *Quarterly Reviews* **8**, 147-167 (1954).
Betts,J. The kinetics of hydrocarbon autoxidation in the liquid phase. *Quarterly Reviews* **25**, 265-288 (1971).
Bolland,J.L. Kinetic studies in the chemistry of rubber and related materials. I. The thermal oxidation of ethyl linoleate. *Proc. Royal Soc.* (London) A186, 218-236 (1946).
Bolland,J.L. Kinetic studies in the chemistry of rubber and related materials. VI. The benzoyl peroxide-catalyzed oxidation of ethyl linoleate. *Trans. Faraday Soc.* **44**, 669-677 (1948).
Bolland,J.L. Kinetics of olefin oxidation. *Quart. Rev. Chem. Soc.* **3**, 1-21 (1949).
Chan,H.W.-S., The mechanism of autoxidation, in *Autoxidation of Unsaturated Lipids*, pp. 1-16, (1987) (edited by H. W.-S. Chan), Academic Press, London.
Cosgrove,J.P., Church,D.F. and Pryor,W.A. The kinetics of the autoxidation of polyunsaturated fatty acids. *Lipids* **22**, 299-304 (1987).
Frankel,E.N. Lipid oxidation. *Prog. Lipid Res.* **19**, 1-22 (1980).
Heaton,F.W. and Uri,N. The aerobic oxidation of unsaturated fatty acids and their esters: cobalt stearate-catalyzed oxidation of linoleic acid. *J. Lipid Res.* **2**, 152-160 (1961).
Holman,R.T., and Elmer,O.C. The rates of oxidation of unsaturated fatty acids and esters. *J. Am. Oil Chem. Soc.* **24**, 127-129 (1947).
Howard,J. A. Homogeneous liquid-phase autoxidations, in *Free Radicals*, Vol. II, pp. 4-62 (1973) (edited by J. K. Kochi), John Wiley & Sons, New York.
Kanner,J., German,J.B., and Kinsella,J.E. Initiation of lipid peroxidation in biological systems. *CRC Critical Reviews Food Science and Nutrition* **25**, 317-364 (1987).
Labuza,T.P. Kinetics of lipid oxidation in foods. *CRC Food Technol.* **2**, 355-405 (1971).

Ragnarsson,J.O., and Labuza,T.P. Accelerated shelf-life testing for oxidative rancidity in foods- a review. *Food Chem.* **2**, 291-308 (1977).

Russell,G.A. Pathways in autoxidation, in *Peroxide Reaction Mechanisms*, pp. 107-128 (1962) (edited by J. O. Edwards), Interscience Publ., New York.

Schaich,K.M. Metals and lipid oxidation. Contemporary issues. *Lipids* **27**, 209-218 (1992).

Scott,G. Initiators, prooxidants and sensitizers, in *Atmospheric Oxidation and Antioxidants*, Vol.I, pp. 83-119 (1993) (edited by G. Scott), Elsevier, Amsterdam, The Netherlands.

Walling,C. *Free Radicals in Solution.* (1957) (John Wiley & Sons., New York).

Zhang,J.R., Cazier,A.R., Lutzke,B.S. and Hall,E.D. HPLC-chemiluminescence and thermospray LC/MS study of hydroperoxides generated from phosphatidylcholine. *Free Radical Biol. Med.* **18**, 1-10 (1995).

CHAPTER 2

HYDROPEROXIDE FORMATION

A. OLEATE AUTOXIDATION

The classical mechanism for the free radical oxidation of methyl oleate involves hydrogen abstraction at the allylic carbon-8 and carbon-11 to produce two delocalized three-carbon allylic radicals (Figure 2-1). According to this mechanism, oxygen attack at the end-carbon positions of these intermediates produces a mixture of four allylic hydroperoxides containing OOH groups on carbons 8, 9, 10, and 11, in equal amounts:

9-hydroperoxy-*trans*-10-octadecenoate (*trans*-9-OOH)
11-hydroperoxy-*cis*-9-octadecenoate (*cis*-11-OOH)
10-hydroperoxy-*trans*-8-octadecenoate (*trans*-10-OOH)
8-hydroperoxy-*cis*-9-octadecenoate (*cis*-8-OOH)

However, studies based on gas-chromatography-mass spectrometry (GC-MS) and on high-performance liquid chromatography (HPLC) showed a small but consistently higher amount of the 8- and 11-OOH (25-29%) than the 9- and 10-OOH (22-25%) (Table 2-1). In the GC-MS method a mass spectrometry-computer summation approach was employed for quantification using authentic synthetic samples of 8-, 9-, 10- and 11-hydroxystearate, as the trimethylsilyl (TMS) ether derivatives. This computer summation approach was developed because the isomeric hydroxy TMS ethers were partially separated by GC and all the mass spectra within the appropriate GC peak were added for proper quantification. The quantitative analyses by the HPLC method, carried out with a normal phase silica column on the hydroxyoctadecanoate derivatives, were in complete agreement with the GC-MS method.

Stereochemical studies based on ^{13}C-nuclear magnetic resonance spectroscopy (^{13}C-NMR) showed the presence of eight *cis* and *trans* allylic hydroperoxides (Table 2-1). To determine the isomeric distribution of allylic hydroxyoleate derivatives, *cis* and *trans* fractions were separated by silver nitrate-thin layer chromatography (TLC), a procedure that separates according to the number, position and geometry of double bonds, and they were hydrogenated prior to GC-MS analyses of the TMS ether derivatives. More recently the six major hydroperoxide isomers of methyl oleate were partially separated by silica HPLC and identified by chemical-ionization mass

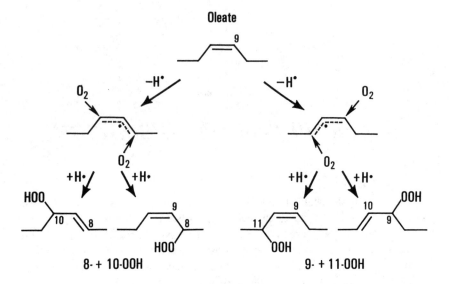

Figure 2-1. Mechanism of oleate autoxidation

spectrometry and ^{1}H NMR (Table 2-1). These hydroperoxide isomers were better separated as the hydroxy oleate derivatives by the same silica HPLC method and reanalysed by GC-MS.

When pure neat methyl oleate was oxidized, with increasing temperature, the relative amounts of cis-8-OOH and cis-11-OOH isomers decreased, the trans-8-OOH and trans-11-OOH isomers as well as the cis-9- and cis-10-isomers increased, while the trans-9- and trans-10-OOH isomers did not change (Table 2-1). When a hexane solution of methyl oleate was oxidized at 30°C in the presence of the free radical di-*tert*-butyl hyponitrite, the distribution of hydroperoxide isomers was similar to that obtained at 25°C, except for greater amounts of trans-11-OOH (plus 3%) and trans-8-OOH isomers (plus 4%) (Table 2-1). These changes in the configuration of hydroperoxides with temperature of oxidation and with the concentration of the oleate substrate can be attributed to both kinetic and thermodynamic factors that determine the stereochemistry of either the allylic carbon or the peroxyl radical intermediates or both (Figure 2-2). The stereochemistry of the allylic hydroperoxides can depend on kinetic and thermodynamic factors according to the conditions of oxidation. The temperature of oxidation has a significant effect on the cis and trans configurations of the initial hydroperoxide formed. Temperature effects reflect significant changes in the 8- and 11-hydroperoxides and very little in the 9- and 10-hydroperoxides isomers (Table 2-1).

A number of mechanisms have been advanced to explain the formation of all eight cis and trans isomers of 8-, 9-, 10- and 11-hydroperoxides in

TABLE 2-1.
Hydroperoxides from autoxidation of methyl oleate (as % of total) [a]

Temp. °C	8-OOH			9-OOH			10-OOH			11-OOH		
	cis	trans	total	cis	trans	total	cis	trans	Total	cis	trans	total
25	14.1	12.3	26.4	1.1	23.1	24.2	1.1	21.7	22.8	13.7	12.9	26.6
30[b]	13	16	29	1	21	22	1	21	22	13	16	29
40	10.6	16.0	26.6	1.6	22.0	23.6	1.7	21.7	23.4	10.1	16.0	26.4
50	8.3	17.8	26.1	2.2	22.5	24.7	2.2	21.3	23.5	8.3	17.4	25.7
75	6.1	19.0	25.1	2.7	22.5	25.1	2.9	22.0	24.9	5.4	19.5	24.9

[a] From Frankel et al. (1984): oxidation with neat methyl oleate; separated products analysed by GC-MS and ^{13}C-NMR.
[b] From Porter et al. (1994): oxidation of methyl oleate in hexane solution initiated with di-*tert*-butyl hyponitrite; products analysed by normal phase HPLC and by ^{1}H-NMR.

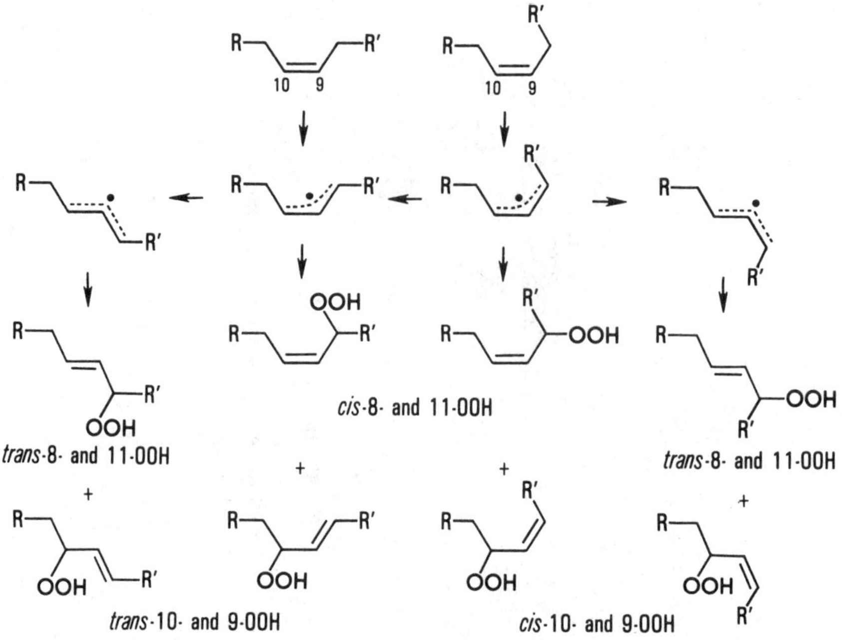

Figure 2-2. Mechanism II of oleate oxidation (Frankel et al., 1984) autoxidized methyl oleate. In one mechanism, the delocalized radicals formed by abstraction of the hydrogens allylic to the double bond of oleate lose their stereochemistry resulting in a change of conformation, especially at elevated temperatures (Figure 2-2). In another mechanism, the stereochemistry of the hydroperoxides is determined by the *cisoid* or *transoid* configurations of the peroxyl radicals rather than those of the allylic carbon radicals, which have a very short lifetime in the presence of oxygen. The conformation of the allylic carbon radicals may thus be established initially by the conformation of the oleate according to the temperature of oxidation. The relative amounts of *cisoid* and *transoid* conformers of the hybrid allylic carbon radical intermediates formed from oleate will depend on the degree of oxidation. The small preference for the reaction of oxygen on carbon-8 and carbon-11 may indicate that some hydrogen abstraction occurs from conformers of non-planar radicals formed from methyl oleate at lower temperatures. Thus, the small preference for the 8- and 11-hydroperoxides may be explained by a reaction of oxygen with radicals before they rotate to a planar conformation.

B. LINOLEATE AUTOXIDATION

Linoleate is 40 times more reactive than oleate (Chapter 1, Table 1-1) because it has an active bis-allylic methylene group on carbon-11 between two double bonds that can lose a hydrogen atom very readily. Hydrogen abstraction

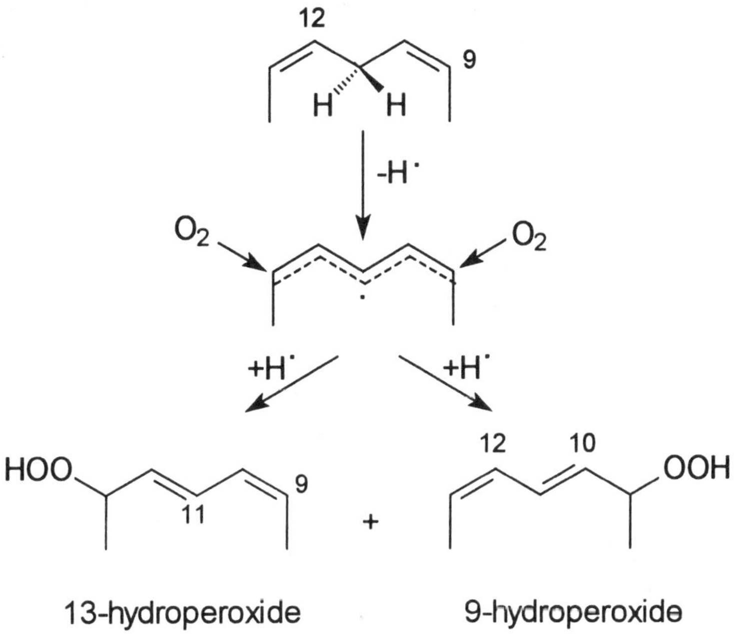

13-hydroperoxide   9-hydroperoxide

Figure 2-3. Mechanism of linoleate autoxidation

at the carbon-11 position of linoleate produces a hybrid pentadienyl radical, which reacts with oxygen at the end carbon-9 and carbon-13 positions to produce a mixture of two conjugated diene 9- and 13-hydroperoxides (Figure 2-3). The greater reactivity of linoleate to autoxidation is due to the formation of a pentadienyl radical intermediate which is more effectively stabilized by resonance, and the resulting dienoic hydroperoxides produced that are stabilized by conjugation. These isomeric conjugated dienoic hydroperoxides are formed in equal concentrations at different levels of autoxidation and a wide range of temperatures.

Stereochemical studies based on HPLC and ^{13}C-NMR showed the formation of a mixture of four *cis,trans* and *trans,trans* conjugated diene hydroperoxides:

9-hydroperoxy-*trans*-10, *cis*-12-octadecadienoate (*cis,trans*-9-OOH)
9-hydroperoxy-*trans*-10, *trans*-12-octadecadienoate (*trans,trans*-9-OOH)
13-hydroperoxy-*cis*-9, *trans*-11-octadecadienoate (*cis,trans*-13-OOH)
13-hydroperoxy-*trans*-9, *trans*-11-octadecadienoate (*trans,trans*-13-OOH)

Initially and at low temperatures, the *cis,trans*-OOH isomers are predominant. The proportion of *trans,trans*-OOH isomers increases with the level and temperature of oxidation (Table 2-2). Each of the *cis,trans*-9- and 13-OOH isomers are readily inter-converted into mixtures containing the corresponding *trans,trans*-OOH isomers.

TABLE 2-2.
Hydroperoxides from autoxidation of methyl linoleate (as % of total). [a]

Temp °C	13-OOH cis,trans	13-OOH trans,trans	9-OOH cis,trans	9-OOH trans,trans	total 13-OOH	total 9-OOH
25	31.0	20.1	29.7	19.2	51.2	48.9
50	19.5	28.0	22.9	29.6	47.5	52.5
65	18.6	33.5	17.5	30.4	52.1	47.9

[a]From Frankel et al. (1990); analyses by HPLC and by ^{13}C-NMR were averaged.

As with oleate, several mechanisms have been advanced to explain the stereochemical formation of the four cis,trans- and trans,trans isomers of the 9- and 13-hydroperoxides in autoxidized methyl linoleate. In one mechanism, the initial pentadienyl radical assumes four conformations before reaction with oxygen (Figure 2-4). An initial cis,cis-pentadienyl radical is formed which is converted into trans,cis/cis,trans-pentadienyl radicals and into a trans,trans radical at higher levels and temperatures of oxidation. Under these conditions the ratio of cis,trans- and trans,trans-hydroperoxides decreases but the proportion of the 9- and 13-hydroperoxides remains the same because the pentadienyl radicals react equally on carbon-9 and carbon-13. These radicals react with oxygen reversibly to form the corresponding cis,trans- and trans,trans-peroxyl radicals. The loss of oxygen leads to a pool of pentadienyl radicals that undergo rearrangement.

In another proposed mechanism, the stereochemistry of linoleate hydroperoxides is determined by the configuration of the peroxyl radicals rather than that of the carbon pentadienyl radicals, which have a very short life-time because they react very rapidly with oxygen (Figure 2-5). Accordingly, the interconversion of cis,trans- and trans,trans-hydroperoxides results from the loss of oxygen from the peroxyl radicals by β-scission to produce carbon pentadienyl radicals of different conformations. The trans,cis/trans,trans product ratio is determined by the relative rate of β-scission (k_β) of the carbon-oxygen bond of the peroxyl radicals and their rate of hydrogen abstraction (KP) to form the corresponding hydroperoxides. The interconversion of cis,trans- and trans,trans-hydroperoxides is inhibited in the presence of antioxidants, which act as hydrogen donors. In the presence of 5% α-tocopherol, the formation of trans,trans-hydroperoxides is completely inhibited during autoxidation of methyl linoleate.

In considering the merits of these mechanisms to explain the stereochemistry of linoleate autoxidation we should appreciate the marked variation in conditions used by various workers to study autoxidation. In some studies conducted in neat viscous lipid systems where autoxidation was initiated in the presence of trace metals and hydroperoxide impurities, hydroperoxidation would be influenced by factors favoring the formation of thermodynamically stable trans,trans- hydroperoxides. In other studies,

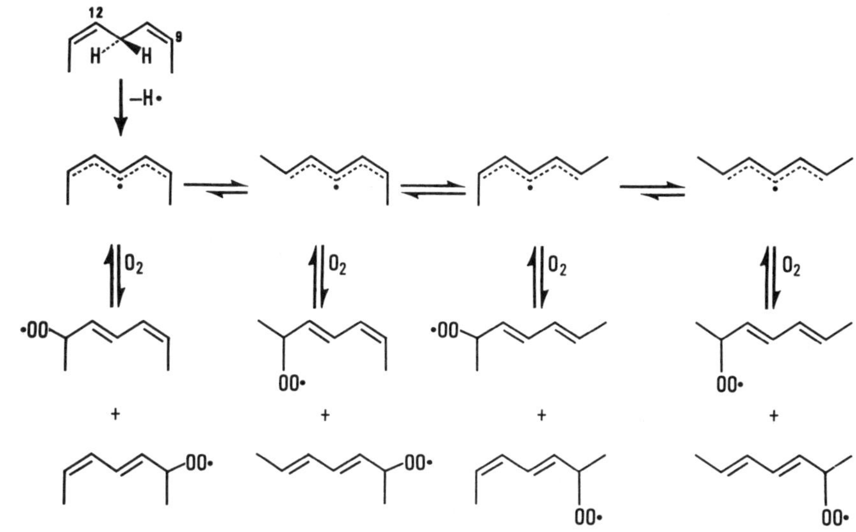

Figure 2-4. Mechanism II of linoleate autoxidation (Frankel, 1979)

thermodynamic factors were minimized by conducting the oxidation in solution in the presence of artificial non-lipid initiators and at very low levels of conversion. Under these conditions, the autoxidation was kinetically controlled and favored a higher *trans,cis/trans,trans* product ratio. In the oxidation of neat lipids or multi-phased emulsion systems, initiated by metal catalysts and oxidized lipid impurities, at elevated temperatures and at high degrees of conversion, the reaction may become more controlled by thermodynamic than by kinetic factors.

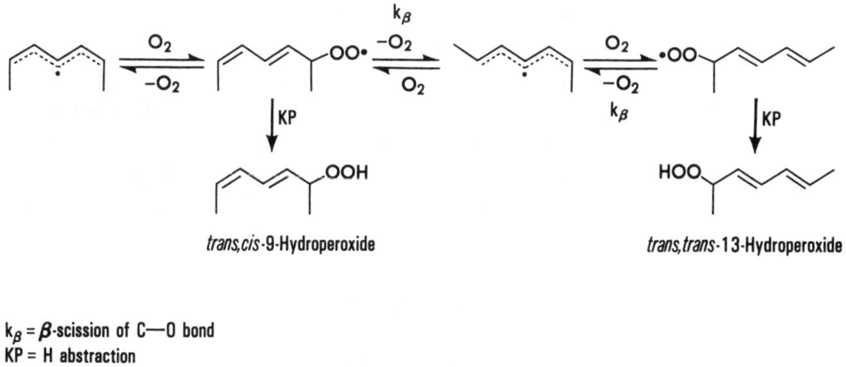

$k_\beta = \beta$-scission of C—O bond
KP = H abstraction

Figure 2-5. Mechanism III of linoleate autoxidation (Porter *et al.*, 1980)

C. LINOLENATE AUTOXIDATION

Methyl linolenate has two bis-allylic methylene groups and reacts twice as fast with oxygen as linoleate (Chapter 1, Table 1-1). Linoleate was 40 times more reactive than oleate, and linolenate was 2.4 times more reactive than

LIPID OXIDATION

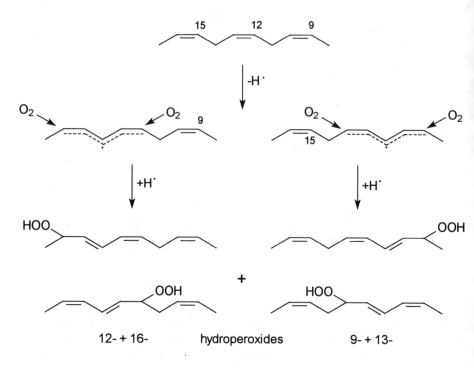

Figure 2-6. Mechanism of linolenate autoxidation (Frankel et al., 1961)

linoleate. When comparing the mono allylic hydrogens of oleate with the bis-allylic hydrogens of linoleate, the rates increased 40 fold compared to 2 fold by introducing the second bis-allylic hydrogens of linolenate. Therefore, the two bis-allylic methylene groups in linolenate act independently and are not activated by each other. By the same mechanism as linoleate, two pentadienyl radicals are formed by hydrogen abstraction on carbon-11 and carbon-14 between the two 1,4-diene systems on carbon-9 and carbon-13, on one hand, and on carbon-12 and carbon-16, on the other hand (Figure 2-6). Reaction with oxygen at the end-carbon positions of each pentadienyl radical produces a mixture of four peroxyl radicals leading to the corresponding conjugated diene 9-, 12-, 13- and 16-hydroperoxides containing a third isolated cis double bond.

9-hydroperoxy-trans-10, cis-12, cis-15-octadecatrienoate
(trans,cis,cis-9-OOH)
13-hydroperoxy-cis-9, trans-11, cis-15-octadecatrienoate
(cis,trans,cis-13-OOH)
12-hydroperoxy-cis-9, trans-13, cis-15-octadecatrienoate
(cis,trans,cis-12-OOH)
16-hydroperoxy-cis-9, cis-12, trans-14-octadecatrienoate
(cis,cis,trans-16-OOH)

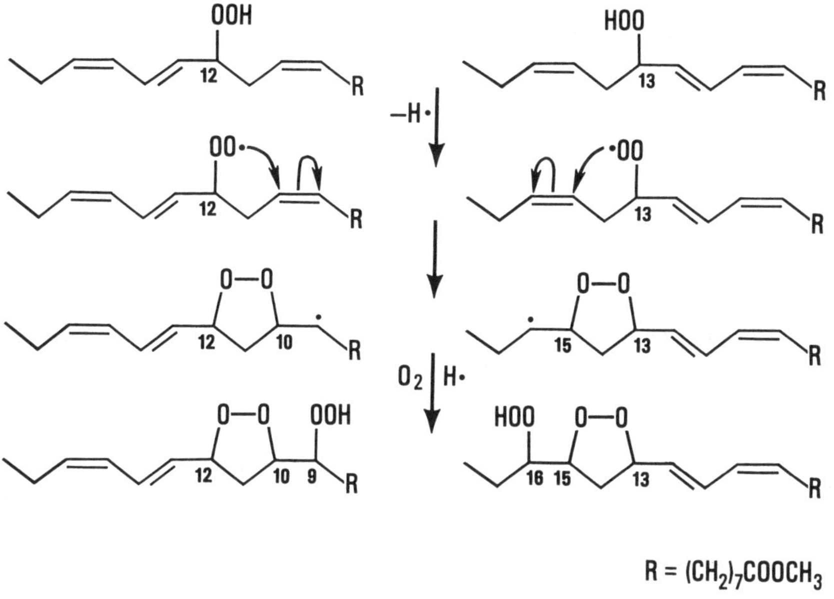

Figure 2-7. Cyclization of 12- and 13-hydroperoxides of linolenate (Neff et al., 1981)

These hydroperoxides can be divided into two distinct groups: the "external" (or "outer") 9- and 16-hydroperoxide isomers, which have the OOH group outside the system of double bonds, and the "internal" (or "inner") 12- and 13-hydroperoxide isomers, which have the OOH group located between a cis,trans-conjugated diene system and a single cis double bond separated by a homoallylic methylene group. The analyses of linolenate autoxidized at a wide range of conversion (2 to 30%) and temperatures (25 to 80°C) showed that the amounts of external 9- and 16-hydroperoxides were about four times higher than the internal 12- and 13-hydroperoxides (Table 2-3). The GC-MS analyses of isomeric distribution of linolenate hydroperoxides are in complete agreement with different analytical approaches based on oxidative cleavage products and on HPLC analyses of hydroxystearate derivatives.

This uneven distribution of isomeric hydroperoxides of linolenate is attributed to the tendency of the 12- and 13-peroxyl radicals to undergo rapid 1,3-cyclization and further oxidation to form 5-membered hydroperoxy epidioxides as major products (25%) (Figure 2-7). The peroxyl radicals of the 12- and 13-hydroperoxides have a cis- double bond homoallylic to the peroxyl group that permits their facile 1,3 cyclization by intramolecular radical addition to the double bond and formation of a new radical that reacts with oxygen. This rapid cyclization accounts for the lower concentrations of the internal 12- and 13-hydroperoxides observed under a wide range of conditions of oxidation (Table 2-3). In contrast to linoleate, during autoxidation, the

TABLE 2-3.
Hydroperoxides from autoxidation of methyl linolenate (as % of total). [a]

Temp. °C	9-OOH	12-OOH	13-OOH	16-OOH
25	30.6	10.1	10.8	48.6
40	33.0	10.8	12.4	43.9
60	29.4	11.4	11.8	47.5
80	34.2	11.1	11.5	43.3

[a]From Frankel et al. (1977); GC-MS analyses of samples autoxidized to different peroxides and temperatures were averaged.

cis,trans-hydroperoxides of linolenate are not readily isomerized to the trans,trans configuration apparently because cyclization is much more favorable and competes with geometric isomerization. The mechanism for cyclization of the internal 12- and 13-hydroperoxyl radicals of linolenate is well demonstrated by the evidence that this cyclization is completely inhibited by the addition of 5% α-tocopherol to linolenate. α-Tocopherol is a good hydrogen donor that inhibits the cyclization resulting in an even distribution of the 9-, 12-, 13- and 16-hydroperoxides. The peroxyl radicals derived from the internal 12- and 13-hydroperoxides of linolenate can also undergo 1,5-cyclization to form two isomeric 9- and 16-hydroperoxy bicycloendoperoxides (Figure 2-8). These hydroperoxy bicycloendoperoxides are characteristically formed from polyunsaturated fatty acids containing three or more double bonds, and are precursors of malonaldehyde and thiobarbituric acid (TBA)-reactive materials (Chapter 4, Table 4-7).

The significantly higher relative proportion of the 16-hydroperoxide than of the 9-hydroperoxide of linolenate (Table 2-3) indicates that oxygen attack favors the terminal carbon-16 position of the 12,16-pentadienyl radical closest to the end of the fatty acid chain of linolenate. The same preference of oxygen attack at the end of the fatty acid molecule was confirmed in the autoxidation of cis-9, cis-15-octadecadienoate and cis-12, cis-15-octadecadienoate. These isomeric dienes are found in partially hydrogenated methyl linolenate. The 9,15-diene has two isolated cis double bonds, and oxidizes like oleate to form four isomeric 8-, 9-, 10- and 11- hydroperoxides by oxidation of the cis-9-double bond, and four isomeric 14-, 15-, 16- and 17-hydroperoxides by oxidation of the cis-15-double bond. The hydroperoxides derived from the 9-double bond are fairly evenly distributed (10-12%), as observed for oleate. However, the hydroperoxides derived from the cis-15 double bond are not evenly distributed and show a significantly higher concentration of the 16- (15%) and 17-hydroperoxides (22%) than the 14- (11%) and 15-hydroperoxides (8%) (Figure 2-9). Oxidation of the terminal 16- and 17-hydroperoxides is thus greatly favored. Decomposition of the hydroperoxides derived from the 9,15-diene produces potent volatile compounds that are responsible for the characteristic so-called "hydrogenation" flavor of partially hydrogenated linolenate-containing oils (Chapter 4).

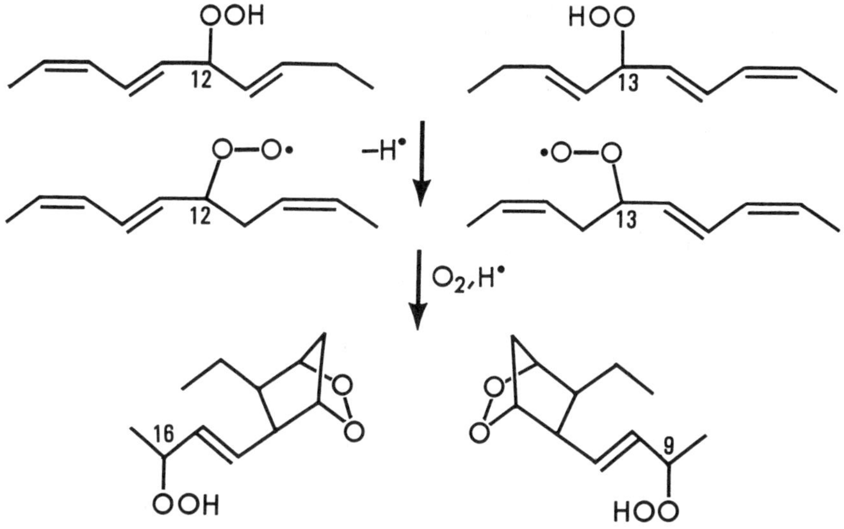

HYDROPEROXIDE FORMATION 33

Figure 2-8. Formation of bicyclo endoperoxides from 12- and 13-hydroperoxides of linolenate

Because the 12,15-diene has a 1,4-diene system it oxidizes like linoleate to form two conjugated dienoic 12- and 16-hydroperoxides. However, in contrast to linoleate, the external 16-hydroperoxide is formed at a higher concentration than the internal 12-hydroperoxide, with (normalized concentrations of 58% and 42%) (Figure 2-9). Therefore, both the 9,15- and 12,15-dienes produce allylic radicals in which the terminal carbons 16 and 17, closest to the end of the fatty acid chain, are the most reactive with oxygen. The same preference of oxygen attack at the terminal double bond position is also observed in other polyunsaturated fatty acids with n-3 double bonds (linolenate, eicosapentaenoic and docosahexaenoic acids), and n-6 double bonds (arachidonic acid). Volatile decomposition products derived from hydroperoxides containing an n-3 double bond are particularly significant for their impact on flavor (Chapter 4).

D. ARACHIDONATE AUTOXIDATION

Methyl arachidonate reacts with oxygen about twice as fast as linolenate (Chapter 1, Table 1-1) because it has three active bis-allylic methylene groups and three 1,4-diene systems, i.e. between carbon-5 and carbon-9, between carbon-8 and carbon-12, and between carbon-11 and carbon-15. By the same mechanism for autoxidation as methyl linoleate and linolenate, hydrogen abstraction on the bis-allylic methylene goups on carbon-7, carbon-10, and carbon-13 of arachidonate produces three pentadienyl radicals between carbons 5 and 9, between carbons 8 and 12 and between carbons 11 and 15. Oxygen attack at either end of these pentadienyl radicals produces a mixture of six isomers with hydroperoxide substitution on carbons 5, 8, 9, 11, 12 and 15

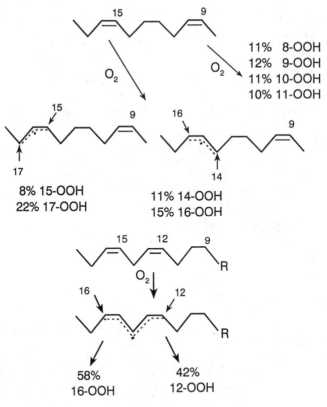

Figure 2-9. Hydroperoxidation of methyl cis-9, cis-15-octadedienoate and methyl cis-12, cis-15-octadedienoate (Frankel et al., 1980)

(Figure 2-10). As with linolenate, the outer 5- and 15-hydroperoxides of arachidonate are formed at higher concentration (52%) than the inner 8-, 9, 11- and 12-hydroperoxides (10-16%). This distribution of hydroperoxides is similar to that of autoxidized methyl linolenate and can be attributed to the tendency of the internal 8-, 9-, 11- and 12-hydroperoxides to cyclize into four hydroperoxy epidioxides and bicycloendoperoxides (Figures 2-7 and 2-8).Also, as observed in linolenate, the 15-hydroperoxide in arachidonate is formed at a higher concentration (33.8%) than the 5-hydroperoxide (18.3%), indicating preferential oxygen attack on the carbon closest to the end of the fatty acid chain. As with methyl linolenate, the addition of 1 to 5% α-tocopherol to arachidonate inhibits the cyclization of the 8-. 9-, 11- and 12-hydroperoxyl radicals and results in an even distribution of the six hydroperoxide isomers of arachidonate.

The hydroperoxy bicycloendoperoxides of linolenic and arachidonic acids have structures that are related to the physiologically important prostaglandins PGG_2 and PGH_2, which promote platelet aggregation (Figure 2-11) (see Chapter 12). The bicycloendoperoxides produced by autoxidation of pure

HYDROPEROXIDE FORMATION

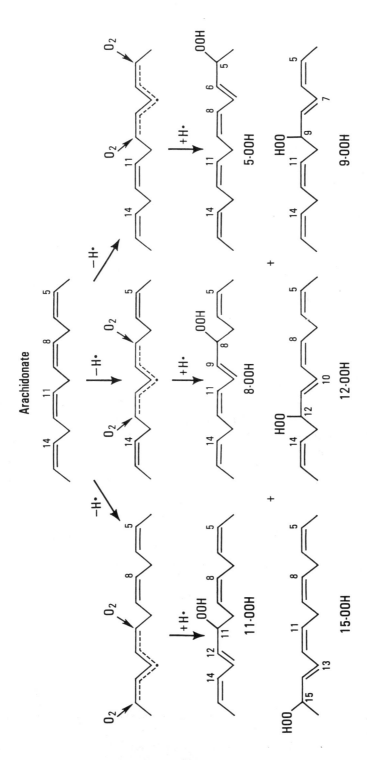

Figure 2-10. Mechanism of arachidonate autoxidation

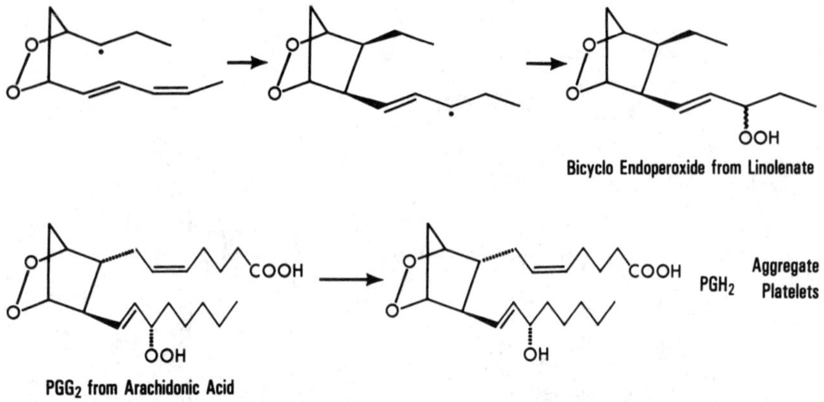

Bicyclo Endoperoxide from Linolenate

PGG$_2$ from Arachidonic Acid

PGH$_2$ Aggregate Platelets

Figure 2-11. Bicyclo endoperoxides from oxidized linolenate and arachidonic acid

13-hydroperoxide of α-linolenate and 9-hydroperoxide of γ-linolenate were purified by HPLC and shown to have mainly *cis* substituents, in contrast to the natural *trans* stereochemistry of prostaglandins derived enzymatically from arachidonic acid. This difference in stereochemistry between the non-enzymatically and enzymatically produced bicycloendoperoxides may be important physiologically.

E. AUTOXIDATION OF POLYUNSATURATED FATTY ACIDS

As the number of double bonds increases in polyunsaturated fatty acids (PUFA), they produce more complex mixtures of hydroperoxides which are easily decomposed and become very difficult to analyse quantitatively. The most important n-3 PUFA found in fish and marine oils include all *cis*-5,8,11,14,17-eicosapentaenoic acid (EPA), and all *cis*-4,7,10,13,16,19-docosahexaenoic acid (DHA). The hydroperoxides produced from EPA and

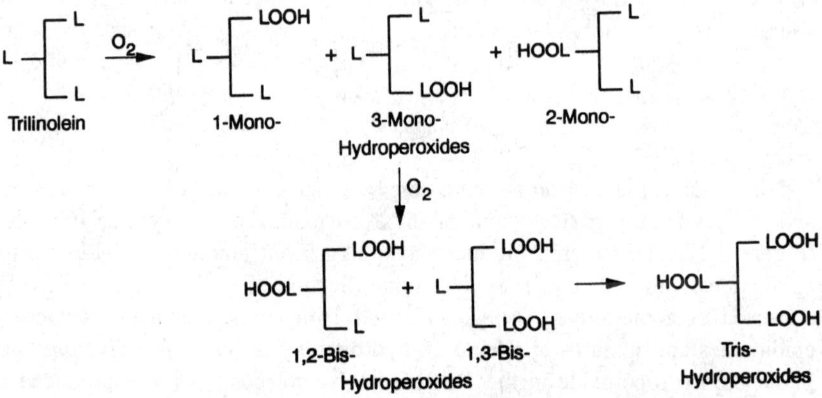

Trilinolein → 1-Mono- → 3-Mono-Hydroperoxides + 2-Mono- → 1,2-Bis-Hydroperoxides + 1,3-Bis-Hydroperoxides → Tris-Hydroperoxides

Figure 2-12. Mechanism of trilinolein autoxidation (L = linoleate) (Neff *et al.*, 1990)

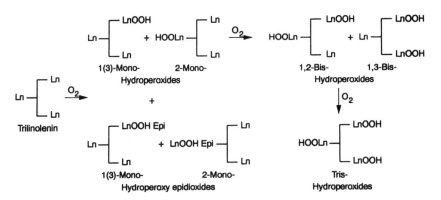

Figure 2-13. Mechanism trilinolenin autoxidation (Ln = linolenate) (Frankel et al., 1990)

DHA have been identified but not quantified. By the same mechanism established for linolenate, EPA produced the eight 5-, 8-, 9-, 11-, 12-, 14-, 15- and 18-hydroperoxides. DHA produced the ten 4-, 7-, 8-, 10-, 11-, 13-, 14-, 16-, 17- and 20-hydroperoxides.

We have seen in Chapter 1 that the oxidizability of 18:2, 18:3, 20:4 and 22:6 was linearly related to the number of bis-allylic positions in the fatty esters. From this relationship, the oxidizability of each PUFA is increased approximately two fold for each active bis-allylic methylene group. Thus, the oxidizability of 22:6 was five times greater than that of 18:2.

F. AUTOXIDATION OF TRIACYLGLYCEROLS

Trilinolein formed monohydroperoxides located on the 1,3- and 2-positions of the glycerol moiety as the main initial products. The mono-hydroperoxides were further oxidized to form a mixture of 1,2- and 1,3-bis-hydroperoxides, which are also oxidized to trishydroperoxides by sequential oxygen addition (Figure 2-12). The hydroperoxides were composed of cis,trans- and trans,trans-9- and 13-isomers, which are located in the 1(3)- and 2-triacylglycerol positions in a ratio of two. Therefore, the oxidation of trilinolein showed no positional preference between the 1(3)- and 2-triacylglycerol positions.

Trilinolenin produced on autoxidation 1(3)- and 2-monohydroperoxides, 1,2- and 1,3-bis-hydroperoxides and tris-hydroperoxides by sequential oxidation (Figure 2-13). Trilinolenin produced also significant amounts of hydroperoxy epidioxides, formed by 1,3-cyclization (Section C). The hydroperoxides consisted of a mixture of 9-, 12-, 13- and 16-isomers, and the hydroperoxy epidioxides are mixtures of 9- and 16-hydroperoxy isomers. The cis,trans- 16-linolenate hydroperoxide in the 1(3)-triacylglycerol positions was produced in a ratio of 2.3 relative to the 2-triacylglycerol position, compared to 1.8 for the corresponding cis,trans 9-linolenate hydroperoxide. These results support a

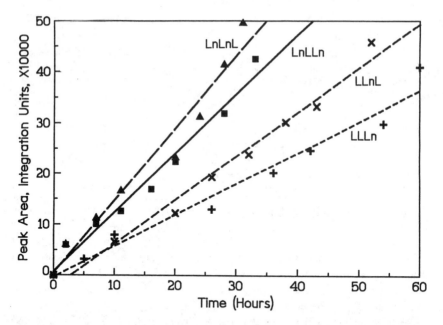

Figure 2-14. Oxidation of four synthetic triacylglycerols containing linoleate (L) and linolenate (Ln) in different specific positions at 40°C (Miyashita et al., 1990). Formation of oxidation products was determined by HPLC as total peak area of hydroperoxides and hydroperoxy epidioxides. With permission of AOCS Press.

small preferential attack of oxygen on the n-3 double bond of linolenate in the 1(3)-triacylglycerol positions. Triacylglycerols containing EPA and palmitic acid also produced monohydroperoxides as major products (70-80%) including hydroperoxy epidioxides and other unidentified secondary products. Tri-EPA were further oxidized to bis- and tris-hydroperoxides. These hydroperoxides were separated by HPLC and identified by liquid chromatography-mass spectrometry (LC-MS) with electrospray ionization.

Four synthetic triacylglycerols containing linoleate (L) and linolenate (Ln) in specific 1,2 or 1,3-positions (LLLn, LLnL, LnLnL and LnLLn) were oxidized to determine the effect of fatty acid position on rates and products of autoxidation. The rates and yields of oxidation products decreased in the order: LnLnL > LnLLn > LLnL > LLLn (Figure 2-14). Therefore, triacylglycerols containing two Ln and one L were more easily oxidized when Ln is in the 1,2- than the 1,3-triacylglycerol position. On the other hand, triacylglycerols containing two L and one Ln were more easily oxidized when L is in the 1,3- than the 1,2-triacylglycerol position. Interactions in the 1,2-positions between either two Ln or one Ln and one L appear to increase the oxidative susceptibility of these triacylglycerols. This interaction in the 1,2-positions was supported by further studies with synthetic triacylglycerols containing linoleate and palmitate (P). Thus, the oxidation rates and yields of triacylgycerol hydroperoxides at 60°C were much greater for LLP than for LPL. In triacylglycerols containing two P and one L, PLP was more easily oxidized than PPL. Thus, when 1,2-interactions between L were eliminated, the triacylglycerols were more easily oxidized when L is in the 2- than in the 1,3-triacylglycerol position. This trend is the same as that observed above for the dilinoleyl-monolinolenyl glycerols (Figure 2-14).

F. AUTOXIDATION OF PHOSPHOLIPIDS

The autoxidation of polyunsaturated phospholipids was shown, by reverse-phase HPLC of the hydroxylinoleate hydroperoxides, to follow a similar course as for the fatty esters. The liposome of oxidized dilinoleyl-phosphatidylcholine produced two hydroxydienes derived from phosphatidylcholine containing one linoleate hydroperoxide, and a phosphatidylcholine containing two linoleate hydroperoxides. Palmitoyl-2-linoleyl-phosphatidylcholine produced by air oxidation the four *cis,trans*- and *trans,trans*-9- and 13-linoleate hydroperoxides. 1-Stearoyl-2-arachidonyl-phosphatidylcholine formed conjugated diene hydroperoxides. In the presence of increasing proportions of saturated dipalmitoyl-lecithin, the oxidation of soybean lecithin produced increasing proportions of hydroxyepoxy fatty acids and their hydrolysis products, trihydroxy acids. Apparently, the saturated lecithin prevents intermolecular oxygen *trans*fer. The yield of *trans, trans* hydroperoxides was much greater than that of the *cis,trans* isomers at all stages of oxidation of soybean phosphatidylcholine liposomes. Soybean phosphatidylcholine is a much poorer H donor than methyl linoleate in solution. The much slower oxidizability of egg lecithin initiated by di-tert-butyl hyponitrile in an aqueous liposome system (0.016 $M^{-\frac{1}{2}}s^{-\frac{1}{2}}$) than in cholorobenzene solution (0.61 $M^{-\frac{1}{2}}s^{-\frac{1}{2}}$) was attributed to the high microviscosity of the lecithin bilayer.

The direct separation of phospholipid hydroperoxides has now been achieved by further advances in HPLC and increased sensitivity by chemiluminescence detection. By reverse-phase HPLC, the hydroperoxides of phosphatidylcholine and phosphatidylethanolamine were separated and determined in stored spray-dried eggs (2.4-6.4%), fresh ground meat (0.2%), and raw fish (0.4%). Silica HPLC with post-column chemiluminescence detection has been used to determine phospholipid hydroperoxides in human blood plasma and in the low-density lipoprotein fraction.

Bibliography

Barclay,L.R.C. and Ingold,K.U. Autoxidation of biological molecules. 2. The autoxidation of a model membrane. A comparison of the autoxidation of egg lecithin phosphatidylcholine in water and in chlorobenzene. *J. Am. Chem. Soc.* **103**, 6478-6485 (1981).

Chan,H.W.-S., and Coxon,D.T. Lipid hydroperoxides, in *Autoxidation of Unsaturated Lipids*, pp. 17-50 (1987) (edited by H.W.-S. Chan), Academic Press, London.

Coxon,D.T., Price,K.R. and Chan,H.W.-S. Formation, isolation and structure determination of methyl linolenate diperoxides. *Chem. Phys. Lipids* **28**, 365-378 (1981).

Crawford,C.G., Plattner,R.D., Sessa,D.J. and Rackis,J.J. Separation of oxidized and unoxidized molecular species of phosphatidylcholine by high-pressure liquid chromatography. *Lipids* **15**, 91-94 (1980).

Endo,Y. Hoshizaki,S. And Fujimoto,K. Autoxidation of synthetic isomers of triacylglycerol containing eicosapentaenoic acid. *J. Am. Oil Chem. Soc.* **74**, 543-548 (1997).

Frankel,E. N., Evans,C.D., McConnell,D.G., Selke,E. and Dutton,H.J. Autoxidation of methyl linolenate. Isolation and characterization of hydroperoxides. *J. Org. Chem.* **26**, 4663-4669 (1961).

Frankel,E.N., Neff,W.E., Rohwedder,W.K., Khambay,B.P.S., Garwood,R.F. and Weedon,B.C.L. Analysis of autoxidized fats by gas chromatography-mass spectrometry: I. Methyl oleate. *Lipids* **12**, 901-907 (1977).

Frankel,E.N., Neff,W.E., Rohwedder,W.K., Khambay,B.P.S., Garwood,R.F. and Weedon,B.C.L. Analysis of autoxidized fats by gas chromatography-mass spectrometry: II. Methyl linoleate. *Lipids* **12**, 908-913 (1977).

Frankel,E.N., Neff,W.E., Rohwedder,W.K., Khambay,B.P.S., Garwood,R.F. and Weedon,B.C.L. Analysis of autoxidized fats by gas chromatography-mass spectrometry: III. Methyl linolenate. *Lipids* **12**, 1055-1061 (1977).

Frankel,E.N. Autoxidation, in *Fatty Acids*, pp. 353-378 (1979) (edited by E. H. Pryde), American Oil Chemists' Society, Champaign, IL.

Frankel,E.N. Lipid oxidation. *Prog. Lipid Res.* **19**, 1-22 (1980).

Frankel,E.N., Dufek,E.J. and Neff,W.E. Analysis of autoxidized fats by gas chromatography-mass spectrometry: VI. Methyl 9,15- and 12,15-octadecadienoate. *Lipids* **15**, 661-667 (1980).

Frankel,E.N., Garwood,R.F., Khambay,B.P.S., Moss,G. and Weedon,B.C.L. Stereochemistry of olefin and fatty acid oxidation. III. The allylic hydroperoxides from the autoxidation of methyl oleate. *J. Chem. Soc., Perkin Trans.* I. 2233-2240 (1984).

Frankel,E.N. Chemistry of free radical and singlet oxidation of lipids. *Prog. Lipid Res.* **23**, 197-221 (1985).

Frankel,E.N., Neff,W.E. and Miyashita,K. Autoxidation of polyunsaturated triacylglycerols. II. Trilinolenoylglycerol. *Lipids* **25**, 40-47 (1990).

Frankel,E.N., Neff,W.E. and Weisleder,D. Determination of lipid hydroperoxides by ^{13}C NMR Spectroscopy. *Methods in Enzymology*, **186**, 380-387 (1990).

Gunstone,F.D. Reaction of oxygen and unsaturated fatty acids. *J. Am. Oil Chem. Soc.* **61**, 441-447 (1984).

Mead,J.F. Membrane lipid peroxidation and its prevention. *J. Am. Oil Chem. Soc.* **57**, 393-397 (1980)

Miyashita,K., Frankel,E.N. and Neff,W.E. Autoxidation of polyunsaturated triacylglycerols. III. Synthetic triacylglycerols containing linoleate and linolenate. *Lipids* **25**, 48-53 (1990).

Miyazawa,T., Fujimoto,K. and Oikawa,S-i. Determination of lipid hydroperoxides in low-density lipoprotein from human plasma using high performance liquid chromatography with chemiluminescence detection. *Biomedical Chromatography* **4**, 131-134 (1990).

Miyazawa,T., Yasuda,K., Fujimoto,K. and Kaneda,T. Presence of phosphatidylcholine hydroperoxide in human plasma. *J. Biochem.* (Tokyo) **103**, 744-746 (1988).

Miyazawa,T., Suzuki,T., Fujimoto,K. and Yasuda,K. Chemiluminescent simultaneous determination of phosphatidylcholine hydroperoxide and phosphatidylethanolamine hydroperoxide in the liver and brain of the rat. *J. Lipid Res.* **33**, 1051-1058 (1992).

Neff,W.E. and El-Agaimy,M. Effect of linoleic acid position in triacylglycerols on their oxidative stability. *Lebensm. Wiss. u. Technol.* **29**, 772-775 (1996).

Neff,W.E., Frankel,E.N. and Weisleder,D. High-pressure liquid chromatography of autoxidized lipids: II. Hydroperoxy-cyclic peroxides and other secondary products from methyl linolenate. *Lipids* **16**, 439-448 (1981).

Neff,W.E., FrankelE.N. and Miyashita,K. Autoxidation of polyunsaturated triacylglycerols. I. Trilinoleoylglycerol. *Lipids* **25**, 33-39 (1990).

O'Connor,D.E., Michelich,E.D. and Coleman,M.C, Stereochemical course of the autoxidative cyclization of lipid hydroperoxides to prostaglandin-like bicyclo-endoperoxides. *J. Am. Chem. Soc.* **106**, 3577-3584 (1984).

Porter,N.A., Wolf,R.A., and Weenen,H. The free radical oxidation of polyunsaturated lecithins. *Lipids* **15**, 163-167 (1980).

Porter,N.A. and Wujek,D.G. Autoxidation of polyunsaturated fatty acids, an expanded mechanistic study. *J. Am. Chem. Soc.* **106**, 2626-2629 (1984).

Porter,N.A. Mechanism of fatty acid and phospholipid autoxidation. *Food Tech.* **38**(3), 49 (1984).
Porter,N.A., Mills,K.A. and Carter,R.L. A mechanistic study of oleate autoxidation: Competing peroxyl H-atom abstraction and rearrangement. *J. Am. Chem. Soc.* **116**, 6690-6696 (1994).
Porter,N.A., Caldwell,S.E. and Mills,K.A. Mechanism of free radical oxidation of unsaturated fatty lipids. *Lipids* **30**, 277-290 (1995).
Pryor,W.A., Stanley,J.P. and Blair,E. Autoxidation of polyunsaturated fatty acids. II. A suggested mechanism for the formation of TBA-reactive materials from prostaglandin-like endoperoxides. *Lipids* **11**, 370-379 (1976).
Terao,J. and Matsushita,S. Analysis of hemoprotein catalyzed peroxidation products of arachidonic acid by gas chromatography-mass spectrometry. *Agric. Biol. Chem.* **45**, 595-599 (1981).
Terao,J. and Matsushita,S. Application of high-performance liquid chromatography for the determination of phospholipid hydroperoxides in foods and biological systems. *Free Radical Biol. Med.* **3**, 345-348 (1987).
Terao,J., Kawanishi,M. and Matsushita,S. Application of high-performance liquid chromatography for the estimation of peroxidized phospholipids in stray-dried egg and muscle foods. *J. Agric. Food Chem.* **35**, 613-617 (1987).
Terao,J., Shibata,S.S. and Matsushita,S. Selective quantification of arachidonic acid hydroperoxides and their hydroxy derivatives by reverse-phase high-performance liquid chromatography. *Anal. Biochem.* **169**, 415-423 (1988).
VanRollins,M., and Murphy,R.C. Autooxidation of docosahexaenoic acid: Analysis of ten isomers of hydroxydocosahexaenoate. *J. Lipid Res.* **25**, 507-517 (1984).
Wu,G-S., Stein,R.A. and Mead,J.F. Autoxidation of phosphatidylcholine liposomes. *Lipids* **17**, 403-413 (1982).
Yamauchi,R., Yamada,T., Kato,K. And Ueno,Y. Monohydroperoxides formed by autoxidation and photosensitized oxidation of methyl eicosapentaenoate. *Agric. Biol. Chem.* **47**, 2897-2902 (1983).

CHAPTER 3

PHOTOOXIDATION OF UNSATURATED FATS

The oxidation of unsaturated fats is accelerated by exposure to light. Direct photooxidation is due to free radicals produced by ultraviolet light irradiation which catalyses the decomposition of hydroperoxides (ROOH) and other compounds such as peroxides (ROOR), carbonyl compounds (RCOR), or other oxygen complexes of unsaturated lipids. This type of oxidation proceeds by normal free radical chain reactions (Chapter 1) and can be inhibited by chain-breaking antioxidants, or by ultraviolet deactivators (*e.g.* o-hydroxy-benzophenone) that absorb irradiation without formation of radicals. Direct photochemical oxidation of lipids is generally of little concern because light absorption at less than 220 nm cannot reach lipids unless they are exposed to direct sunlight or fluorescent light without suitable protection by glass or opaque plastic containers. However, in the presence of metals or metal complexes oxygen can become activated and can initiate oxidation either by formation of free radicals or singlet oxygen.

A. PHOTOSENSITIZED OXIDATION

Photooxidation provides an important way to produce hydroperoxides from unsaturated fatty acids and esters in the presence of oxygen, light energy and a photosensitizer. Pigments in foods can serve as photosensitizers by absorbing visible or near-UV light to become electronically excited. Sensitizers have two excited states, the singlet (^{1}Sens) and the triplet (^{3}Sens), which has a longer life-time and initiates photosensitized oxidation. Pigments initiating photosensitized oxidation in foods include chlorophyll, hemeproteins, and riboflavin.

Two types of sensitizers are recognized for photosensitized oxidation. These sensitizers can proceed by two types of processes. A Type I sensitizer serves as a photochemically activated free radical initiator. The sensitizer in the triplet state reacts with the lipid substrate by hydrogen atom or electron transfer to form radicals, which can react with oxygen (1). The hydroperoxides produced from the conjugated lipid radical are the same as those from free radical autoxidation. However, in contrast to the free radical oxidation, the photosensitized reaction is not inhibited by chain-breaking antioxidants (Chapter 8).

$$^3\text{Sens} + \text{LH} \xrightarrow{h\nu} [\text{intermediate}] + O_2 \qquad (1)$$

$$\downarrow$$

$$\text{hydroperoxides} + \text{Sens}$$

Riboflavin reacts with unsaturated fatty acid esters by type I photosensitized oxidation to produce the same isomeric hydroperoxides as free radical autoxidation.

The type II sensitizer in the triplet state interacts with oxygen by energy transfer to give singlet oxygen ($^1O_2$), which reacts further with unsaturated lipids (2). This type of photosensitized oxidation is also not inhibited by chain-breaking antioxidants.

$$^3\text{Sens} + O_2 \xrightarrow{h\nu} [\text{intermediate}] + {}^1\text{Sens} \longrightarrow {}^1O_2$$

$$\downarrow \text{LH} \qquad (2)$$

$$\text{hydroperoxides}$$

Chlorophyll, methylene blue and erythrosine react with unsaturated fatty esters by type II photosensitized oxidation in which singlet oxygen produces hydroperoxides by an entirely different mechanism from free radical autoxidation. In marked contrast to autoxidation, the distribution of hydroperoxides produced from oleate, linoleate and linolenate in the presence of singlet oxygen is very different and is discussed in Section B below.

Singlet oxygen can also be generated directly by non-photochemical processes, including the reaction of hydrogen peroxide and hypochlorous acid, by decomposition of triphenyl phosphite ozonides, and by radio-frequency gas discharge. Other sources of singlet oxygen may include the termination of peroxyl radicals via a tetroxide intermediate, known as the Russell mechanism (Chapter 1). The oxygen thus produced is activated in the singlet state. Another mechanism of singlet oxygen production is assumed in the decomposition of hydroperoxides by hemeproteins.

B. HYDROPEROXIDE FORMATION BY SINGLET OXYGEN

Oxygen in the triplet state becomes activated by electronic excitation to the singlet state. Because singlet oxygen is highly electrophilic, it reacts very rapidly with unsaturated fatty acids by a different mechanism than free radical autoxidation. Singlet oxygen reacts with methyl linoleate at a rate of about 1500 times faster than normal triplet oxygen. Because of this high reactivity, the hydroperoxides formed by singlet oxygen may play an important part in the initiation of free radical oxidation.

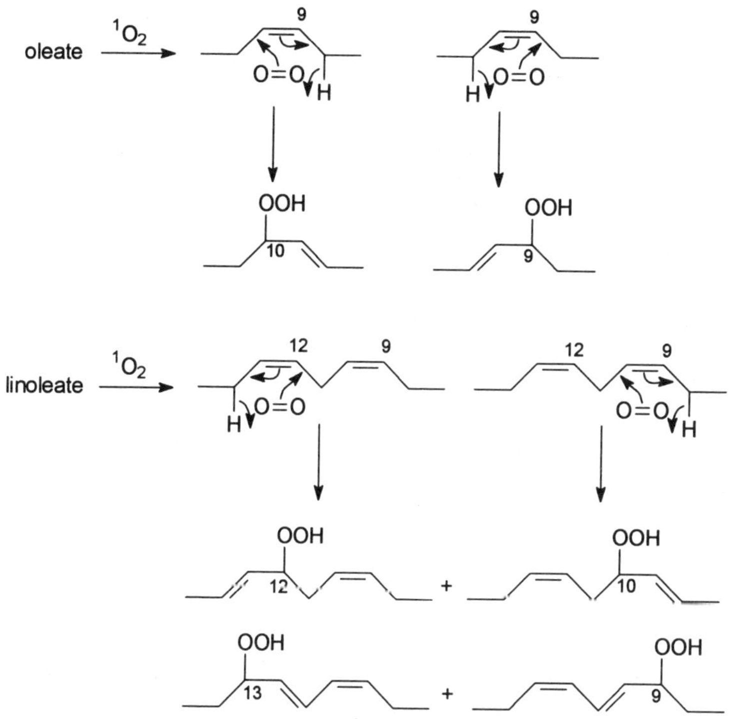

Figure 3-1. Oxidation of oleate and linoleate by singlet oxygen

Singlet oxygen reacts directly with the double bonds of unsaturated fatty acids by a concerted "ene" addition mechanism. Accordingly, oxygen is inserted at either end carbon of a double bond, which is shifted to an allylic position in the *trans* configuration. The resulting hydroperoxides have an allylic *trans* double bond (Figure 3-1). According to this mechanism, the hydroperoxide distribution is different from that of free-radical autoxidation (Chapter 2). Oleate produces a mixture of 9- and 10-hydroperoxides, linoleate a mixture of 9-, 10-, 12- and 13-hydroperoxides, and linolenate a mixture of 9-, 10-, 12-, 13-, 15- and 16-hydroperoxides. Oleate produces an equal mixture of 9- and 10-hydroperoxides as expected by the ene-addition mechanism (Table 3-1).

Linoleate treated with singlet oxygen produces a mixture of four hydroperoxide isomers:

9-hydroperoxy-*trans*-10-*cis*-12-octadecadienoate (*trans,cis*-9-OOH)
13-hydroperoxy-*cis*-9-*trans*-11-octadecadienoate (*cis,trans*-13-OOH)
10-hydroperoxiy-*trans*-8-*cis*-12-octadecadienoate (*trans,cis*-10-OOH)
12-hydroperoxy-*cis*-9-*trans*-13-octadecadienoate (*cis,trans*-12-OOH)

The *trans,cis*-9-OOH and the *cis,trans*-13-OOH are conjugated and have the same structures as those formed by free radical autoxidation (Chapter 2). However, the *trans,cis*-10-OOH and *cis,trans*-12-OOH are unconjugated and

TABLE 3-1.
Isomeric Hydroperoxides in Unsaturated Fatty Acid Esters Oxidized by Singlet Oxygen.

Hydroperoxides	Me Oleate	Me Linoleate	Me Linolenate
trans-9-OOH	48-51		
trans-10-OOH	49-52		
trans,cis-9-OOH		28	
cis,trans-13-OOH		34	
trans,cis-10-OOH		18	
cis,trans-12-OOH		21	
trans,cis,cis-9-OOH			20-23
trans,cis,cis-10-OOH			13
cis,trans,cis-12-OOH			12-14
cis,trans,cis-13-OOH			14-15
cis,cis,trans-15-OOH			12-13
cis,cis,trans-16-OOH			25-26

Sources: Frankel et al. (1980,1982) and Neff et al. (1980).

unique to singlet oxygenation. The external conjugated 9- and 13-hydroperoxides are formed in a ratio of 2 to 1 relative to the unconjugated internal 10- and 12-hydroperoxides. (Table 3-1). The lower concentration of the internal 10- and 12-hydroperoxides is attributed to the tendency of their peroxyl radicals to readily cyclize into a mixture of hydroperoxy epidioxides (Figure 3-2). The internal 10- and 12-hydroperoxides have the same *cis* homoallylic structure (where the hydroperoxide and the double bond are separated by a methylene group) as the internal 12- and 13-hydroperoxides of linolenate, and undergo rapid 1,3-cyclisation (Figure 2-7). Because cyclization of the pure hydroperoxides, prepared by photosensitized oxidation of linoleate, occurs rapidly in the dark at 0-5°C, this is evidently a free radical side reaction that is not photosensitized. The formation of hydroperoxy epidioxides is made possible as a consequence of the unique homoallylic structures of the 10- and 12-hydroperoxides of linoleate. In the presence of hydrogen donor antioxidants such as α-tocopherol, this free radical cyclization is inhibited and an even distribution is produced of the four 9-, 10-, 12- and 13-hydroperoxides.

Under conditions of photosensitized oxidation, the 9- and 13-hydroperoxides prepared by autoxidation of linoleate, produced six-membered cyclic peroxides by 1,4-addition of singlet oxygen to their conjugated diene systems (Figure 3-3). This reaction proceeds by photosensitized isomerisation of the *cis,trans*-conjugated diene hydroperoxides to the *trans,trans* configuration which can then assume a *cisoid conformation* that is required for 1,4-cycloaddition of singlet oxygen to the conjugated diene system of linoleate hydroperoxides.

Linolenate treated with singlet oxygen produces a mixture of six hydroperoxide isomers:

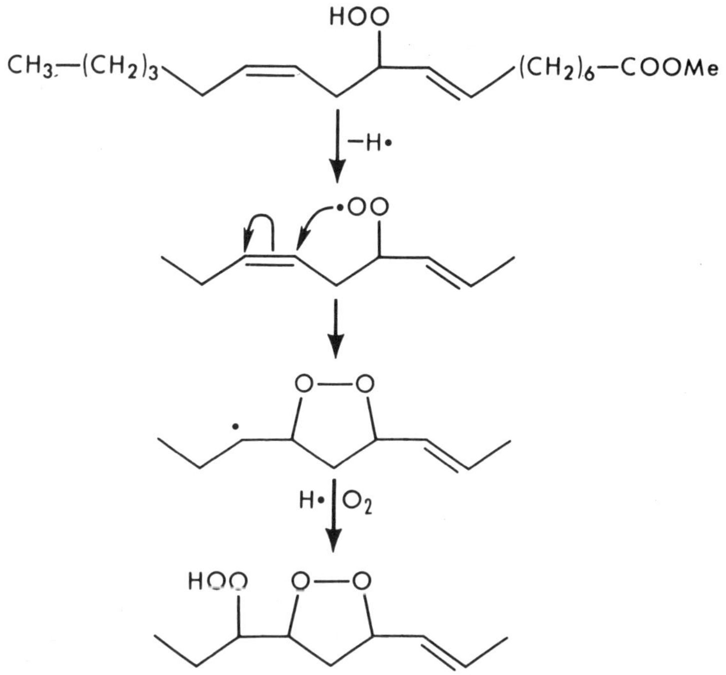

Figure 3-2. Cyclization of trans-8, cis-12-10-hydroperoxyoctadecadienoate produced by singlet oxidation of methyl linoleate (Frankel et al., 1982).

9-hydroperoxy-trans-10, cis-12, cis-15-octadecatrienoate
(trans,cis,cis-9-OOH)
10-hydroperoxy-trans-8, cis-12, cis-15-octadecatrienoate
(trans,cis,cis-10-OOH)
12-hydroperoxy-cis-9, trans-13, cis-15-octadecatrienoate
(cis,trans,cis-12-OOH)
13-hydroperoxy-cis-9, trans-11,cis-15-octadecatrienoate
(cis,trans,cis-13-OOH)
15-hydroperoxy-cis-9, cis-12, trans-16-octadecatrienoate
(cis,cis,trans-15-OOH)
16-hydroperoxy-cis-9, cis-12, trans-14-octadecatrienoate
(cis,cis,trans-16-OOH)

As with linoleate, the outer 9- and 16-hydroperoxides are formed in significantly higher proportion than the inner 10-, 12-, 13- and 15-hydroperoxides (Table 3-1). The inner hydroperoxide isomers have cis-homoallylic structures that lead to 1,3-cyclisation of their peroxyl radicals. The major hydroperoxy epidioxides formed are the same as those produced from the 12- and 13-hydroperoxyl radicals from autoxidized methyl linolenate (Figure 2-7). The 10- and 15-hydroperoxyl radicals have homoallylic unsaturation and undergo serial cyclizations to produce a mixture of

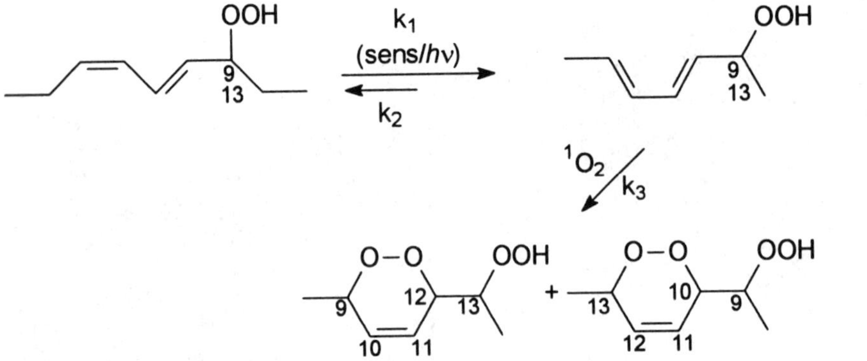

Figure 3-3. Formation of six-membered cyclic peroxides by photosensitized oxidation of 9- and 13-hydroperoxides of methyl linoleate (Neff et al., 1983). With permission of AOCS Press

bis-epidioxides (Figure 3-4). The formation of hydroperoxy bis-epidioxides is a free radical cyclization that occurs as a rapid side-reaction, only because of the unique homoallylic structures of the 10- and 15-hydroperoxides produced from linolenate treated with singlet oxygen. The photosensitized oxidation of methyl linolenate produces also small amounts of hydroperoxy bicyclo-endoperoxides (with the OOH group on either carbon-9 or carbon-16) (Figure 2-11), and isomeric mixtures of dihydroperoxides (9,12-, 10,12-, 13,15-, 13,16- 10,16-, 9,15- and 9,16-) (Chapter 4, Figure 4-5).

Under conditions of photosensitized oxidation, the mixture of 9-, 12-, 13- and 16-hydroperoxides prepared by autoxidation of methyl linolenate produced cyclic peroxides and hydroperoxy bis-cyclic peroxides containing five- and six-membered rings. The six-membered rings are formed by 1,4-cycloaddition of singlet oxygen to the conjugated 1,3-diene systems of the terminal 9- and 16-hydroperoxides after *trans,trans* isomerization (Figure 3-5). The *trans,cis*-diene hydroperoxides undergo photosensitized isomerization to the *trans,trans* configuration which assumes a *cisoid* conformation as required for cyclo-addition with singlet oxygen. The 1,4-cyclization of the 9- and 16-hydroperoxides is less favorable than the free radical cyclization of the 12- and 13-hydroperoxides of linolenate because the five-membered hydroperoxy mono-epidioxides are formed in higher proportion than the six-membered hydroperoxy mono-epidioxides. The hydroperoxy bis-epidioxides are formed by cyclo-addition of singlet oxygen to the conjugated diene systems of the hydroperoxy mono-epidioxides derived from the 12- and 13-hydroperoxides after *trans,trans* isomerization (Figure 3-6).

C. INHIBITION OF PHOTOSENSITIZED OXIDATION

Plants containing chlorophyll are well protected by carotenoids against the atmospheric damage of photosensitized oxidation. The same mechanism of protection is generally invoked for the inhibition of photosensitized lipid

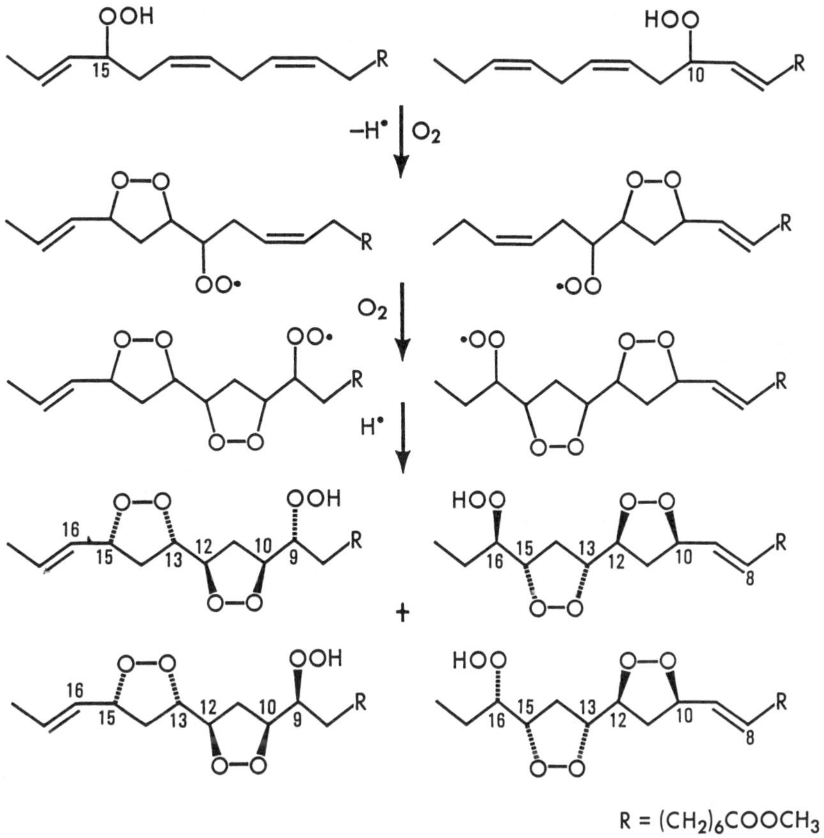

Figure 3-4. Serial cyclization of 10- and 15-hydroperoxides from photosensitized oxidized methyl linolenate (Neff et al., 1982).

oxidation by carotenoids. Carotenoids protect unsaturated lipids against photosensitized oxidation by interfering with the activation of triplet oxygen into singlet oxygen. This quenching effect takes place by an energy transfer mechanism from singlet oxygen to carotene. By a similar energy transfer mechanism, carotenoids also react with the triplet state of the excited sensitizers (Figure 3-7). α-Tocopherol is highly reactive toward singlet oxygen and inhibits photosensitized oxidation by quenching singlet oxygen and by forming stable addition products.

The *chemical quenching* by the reaction with singlet oxygen to produce stable products, can be distinguished from the *physical quenching* without undergoing chemical changes. The rate of physical quenching by β-carotene approaches diffusion rate (1.5×10^{10} M^{-1} sec^{-1}) and is much greater than that for α-tocopherol (Table 3-2). On the other hand, the rate of chemical quenching is much greater for α-tocopherol than that for β-carotene. γ-Tocopherol and δ-tocopherol, present in soybean oil at much greater

Figure 3-5. Formation of six-membered cyclic peroxides by photosensitized oxidation of 9- and 16-hydroperoxides of methyl linolenate (Neff and Frankel, 1984). With permission of AOCS Press.

Figure 3-6. Formation of five- and six-membered cyclic peroxides by photosensitized oxidation of 9- and 16-hydroperoxy epidioxides of methyl linolenate (Neff and Frankel, 1984). With permission of AOCS Press.

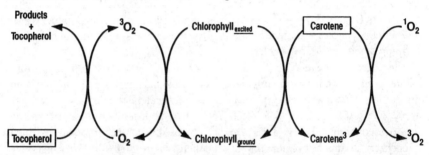

Figure 3-7. Inhibition of photosensitized oxidation by carotene and tocopherol.

TABLE 3-2.
Rates of quenching of singlet oxygen by tocopherols and β-carotene (in $M^{-1}sec^{-1}$).

Type of quenching	α-Tocopherol	γ-Tocopherol	δ-Tocopherol	β-Carotene
Physical	2.6×10^8	1.8×10^8	1.0×10^8	1.5×10^{10}
Chemical	6.6×10^6	2.6×10^6	0.7×10^6	0.8×10^2

Source: Matsushita and Terao (1980)

concentrations than α-tocopherol (Chapter 8), are less efficient singlet oxygen quenchers than α-tocopherol.

The inhibition of photosensitized oxidation by β-carotene is complicated because it is highly susceptible to autoxidation and is quickly destroyed in the presence of radicals or hydroperoxides. To be effective in unsaturated lipids exposed to light irradiation, β-carotene must be protected by an antioxidant. In soybean oil containing natural tocopherols, β-carotene was effective in protecting against light oxidation (Table 3-3). With no citric acid, 20 ppm of β-carotene improved the flavor and oxidative stability of soybean oil; with citric acid, 5-10 ppm of β-carotene was sufficient to show an improvement. On the other hand, in distilled soybean methyl esters containing no natural tocopherols, a larger amount of β-carotene (1000 ppm) was necessary to show a protective effect against light oxidation (Figure 3-8). α-Tocopherol and β-carotene were effective singlet oxygen quenchers, but α-tocopherol was more effective in protecting soybean esters against light oxidation. Because β-carotene was rapidly bleached, its effect was confounded by its depletion

TABLE 3-3.
Effect of β-carotene on flavor and oxidative stability of light-exposed soybean oil.

β-carotene ppm	Flavor score 8 hr	Flavor score 16 hr	Peroxide value 8 hr	Peroxide value 16 hr
no citric acid				
0	5.0	4.3	1.3	3.0
1	5.2	4.2	1.0	2.0
5	5.5	4.5	0.6	1.8
10	5.7	4.3	1.0	2.2
15	5.5	4.7	1.1	1.4
20	6.5	6.0	1.1	2.1
+100 ppm citric acid				
0	5.4	5.4	2.2	3.5
1	6.2	5.9	0.9	3.3
5	6.8	6.3	0.4	2.9
10	6.7	6.3	1.1	0.7
15	6.2	6.4	0.9	1.3
20	6.2	6.4	1.0	2.3

Source: Warner and Frankel (1987); samples exposed to fluorescent lighting of 7535 lux; flavor scores are based on a 1-10 intensity scale, with 10 = bland and 1 = strong; peroxide values in meq/kg

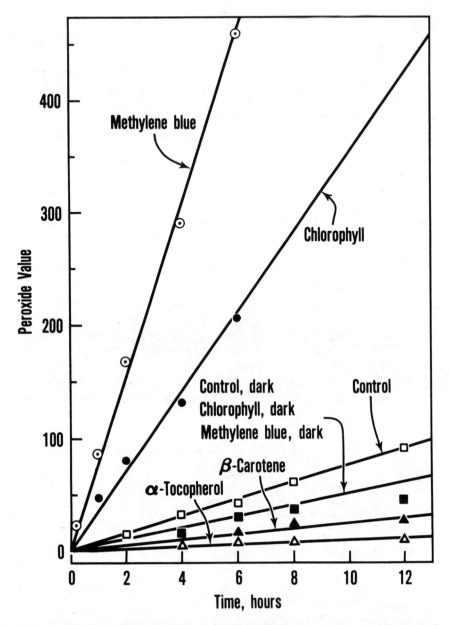

Figure 3-8. Effect of methylene blue and chlorophyll as photosensitizers, and β-carotene and α-tocopherol as singlet oxygen quenchers, on the photo-oxidation of distilled soybean esters at 35°C (Frankel et al., 1979). With permission of AOCS Press.

during oxidation. When soybean oil was stored in the dark, β-carotene at high concentrations promoted autoxidation, apparently due to its oxidation products.

In soybean oil and rapeseed oils chromatographed to remove the natural tocopherols, β-carotene by itself acted as a prooxidant by promoting oxidation. However, β-carotene effectively inhibited light oxidation in the presence of added δ- or γ-tocopherols. The mixture of β-carotene and γ-tocopherol was more effective than γ-tocopherol alone in inhibiting light oxidation of chromatographed rapeseed oil. Therefore, β-carotene is an effective inhibitor of photosensitized oxidation in vegetable oils only when it is protected against oxidation or reinforced by natural tocopherols.

In addition to the singlet oxygen quenching property of carotenoids, other mechanisms have been invoked for their inhibition of photosensitized oxidation. Carotenoids may either act as built-in filters for light of short wavelengths, or by being preferentially oxidized they may exert a sparing effect until they are completely oxidized. There is good evidence, however, that the main effect of carotenoids is their singlet oxygen quenching, because the rate of this quenching is a function of the number of conjugated double bonds in their chain. Carotenoids with nine or more double bonds are much more efficient than those with seven or less double bonds. This is the range in which the carotenoids are most effective in inhibiting photosensitized oxidation.

Cold pressed virgin olive oil and unprocessed vegetable oils obtain their color from natural carotenoids. However, carotenoids are completely removed by bleaching and deodorization during processing of vegetable oils. Therefore, for protection against light oxidation β-carotene must be added after deodorization in oils containing natural tocopherols, at concentrations below 20 ppm provided the resulting color is not objectionable to the consumer.

BIBLIOGRAPHY

Chan,H.W.-S. Photo-sensitized oxidation of unsaturated fatty acid methyl esters. The identification of different pathways. *J. Am. Oil Chem. Soc.* **54**, 100-104 (1977).
Chan,H.W.-S. and Coxon, D.T. Lipid hydroperoxides, in *Autoxidation of Unsaturated Lipids*, pp. 30-31 (1987) (Edited by H.W.-S. Chan), Academic Press, London.
Clements,A.H., Van Den Engh,R.H., Frost,D.J., Hoogenhaut,K. and Nooi,J.R. Participation of singlet oxygen in the photosensitized oxidation of 1,4-dienoic acid systems and photooxidation of soybean oil. *J.Am. Oil Chem. Soc.* **50**, 325-330 (1973).
Foote,C.S. Photosensitized oxygenation and the role of singlet oxygen. *Accounts Chem. Revs.* **1**, 104-110 (1968).
Foote,C.S. Photosensitized oxidation and singlet oxygen: Consequences in biological systems, in *Free Radicals in Biology*, Vol. II, pp. 85-133 (1976) (Edited by W.A. Pryor), Academic Press, New York.
Foote,C.S. Mechanisms of photo-oxidation, in *Singlet Oxygen. Reactions with Organic Compounds & Polymers*, pp. 135-146 (1978) (Edited by B. Rånby and J.F. Rabek), John Wiley & Sons, Chichester.
Frankel,E.N., Neff,W.E. and Bessler,T.R. Analysis of autoxidized fats by gas chromatography-mass spectrometry: V. Photosensitized oxidation. *Lipids* **14**, 961-967 (1979).
Frankel,E.N. Lipid oxidation. Prog. *Lipid Res.* **19**, 1-22 (1980).
Frankel,E.N., Neff,W.E., Selke,E. and Weisleder,D. Photosensitized oxidation of methyl linoleate. Secondary and volatile thermal decomposition products. *Lipids* **17**, 11-18 (1982).
Frankel,E.N. Chemistry of free radical and singlet oxidation of lipids. Prog. *Lipid Res.* **23**, 197-221 (1985).
Frankel,E.N. The antioxidant and nutritional effects of tocopherols, ascorbic acid and beta-carotene in relation to processing of edible oils. *Bibliotheca Nutritio et Dieta. Basel, Krager* **43**, 297-312 (1989).

Frimer,A.A. Singlet oxygen in peroxide chemistry, in *The Chemistry of Functional Groups, Peroxides*, pp. 201-234 (1983) (edited by S. Patai), John Wiley & Sons Ltd., New York.

Haila,K. And Heinonen,M. Action of β-carotene on purified rapeseed oil during light storage. *Lebensm.-Wiss. u. Technol.* **27**, 573-577 (1994).

Hawco,F.J., O'Brien,C.R. and O'Brien,P.J. Singlet oxygen during hemoprotein catalyzed lipid peroxide decomposition. *Biochem. Biophys. Res. Commun.* **76**, 354-361 (1977).

Korycka-Dahl,M.B. and Richardson,T. Activated oxygen species and oxidation of food constituents. *CRC Critical Revs. Food Sci. & Nutrition* **10**, 209-241 (1978).

Lee,S-H. and Min,D.B. Effects, quenching mechanisms, and kinetics of carotenoids in chlorophyll-sensitized photooxidation of soybean oil. *J. Agric. Food Chem.* **38**, 1630-1634 (1990).

Matsushita,S. and Terao,J. Singlet oxygen-initiated photooxidation of unsaturated fatty acid esters and inhibitory effects of tocopherols and β-carotene, in *Autoxidation in Food and Biological Systems*, pp. 27-44 (1980) (Edited by M. Simic and M. Karel), Plenum Press, New York.

Neff,W.E. and Frankel,E.N. Quantitative analyses of hydroxystearate isomers from hydroperoxides by high-pressure liquid chromatography of autoxidized and photosensitized-oxidized fatty esters. *Lipids* **15**, 587-590 (1980).

Neff,W.E., Frankel,E.N. and Weisleder,D. Photosensitized oxidation of methyl linoleate. Secondary products. *Lipids* **17**, 780-790 (1982).

Neff,W.E., Frankel,E.N. and Weisleder,D. Photosensitized oxidation of methyl linoleate monohydroperoxides: Hydroperoxy cyclic peroxides, dihydroperoxides, keto esters and volatile thermal decomposition products. *Lipids* **18**, 868-876 (1983).

Neff,W.E. and Frankel,E.N. Photosensitized oxidation of methyl linolenate monohydroperoxides: Hydroperoxy cyclic peroxides, dihydroperoxides and hydroperoxy bis-cyclic peroxides. *Lipids* **19**, 952-957 (1984).

Rawls,H.R. and van Santen,P.J. A possible role of singlet oxygen in the initiation of fatty acid autoxidation. *J. Am. Oil Chem. Soc.* **47**, 121-125 (1970).

Sattar,A., deMan,J.M. and Akexander,J.C. Light-induced oxidation of edible oils and fats. *Lebensm.-Wiss. u. Technol.* **9**, 149-152 (1976).

Terao,J. and Matsushita,S. J. The isomeric compositions of monohydroperoxides produced by oxidation of unsaturated fatty acid esters with singlet oxygen. *Food Process. Preserv.* **3**, 329-337 (1980).

Warner,K. and Frankel,E.N. Effect of β-carotene on light stability of soybean oil. *J. Am. Oil Chem. Soc.* **64**, 213-218 (1987).

CHAPTER 4

HYDROPEROXIDE DECOMPOSITION

A complex mixture of monomeric, polymeric and small molecular weight volatile materials is produced when lipid oxidation is carried out to higher levels than 5-10% especially with polyunsaturated fats containing more than two double bonds. Complicated pathways are recognized for hydroperoxide decomposition. However, the extensive literature on the nature of secondary reactions must be interpreted with care because the conditions used in many studies are either too drastic or artificial and not relevant to real food lipid applications. In complex foods, the interaction of lipid hydroperoxides and secondary oxidation products with proteins and other components has a significant impact on oxidative and flavor stability and texture during processing, cooking and storage. The volatile decomposition products contribute to the flavor and may affect the safety of lipid-containing foods by causing damage to proteins and enzymes.

A. MONOMERIC PRODUCTS

On prolonged oxidation oleate, linoleate and linolenate produce a multitude of secondary oxidation products.

1. Oleate

Methyl oleate produces small amounts of allylic keto-oleates (with CO on carbons 8-, 9-,10- and 11), epoxy-stearate or epoxy-oleates (8,9-, 9,10- and 10,11-epoxy), dihydroxy-oleates (8,9- 9,10-, and 10,11-diOH) and dihydroxy-stearates (between carbon-9 and carbon-11). The allylic keto-oleates may be derived by dehydration of the corresponding hydroperoxides. 9,10-Epoxy-stearate may be produced by the reaction of oleate and the hydroperoxides. The other epoxy products can be formed by cyclization of an alkoxy radical formed from the corresponding hydroperoxides of oleate (Figure 4-1). Accordingly, the 11-hydroperoxide forms the 10,11-epoxy ester, the 8-hydroperoxide forms the 8,9-epoxy ester, and the 9- and 10-hydroperoxides form the 9,10-epoxy ester. The 1,2- and 1,4-dihydroxy esters may be formed from a similar alkoxyl radical that undergoes hydroxyl and hydrogen radical substitution via an allylic hydroxy ester radical (Figure 4-1).

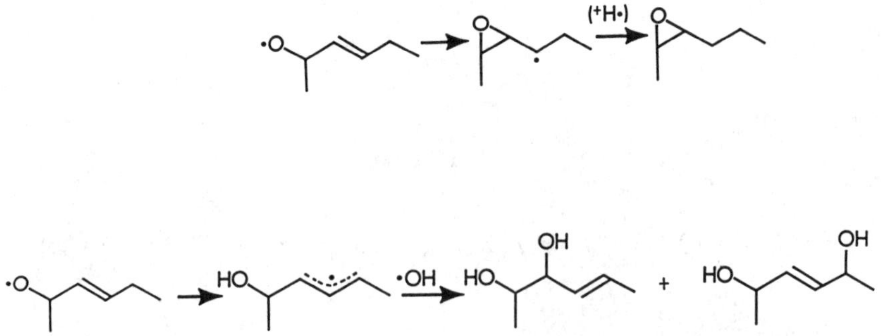

Figure 4-1. Formation of epoxystearate from and dihydroxy oleate from oleate hydroperoxides.

2. Linoleate

A large number of compounds have been identified from autoxidized methyl linoleate and further oxidation of linoleate hydroperoxides, including keto-linoleate (with CO on carbon-9 and carbon-13), epoxyhydroxy-oleate (9,10-12,13-epoxy-11-hydroxy and 9,13-hydroxy-12,13-/9,10-epoxy), 9,13-dihydroxy, and trihydroxy (9,10,13- and 9,12,13-) stearate. The epoxy allylic hydroxy, hydroperoxy and keto products can be formed by cyclization of an alkoxy radical formed from the corresponding hydroperoxides of linoleate (Figure 4-2). The availability of the almost pure 13-hydroperoxide isomer produced from linoleic acid by soybean lipoxygenase has made it possible to clarify the mechanism of decomposition of linoleate hydroperoxides under different conditions. Decomposition of the 13-hydroperoxide isomer of linoleic acid in the presence of various free radical initiators or decomposers (*e.g.* hemoglobin, iron compounds, di-alkyl peroxy oxalate) produces 13-keto and 13-hydroxy dienes and mixtures of 11-hydroxy-12,13-epoxy-9-oleate and 9-hydroxy-12,13-epoxy-10-oleate isomers. The same hydroxy epoxy compounds are produced from autoxidized linoleic acid containing a mixture of 9- and 13-hydroperoxide isomers. Other polyoxygenated compounds formed by thermal decomposition of linoleate hydroperoxides include di- and tri-hydroxy esters, which are apparently formed from the corresponding dihydroperoxides, keto-hydroperoxides and hydroxy-hydroperoxides.

3. Linolenate

The formation of hydroperoxy epidioxides in significant yields during autoxidation of methyl linolenate was discussed previously (Chapter 2, Figure 2-7). The peroxyl radicals leading to these cyclic peroxides can undergo a second cyclization to produce bicyclo-endoperoxides (Figure 2-8). These cyclic peroxides are precursors of malonaldehyde, a volatile decomposition product characteristic of oxidation of polyunsaturated lipids containing three or more double bonds (Chapter 2).

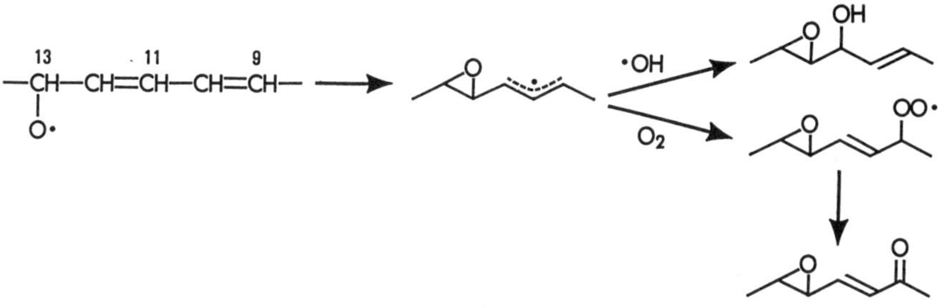

Figure 4-2. Formation of epoxy allylic hydroxy, epoxy allylic hydroperoxy and epoxy allylic keto oleate from linoleate hydroperoxides.

Many precursors of malonaldehyde have been suggested to explain what is measured by the thiobarbituric acid (TBA) test, a popular method to measure rancidity in polyunsaturated lipids (Chapter 5). The cleavage of monocyclic peroxides of linolenate and arachidonate on each side of the endoperoxide ring to form malonaldehyde was proposed as the main mechanism for the formation of the red colored TBA-malonaldehyde adduct (Figure 4-3). The formation of pure malonaldehyde was tested by a GC procedure measuring the tetramethyl acetal derivative, which is stable under the GC conditions. Several purified oxidation products of methyl linoleate and linolenate subjected to either free radical autoxidation or photosensitized oxidation were tested as precursors of malonaldehyde under the acidic methanol conditions used for

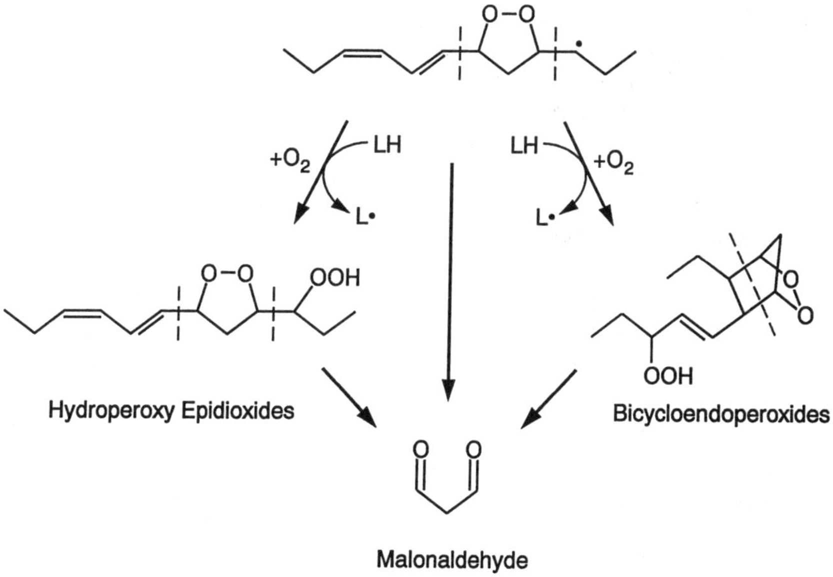

Figure 4-3. Formation of malonaldehyde from hydroperoxy epidioxides and Bicyclo-endoperoxides of methyl linolenate (Pryor et al., 1976).

TABLE 4-1.
Lipid oxidation precursors of malonaldehyde.

Oxidation products	Malonaldehyde (MDA), mol %	TBARS as MDA, mol %
Monohydroperoxides		
Linoleate, $^3O_2$	0.18	0.53
Linoleate, $^1O_2$	0.26	0.18
Linolenate, $^3O_2$	0.22	1.85
Linolenate, $^1O_2$	0.44	1.41
Hydroperoxy epidioxides		
Linoleate, $^1O_2$	24.2	1.64
Linolenate, $^3O_2$	2.34	4.02
Linolenate, $^1O_2$	2.20	2.49
Hydroperoxy bis-epidioxides[a]	5.65	10.5
Hydroperoxy bicycloendoperoxides[a]	4.14	3.41
10,12-dihydroperoxides[a]	4.76	3.77
9,12- & 13,16-dihydroperoxides[a]	0.81	6.29
9,16-dihydroperoxides[a]	0.0	7.63
10,16-dihydroperoxides[a]	0.0	2.74

From: Frankel and Neff (1983).
$^3O_2$: produced by autoxidation; $^1O_2$: produced by oxidation with singlet oxygen.
[a] Products of oxidized methyl linolenate.

acetalation (Table 4-1). Under these acidic decomposition conditions, the hydroperoxy epidioxides of linoleate were the most significant precursors of malonaldehyde followed by the hydroperoxy bis-epidioxides of linolenate, 1,3-dihydroperoxides of linolenate (see below), hydroperoxy bicyclo-endoperoxides of linolenate, and the hydroperoxy epidioxides of linolenate. The monohydroperoxides of linoleate and linolenate were not significant sources of malonaldehyde. Completely different trends of malonaldehyde formation were observed by the TBA test.

A mixture of 9,12-, 9,16- and 13,16-dihydroperoxides is formed in smaller yields than the cyclic peroxides by secondary oxidation of the pentadienyl systems of the 9- and 16-hydroperoxides of methyl linolenate. The 12,15-pentadienyl radical produced by hydrogen abstraction on carbon-14 of the 9-hydroperoxide reacts with oxygen at the end carbon positions to form the 9,12- and 9,16-dihydroperoxides (Figure 4-4). The corresponding pentadienyl radical produced by hydrogen abstraction on carbon-11 of the 16-hydroperoxide of linolenate reacts with oxygen to form the 9,16- and 13,16-dihydroperoxides. The 9,12- and 13,16-dihydroperoxides have a *cis,trans* conjugated diene system, while the 9,16-dihydroperoxide has a *trans,cis,trans* conjugated triene system. In the presence of 10% α-tocopherol, the 9,16-dihydroperoxide is produced selectively from methyl linolenate. As discussed previously (Chapter 3), the hydrogen donating ability of α-tocopherol inhibits the cyclization reaction, thereby promoting dihydroperoxidation. Minor secondary oxidation products of methyl linolenate include epoxy-hydroxydienes and epoxy-hydroperoxydienes.

OOH
16
11
-H•
OOH
16
OOH OOH
16
13
O2 + H•
OOH
16
OOH
9
OOH
9
14
-H•
OOH
9
OOH OOH
12
9

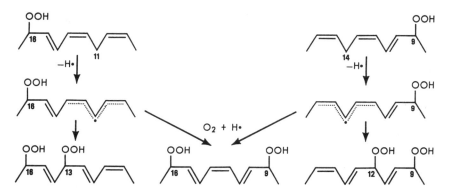

Figure 4-4. Formation of dihydroperoxides from autoxidized methyl linolenate (Neff *et al.*, 1981).

B. OLIGOMERIC PRODUCTS OF METHYL LINOLEATE AND METHYL LINOLENATE

Dimeric compounds are formed during the autoxidation of methyl linoleate and linoleate hydroperoxides under ambient conditions. Dimers formed from methyl linoleate hydroperoxides are composed of unsaturated fatty esters cross-linked through either peroxide or ether linkages and containing hydroperoxy, hydroxy or oxo substituents (Figure 4-5).

Peroxy-linked dimers are also formed from linoleate hydroperoxides in the presence of free radical initiators and copper palmitate, and carbon-carbon linked dimers in the presence of copper catalysts. Decomposition of methyl linoleate hydroperoxides at 210°C under nitrogen produces mainly carbon-carbon linked dimers (82%), monomers with loss of diene conjugation, volatile compounds (4-5%) and water. The resulting dimers contain carbonyl and hydroxyl groups and double bonds scattered between carbon 8 and carbon 10. Linoleate hydroperoxides can dimerize by one of the termination reactions discussed in Chapter 1. The termination reactions involving combination of either alkyl, alkoxyl, or peroxyl radical intermediates produce dimers with

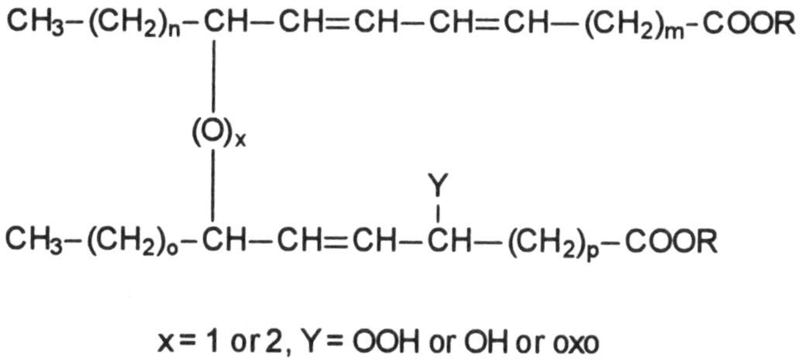

CH3-(CH2)n-CH-CH=CH-CH=CH-(CH2)m-COOR
(O)x
Y
CH3-(CH2)o-CH-CH=CH-CH-(CH2)p-COOR
x = 1 or 2, Y = OOH or OH or oxo

Figure 4-5. Dimers from methyl linoleate hydroperoxides.

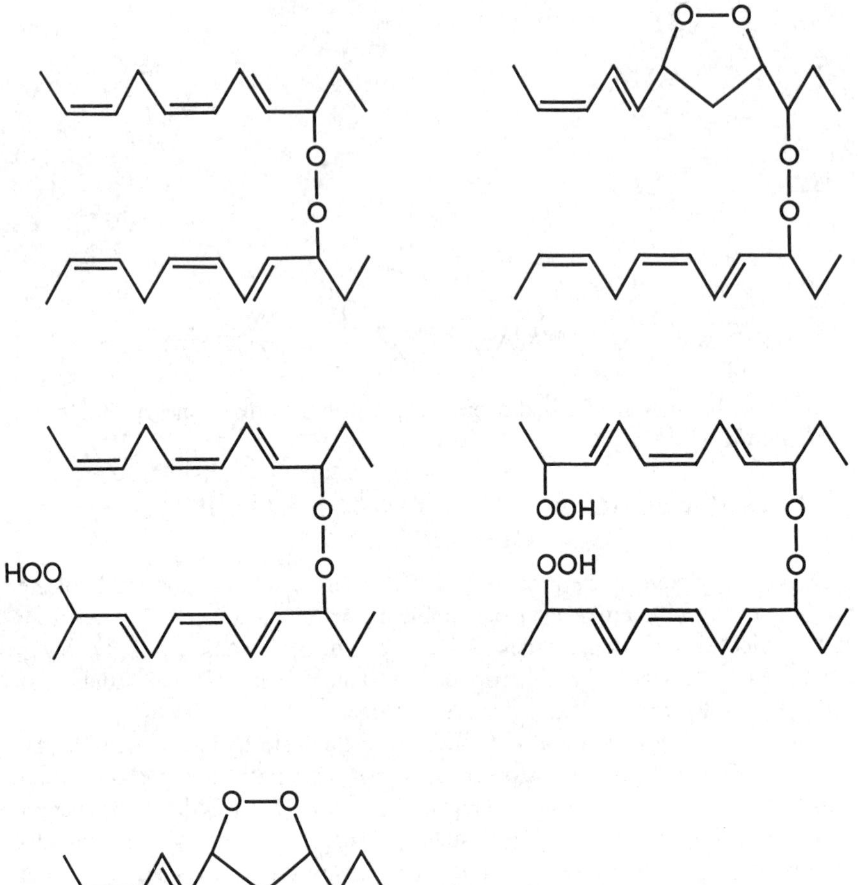

Figure 4-6. Dimers from autoxidized methyl linolenate (Neff *et al.*, 1988).

either carbon-carbon, carbon ether, or peroxy links. The carbon-carbon and carbon-oxygen linked dimers are favored at elevated temperatures and the peroxy-linked dimers at ambient temperatures. The peroxy-linked dimers may also decompose to the ether-linked and carbon-carbon linked dimers via the corresponding alkyl and alkoxyl radical intermediates.

Dimers and oligomers are significant secondary products occurring in methyl linolenate highly oxidized under mild conditions (25-40°C). These secondary products include conjugated diene-triene, dihydroperoxides or hydroperoxy epidioxide units cross-linked with peroxide groups (Figure 4-6). Dimers are produced in yields of 6-8% from methyl linolenate autoxidized at 40°C. These dimers are derived from hydroperoxides (80%) and from hydroperoxy epidioxides (20%), and are 90% peroxide-linked and 10% ether- and/or carbon-linked. Oligomers are the principal products formed by decomposition of pure hydroperoxides of methyl linolenate in the presence of ferric chloride and ascorbic acid (73%, molecular weight of 625), and thermally at 150°C (62%, molecular weight of 834). These oligomers contain monomeric units linked by either ether or carbon-carbon bonds (57%) or by peroxide bonds (43%). The peroxide-linked dimers can be easily cleaved thermally and produce volatile compounds contributing to flavor deterioration of linolenate-containing oils (Section D).

C. OLIGOMERIC PRODUCTS OF TRIACYLGLYCEROLS

In contrast to methyl linoleate and its hydroperoxides, which tend to dimerize readily under ambient conditions, highly oxidized trilinoleoylglycerol or the mono-hydroperoxides from trilinoleoylglycerol (Chapter 2) do not dimerize under the same conditions. Dimerization at ambient conditions is evidently significant only in the methyl esters of unsaturated fatty acids, because intermolecular condensations of peroxyl radicals are favored. On the other hand, further oxidation of the monohydroperoxides of trilinoleoylglycerol to bis- and tris-hydroperoxides appears to be the preferred reaction (Chapter 2, Section F). Intramolecular hydrogen abstraction from the linoleoyl residues can evidently occur more favorably than intermolecular condensation of the peroxyl radicals to form dimers. There is also no evidence for dimerization of tris-hydroperoxides of trilinoleoylglycerol. Alternate pathways of decomposition of oxidized triacylglycerols may be more favorable. Therefore, the simple esters of unsaturated fatty acids do not necessarily provide valid models for the oxidative dimerization of unsaturated triacylglycerols. The extensive use of methyl linoleate as a model lipid in decomposition studies may not apply to triacylglycerols and may not be relevant to fats and oils in general.

Although highly oxidized polyunsaturated triacylglycerols do not undergo dimerization and oligomerization under mild temperatures, these reactions become significant in oxidized soybean oil and other vegetable oils at the elevated temperatures of frying (Chapter 11), and of deodorization at 210°C and above. Column chromatographic analyses of refined-bleached and deodorized soybean oils show the presence of 1 to 3% high-molecular weight compounds consisting of both unpolar non-oxygenated and polar oxygenated fractions with average molecular weight of dimers. The non-oxygenated dimers, referred to as "thermal dimers", and the oxygenated dimers, referred to as "oxidative dimers", are complex high-molecular-weight decomposition mixtures that are not well characterized. Partial separation and analyses by size-exclusion chromatography show triacylglycerols dimeric compounds ranging from 0.5 to 2.8% in vegetable oils oxidized to high peroxide values (100-400). These high molecular-weight compounds found in oxidized-deodorized vegetable and fish oils contain significant amounts of carbonyl function that can also can be estimated by the anisidine value, which provides another measure of non-volatile carbonyl compounds (Chapter 5).

The presence of increasing levels of high-molecular weight compounds in vegetable oils is related to the degree of oxidation prior to deodorization. This type of oxidation occurring during processing has been referred to as "hidden oxidation" because the hydroperoxides are completely decomposed at the elevated temperatures of deodorization into dimers and the resulting deodorized oils have a peroxide value of zero. The high-molecular-weight materials having an average molecular weight of "dimers" decrease the oxidative stability of the oils upon further storage. Dimers and oligomers can be measured in these oils by either silicic acid column chromatography or by size exclusion chromatography to obtain information on how much oxidative damage occurs during processing. These measurements can be used to estimate the future oxidative stability of fats and oils (Chapters 5 and 6).

D. VOLATILE PRODUCTS

Flavor deterioration of food lipids is caused mainly by the presence of volatile lipid oxidation products that have an impact on flavor at extremely low concentrations, often below 1 ppm. An understanding of the sources of volatile oxidation products provides the basis for improved methods to control and evaluate flavor deterioration. The decomposition of lipid hydroperoxides produces carbonyl compounds, alcohols and hydrocarbons under various conditions of elevated temperatures and in the presence of metal catalysts.

Thermal decomposition of monohydroperoxides produces alkoxyl radicals undergoing homolytic β-scission to form aldehydes, alkyl and olefinic radicals (Figure 4-7). The alkyl radicals form either a hydrocarbon by hydrogen abstraction or an alcohol by reacting with a hydroxyl radical. Similarly, the olefinic radical reacts either with a hydrogen radical to form an olefin or reacts with a hydroxyl radical to produce 1-enols which tautomerize into saturated aldehydes. The formation of lower aldehydes proceeds by reaction of the alkyl radicals with oxygen to produce primary hydroperoxides, which decompose further to produce aldehydes via an alkoxyl radical or cleave to produce formaldehyde and a lower alkyl radical containing one less carbon.

The main volatile decomposition products formed from oleate, linoleate and linolenate are those expected from the cleavage of the alkoxyl radicals formed from the hydroperoxides of autoxidized and photosensitized oxidized fatty esters (Figure 4-7). Gas chromatography (GC) by direct injection (Chapter 5) is a useful technique used to study the volatile decomposition products of hydroperoxides that can be identified by GC-MS. This technique of anaerobic pyrolysis eliminates or minimizes secondary reactions between hydroperoxides

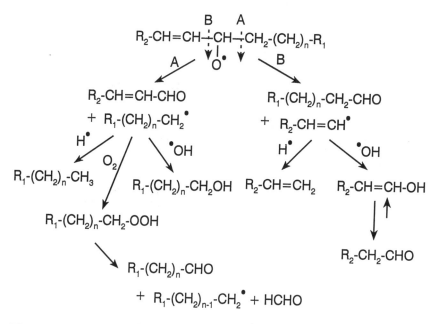

Figure 4-7. Homolytic β-scission of fatty ester monohydroperoxides.

and unsaturated fatty esters that would confound the interpretation of results. However, the volatile compounds analysed are those best amenable to the GC-MS technique used for analysis. The quantitative analyses are also affected by the thermal stability of the volatile aldehydes produced. A further complication arises from the ready interconversion of isomeric hydroperoxides that occurs as a side-reaction during thermal decomposition under the GC conditions.

With autoxidized oleate, decanal and 2-undecenal are produced by homolytic cleavages A and B on either side of the alkoxyl intermediate group from the 8-hydroperoxide, nonanal from either the 9- or 10-hydroperoxide, octane from the 10-hydroperoxide, and octanal and heptane from the 11-hydroperoxide (Figure 4-8). A mixture of 8-, 9-, 10- and 11-oxo esters is produced by cleavage A on the ester side of the hydroperoxides. With the monohydroperoxides of triacylglycerols containing oleate, the corresponding oxo glycerides are formed; these high-molecular weight nonvolatile aldehydo-glycerides, referred also as "core" aldehydes, are useful predictors of quality of edible oils (Chapter 5).

The hydroperoxides of oleate subjected to photosensitized oxidation not only produced the volatiles expected from the 9- and 10-isomers but also those derived from the 8- and 11-isomers. These results are explained by the interconversion between the 9- and 10-hydroperoxides of oleate formed by photosensitized oxidation into a mixture of 8-, 9-, 10-, and 11-hydroperoxides (Figure 4-9). Although both types of hydroperoxide produce the same major volatiles, the photosensitized oxidation-derived hydroperoxides produced more of the volatiles derived from the 9- and 10-hydroperoxide isomers (octane, 1-

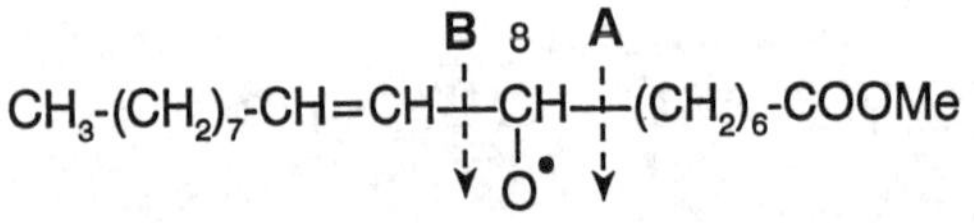

$$CH_3\text{-}(CH_2)_7\text{-}CH=CH\underset{\downarrow}{\overset{B\ 8}{|}}CH\underset{O^\bullet}{\overset{A}{|}}(CH_2)_6\text{-}COOMe \quad \text{8-OOH}$$

Cleavage A: 2-undecenal (2%, 7%) + Me heptanoate (2%, 5%)
Cleavage B: decanal (4%, 2%) + Me 8-oxooctanoate (4%, 3%)

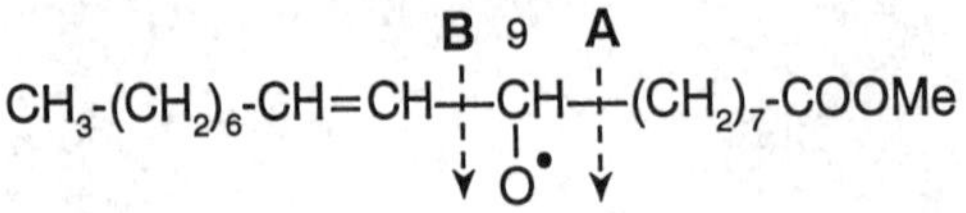

$$CH_3\text{-}(CH_2)_6\text{-}CH=CH\underset{\downarrow}{\overset{B\ 9}{|}}CH\underset{O^\bullet}{\overset{A}{|}}(CH_2)_7\text{-}COOMe \quad \text{9-OOH}$$

Cleavage A: 2-decenal (5%, 12%) + Me octanoate (5%, 10%)
Cleavage B: nonanal (15%, 10%) + Me 9-oxononanoate (15%, 11%)

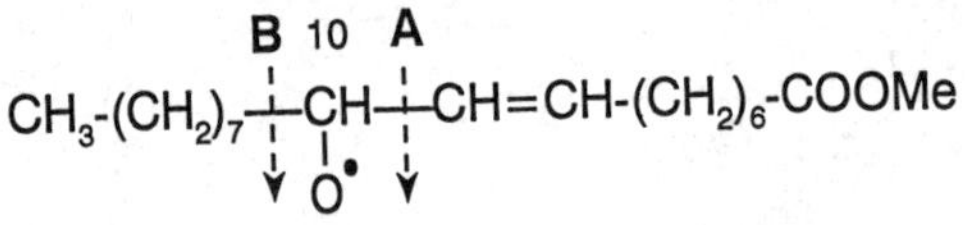

$$CH_3\text{-}(CH_2)_7\underset{\downarrow}{\overset{B\ 10}{|}}CH\underset{O^\bullet}{\overset{A}{|}}CH=CH\text{-}(CH_2)_6\text{-}COOMe \quad \text{10-OOH}$$

Cleavage A: nonanal + Me 9-oxononanoate
Cleavage B: octane (3%, 10%) + Me 10-oxo-8-decenoate (3%, 5%)

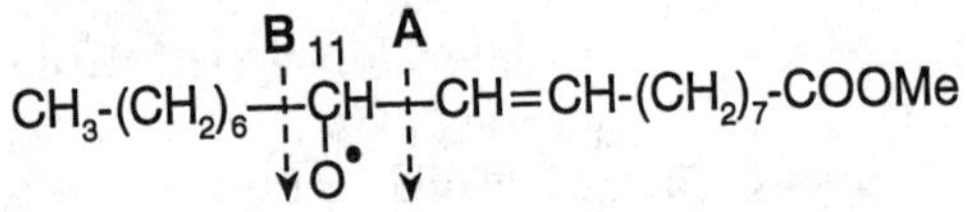

$$CH_3\text{-}(CH_2)_6\underset{\downarrow}{\overset{B\ 11}{|}}CH\underset{O^\bullet}{\overset{A}{|}}CH=CH\text{-}(CH_2)_7\text{-}COOMe \quad \text{11-OOH}$$

Cleavage A: octanal (11%, 4%) + Me 10-oxodecanoate (12%, 2%)
Cleavage B: heptane (4%, 5%) + Me 11-oxo-9-undecenoate (6%, 5%)

Figure 4-8. Cleavage products from methyl oleate hydroperoxides (Frankel et al., 1981). Relative percentage values shown in parentheses are for autoxidation and photosensitized oxidation, respectively.

octanol, 2-decenal, and methyl octanoate). A small amounts of volatile decomposition compounds that are formed from oleate hydroperoxides cannot be explained by the classical A and B cleavage mechanism, including heptanal, 2-nonenal, and methyl nonanoate.

With autoxidized linoleate, 2,4-decadienal and methyl octanoate are produced by homolytic cleavage A on the 9-hydroperoxide, and 3-nonenal and 9-oxononanoate by cleavage B; hexanal, pentane, 1-pentanol and pentanal are produced from the 13-hydroperoxide (Figure 4-10). With linoleate subjected to photosensitized oxidation, the 9- and 13-hydroperoxides produced less 2,4-decadienal, methyl octanoate and pentane, and similar amounts of hexanal. The unique unconjugated 10-hydroperoxide produced 9-oxononanoate and methyl 10-oxo-8-decenoate, and the 12-hydroperoxide produced a significant amount of 2-heptenal (Figure 4-10).

As with oleate some volatile decomposition compounds are formed from

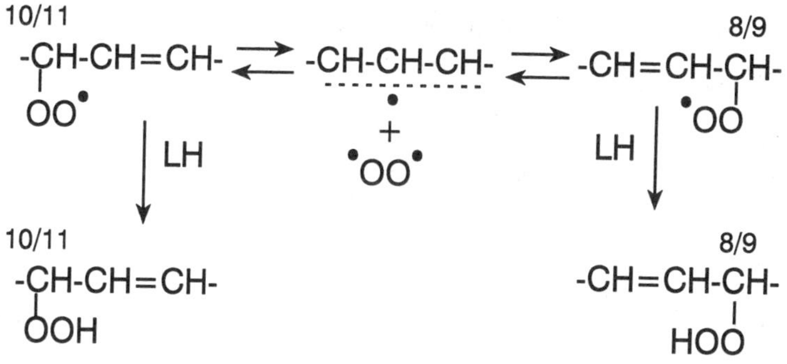

```
  10/11                                                    8/9
-CH-CH=CH-  ⇌  -CH-CH-CH-  ⇌  -CH=CH-CH-
  |                   •                      |
  OO•                 +                     •OO
        | LH        •OO•         LH |
        ↓                           ↓
  10/11                                                    8/9
-CH-CH=CH-                        -CH=CH-CH-
  |                                          |
  OOH                                       HOO
```

Figure 4-9. Thermal rearrangement of oleate hydroperoxides via allylic 3-carbon intermediates.

linoleate hydroperoxides that cannot be explained by the classical A and B cleavage mechanism, including acetaldehyde, 2-pentylfuran, methyl heptanoate, 2-octenal, 2,4-nonadienal, methyl 8-oxooctanoate, and methyl 10-oxodecanoate. Although hydroperoxides derived from both autoxidation and from photosensitized oxidation of linoleate form the same volatiles decomposition products, significant quantitative differences are noted. The autoxidized linoleate hydroperoxides produced more pentane, 2-pentyl furan, 2,4-decadienal and methyl octanoate, and less 10-oxo-8-decenoate and 2-heptenal than the photooxidized linoleate hydroperoxides.

With autoxidized methyl linolenate, decatrienal and methyl octanoate are

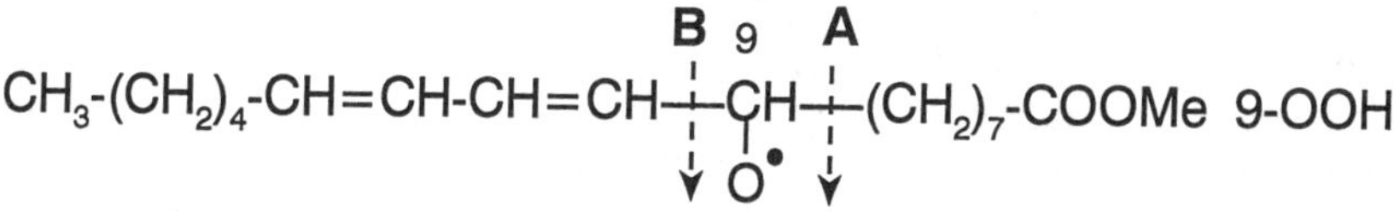

```
                            B  9    A
CH₃-(CH₂)₄-CH=CH-CH=CH─┼─CH─┼─(CH₂)₇-COOMe   9-OOH
                        ↓  |  ↓
                           O•
```

Cleavage A: 2,4-decadienal (14%, 4%) + Me octanoate (15%, 8%)
Cleavage B: 3-nonenal (1%, 2%) + Me 9-oxononanoate (19%, 22%)

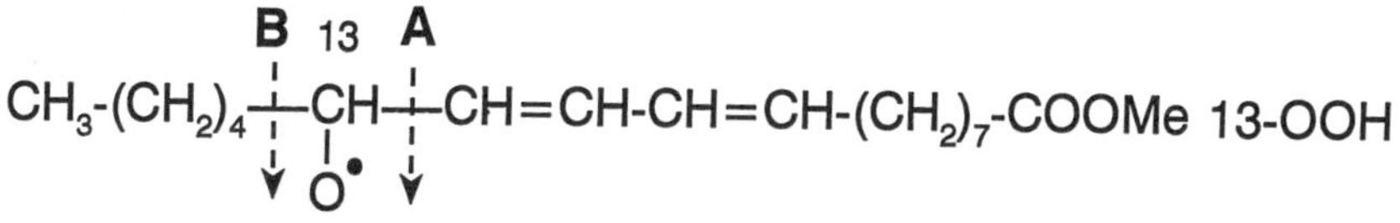

```
              B  13   A
CH₃-(CH₂)₄─┼─CH─┼─CH=CH-CH=CH-(CH₂)₇-COOMe   13-OOH
            ↓  |  ↓
               O•
```

Cleavage A: hexanal (15%, 17%) + Me 12-oxo-9-dodecenoate (?)
Cleavage B: pentane (10%, 4%) + 1-pentanol (1%, 0.3%) +
pentanal (1%, 0.3%) + Me 13-oxo-9,11-tridecadienoate (?)

Figure 4-10. Cleavage products from 9- and 13-hydroperoxides formed by autoxidation of methyl linoleate (Frankel *et al.* 1981). Relative percentage values shown in parentheses are for autoxidation and photosensitized oxidation, respectively. ? = not detected.

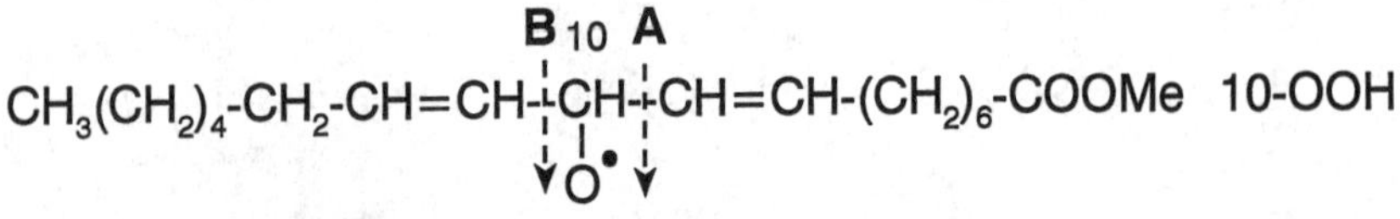

$$B_{10} \quad A$$
$$CH_3(CH_2)_4\text{-}CH_2\text{-}CH=CH\text{-}CH\text{-}CH=CH\text{-}(CH_2)_6\text{-}COOMe \quad 10\text{-}OOH$$
$$\downarrow O^\bullet \downarrow$$

Cleavage A: 3-nonenal (1.4%, 1.6%) + Me 9-oxononanoate (19%, 22%)
Cleavage B: 2-octene + 2-octen-1-ol (tr, 1.9%) +
Me 10-oxo-8-decenoate (5%, 14%)

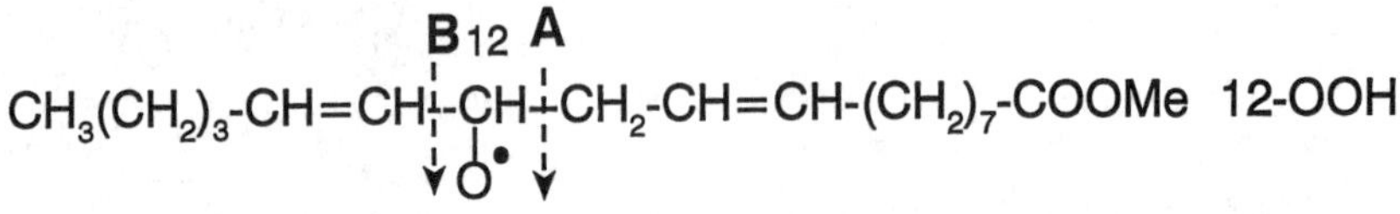

$$B_{12} \quad A$$
$$CH_3(CH_2)_3\text{-}CH=CH\text{-}CH\text{-}CH_2\text{-}CH=CH\text{-}(CH_2)_7\text{-}COOMe \quad 12\text{-}OOH$$
$$\downarrow O^\bullet \downarrow$$

Cleavage A: 2-heptenal (tr, 10%) + Me 9-undecenoate (?)
Cleavage B: hexanal (15%, 17%) + Me 12-oxo-9-dodecenoate (?)

Figure 4-11. Cleavage products from 10- and 12-hydroperoxides formed by photosensitized oxidation of methyl linoleate (Frankel *et al.* 1981).

derived from the 9-hydroperoxide; 2,4-heptadienal from the 12-hydroperoxide; 3-hexenal and 2-pentenal from the 13-hydroperoxide; propanal and ethane from the 16-hydroperoxide (Figure 4-12). As with oleate and linoleate hydroperoxides, common volatiles are produced from both types of hydroperoxides derived by autoxidation and photosensitized oxidation. The autoxidation hydroperoxides of linolenate produced more decatrienal, methyl octanoate and ethane. With linolenate subjected to photosensitized oxidation, 9-oxononanoate and methyl 10-oxo-8-decenoate are produced from the 10-hydroperoxide, and a significant amount of 2-butenal is produced from the 15-hydroperoxide (Figure 4-13).

As with oleate and linoleate some volatile decomposition compounds are formed from linolenate hydroperoxides that cannot be explained by the classical A and B cleavage mechanisms, including acetaldehyde, butanal, 2-butyl furan, methyl heptanoate, 4,5-epoxyhepta-2-enal, methyl nonanoate, and methyl 8-oxooctanoate, and methyl 10-oxo-8-decenoate. Some of these minor volatile oxidation products can be attributed to further oxidation of unsaturated aldehydes. Other factors contribute to the complexity of volatile products formed from hydroperoxides, including temperature of oxidation, metal catalysts, stability of volatile products and competing secondary reactions including dimerization, cyclization, epoxidation and dihydroperoxidation (Section E).

In contrast to thermal and metal-catalysed decomposition of hydroperoxides, which proceeds by homolytic cleavage, decomposition under acid conditions proceeds by a different mechanism involving heterolytic cleavage producing ionic ether intermediates, also referred to as carbocations. The resulting

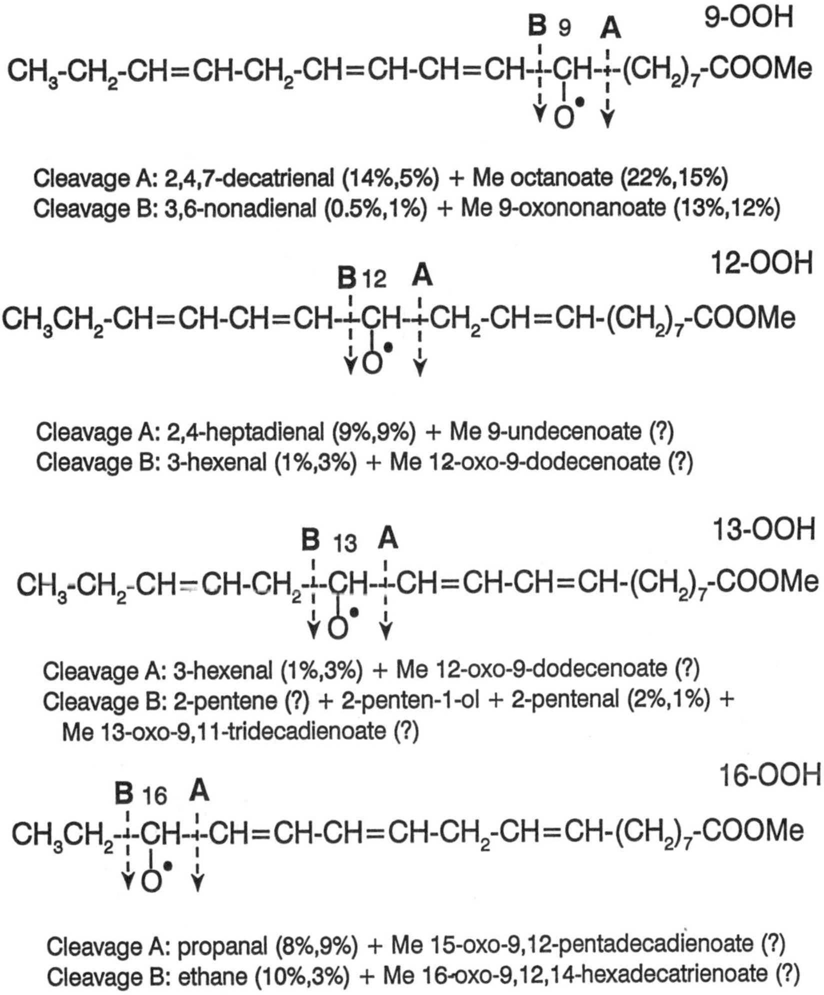

$$\text{B 9 A} \quad \text{9-OOH}$$
$$CH_3\text{-}CH_2\text{-}CH=CH\text{-}CH_2\text{-}CH=CH\text{-}CH=CH\text{-}\overset{|}{\underset{\downarrow}{C}}H\text{-}\underset{O^\bullet}{\overset{|}{\underset{\downarrow}{|}}}\text{-}(CH_2)_7\text{-}COOMe$$

Cleavage A: 2,4,7-decatrienal (14%,5%) + Me octanoate (22%,15%)
Cleavage B: 3,6-nonadienal (0.5%,1%) + Me 9-oxononanoate (13%,12%)

$$\text{B 12 A} \quad \text{12-OOH}$$
$$CH_3CH_2\text{-}CH=CH\text{-}CH=CH\text{-}CH\text{-}CH_2\text{-}CH=CH\text{-}(CH_2)_7\text{-}COOMe$$

Cleavage A: 2,4-heptadienal (9%,9%) + Me 9-undecenoate (?)
Cleavage B: 3-hexenal (1%,3%) + Me 12-oxo-9-dodecenoate (?)

$$\text{B 13 A} \quad \text{13-OOH}$$
$$CH_3\text{-}CH_2\text{-}CH=CH\text{-}CH_2\text{-}CH\text{-}CH=CH\text{-}CH=CH\text{-}(CH_2)_7\text{-}COOMe$$

Cleavage A: 3-hexenal (1%,3%) + Me 12-oxo-9-dodecenoate (?)
Cleavage B: 2-pentene (?) + 2-penten-1-ol + 2-pentenal (2%,1%) +
Me 13-oxo-9,11-tridecadienoate (?)

$$\text{B 16 A} \quad \text{16-OOH}$$
$$CH_3CH_2\text{-}CH\text{-}CH=CH\text{-}CH=CH\text{-}CH_2\text{-}CH=CH\text{-}(CH_2)_7\text{-}COOMe$$

Cleavage A: propanal (8%,9%) + Me 15-oxo-9,12-pentadecadienoate (?)
Cleavage B: ethane (10%,3%) + Me 16-oxo-9,12,14-hexadecatrienoate (?)

Figure 4-12. Cleavage products from 9-, 12-, 13- and 16-hydroperoxides formed by autoxidation of methyl linolenate (Frankel *et al.* 1981).

products are simpler and consist mainly of aldehydes. The acid decomposition products of linoleate hydroperoxides are those expected by selective heterolytic cleavage between the hydroperoxide group and the allylic double bond (Figure 4-14). Acid decomposition with hydrochloric acid-methanol produced the dimethyl acetal derivatives that are stable under conditions of GC-MS. Thus, methyl 9-oxononanoate and 3-nonenal are formed from the 9-hydroperoxide of linoleate; and hexanal and methyl 12-oxo-9-dodecenoate are formed from the 13-hydroperoxide of linoleate (Table 4-2). Methyl 9-oxononanoate and 3,6-nonadienal are formed from the 9-hydroperoxide of linolenate; methyl 12-oxo-9-dodecenoate and 3-hexenal are formed from the 12-hydroperoxide and

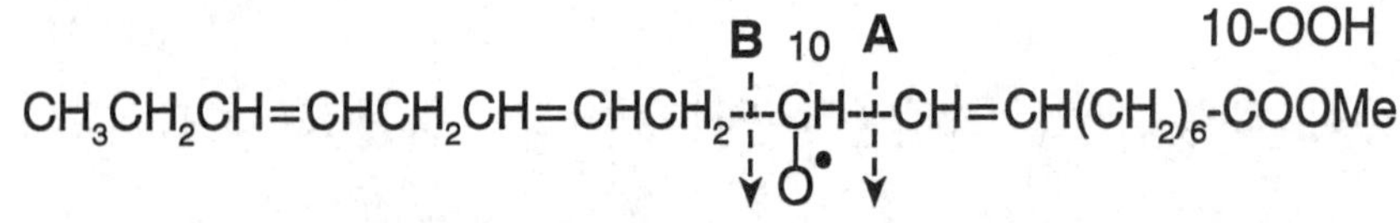

68   LIPID OXIDATION

$$\text{B}_{10} \quad \text{A} \qquad \text{10-OOH}$$
$$CH_3CH_2CH=CHCH_2CH=CHCH_2\text{-}CH\text{-}CH=CH(CH_2)_6\text{-}COOMe$$
$$\downarrow O^\bullet \downarrow$$

Cleavage A: 3,6-nonadienal (0.5%,1%) + Me 9-oxononanoate (13%,12%)
Cleavage B: 2,5-octadiene (?) + 2,5-octadien-1-ol (?) +
Me 10-oxo-8-decenoate (4%,13%)

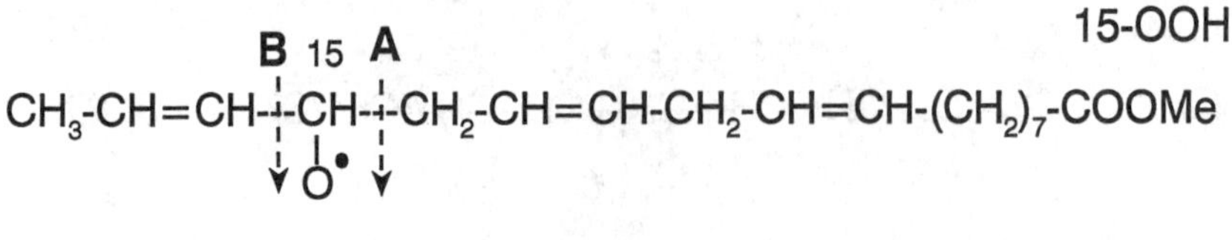

$$\text{B}_{15} \quad \text{A} \qquad \text{15-OOH}$$
$$CH_3\text{-}CH=CH\text{-}CH\text{-}CH_2\text{-}CH=CH\text{-}CH_2\text{-}CH=CH\text{-}(CH_2)_7\text{-}COOMe$$
$$\downarrow O^\bullet \downarrow$$

Cleavage A: 2-butenal (0.5%,11%) + Me 9,12-butadecadienoate (?)
Cleavage B: propanal (8%,9%) + Me 15-oxo-9,12-pentadecadienoate (?)

Figure 4-13. Cleavage products from 10- and 15-hydroperoxides formed by photosensitized oxidation of methyl linolenate (Frankel *et al.* 1981).

13-hydroperoxide of linolenate; and propanal and methyl 15-oxo-9,12-pentadecadienoate are formed from the 16-hydroperoxide of linolenate (Table 4-2).

The classical homolytic cleavage mechanism (Figure 4-7) has been criticized because β-scission on one side of the hydroperoxide group leads to stable alkyl radicals (R·), and on the other side of the hydroperoxide group it leads to vinyl radicals (R'-CH=CH·), which is unstable and unlikely. For this reason, a homolytic-heterolytic mechanism was postulated to avoid the formation of unfavorable vinylic radical intermediates (Figure 4-15). By this "mixed" mechanism, cleavage occurs homolytically on the alkyl side of the hydroperoxide to produce 2-alkenals, and heterolytically on the unsaturated

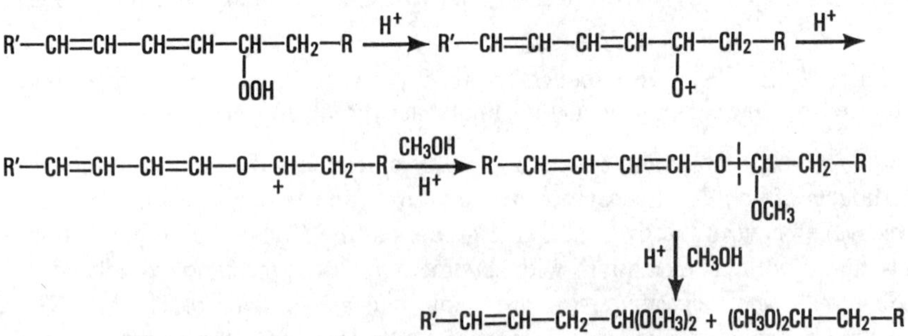

$$R'\text{—}CH=CH\text{—}CH=CH\text{—}\underset{\underset{OOH}{|}}{CH}\text{—}CH_2\text{—}R \xrightarrow{H^+} R'\text{—}CH=CH\text{—}CH=CH\text{—}\underset{\underset{O+}{|}}{CH}\text{—}CH_2\text{—}R \xrightarrow{H^+}$$

$$R'\text{—}CH=CH\text{—}CH=CH\text{—}O\text{—}\underset{+}{CH}\text{—}CH_2\text{—}R \xrightarrow[H^+]{CH_3OH} R'\text{—}CH=CH\text{—}CH=CH\text{—}O\text{—}\underset{\underset{OCH_3}{|}}{CH}\text{—}CH_2\text{—}R$$

$$\downarrow H^+ \mid CH_3OH$$

$$R'\text{—}CH=CH\text{—}CH_2\text{—}CH(OCH_3)_2 + (CH_3O)_2CH\text{—}CH_2\text{—}R$$

Figure 4-14. Acid heterolytic decomposition of linoleate hydroperoxdes (Frankel et al 1984). Stable dimethyl acetal derivatives are produced in the presence of hydrogen chloride-methanol.

TABLE 4-2.
Aldehydes produced by acid decomposition of hydroperoxides of methyl linoleate and linolenate.[a]

Aldehydes as (dimethyl acetals)	Methyl linoleate hydroperoxides		Methyl linolenate hydroperoxides	
	Relative %	Precursors	Relative %	Precursors
Hexanal	21	13-OOH		
3-Nonenal	2	9-OOH		
Me 9-oxononanoate	51	9-OOH		
Me 12-oxo-9-dodecenoate	26	13-OOH		
Propanal			22	16-OOH
3-Hexenal			1	12- + 13-OOH
3,6-Nonadienal			5	9-OOH
Me 9-oxononanoate			22	9-OOH
Me 12-oxo-9-dodecenoate			26	12- + 13-OOH
Me 15-oxo-9,12-pentadecadienoate			24	16-OOH

[a] From: Frankel et al. (1984); relative % values were normalized to 100%; OOH = hydroperoxide

side of the hydroperoxide to produce alkanals. This heterolysis may be favored because hydroperoxides are known to be weakly acidic.

E. VOLATILES FROM SECONDARY OXIDATION PRODUCTS

In addition to the monohydroperoxides, unsaturated aldehydes and ketones undergo autoxidation and provide additional sources of volatile compounds.

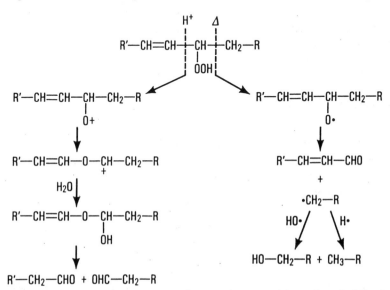

Figure 4-15. Homolytic mechanism for the decomposition of methyl linoleate hydroperoxides (Frankel et al. 1984). With permission of AOCS Press.

Nonvolatile secondary products that undergo further decomposition into volatile products include dimers and oligomers, hydroperoxy epoxides, hydroperoxy epidioxides, and dihydroperoxides. These secondary products contain one or more hydroperoxide groups and decompose the same way as monohydroperoxides to produce similar volatile materials. As with the monohydroperoxides, the multitude of volatile compounds formed from secondary products are important contributors to the flavor quality of lipid-containing foods.

1. Aldehydes

The oxidation of unsaturated aldehydes provides an important source of additional aldehydes from the decomposition of hydroperoxides. The oxidation products of 2-nonenal include alkanals, glyoxal, and mixtures of α-keto aldehydes. The same products are formed from the oxidation of 2,4-heptadienal, which also produces dialdehydes (cis-2-butene-1,4-dial and malonaldehyde). These volatile products are speculated to be produced by the formation of α-hydroperoxide aldehydes which can either undergo cleavage into smaller chain aldehydes and malonaldehyde, or produce α-keto aldehydes (Figure 4-16). By this mechanism, enals produce the next lower alkanal, and 2,4-dienals give the next lower enal and alkanal with four carbons less than the parent aldehyde, together with dialdehydes.

In mixtures, 2-nonenal and 2,4-heptadienal are more rapidly oxidized than methyl linoleate and linolenate, and these fatty esters are more readily oxidized than nonanal. Therefore, during the more advanced stages of oxidation the saturated aldehydes accumulate and the concentration of unsaturated aldehydes are further oxidized to lower aldehydes and dialdehydes. At elevated temperatures, saturated aldehydes and dialdehydes are further oxidized into mono- and dibasic acids. The formation of low-molecular weight volatile acids is used as a measure of thermal oxidation of fats, and is the basis for detection in some high-temperature stability tests (Chapter 6).

2. Dimers and Oligomers

Mixtures of unsaturated dimers and oligomers isolated from thermally decomposed linoleate hydroperoxides and oxidized linoleate produce hydroperoxides. Cleavage of these hydroperoxides produce the same volatile products as the monohydroperoxides. The formation of pentane seems to be favored. High-molecular weight aldehydo-glycerides (or oxo-glycerides) found in oxidized fats are formed by the same cleavage mechanism producing aldehyde esters from the hydroperoxide esters. Aldehydo-glycerides are also referred to as "core" aldehydes.

Dimers isolated from autoxidized linolenate, monohydroperoxides, hydroperoxy epidioxides and dihydroperoxides produce volatile decomposition products that are similar to those formed from the corresponding monohydroperoxides, but their relative concentrations are different. Dimers

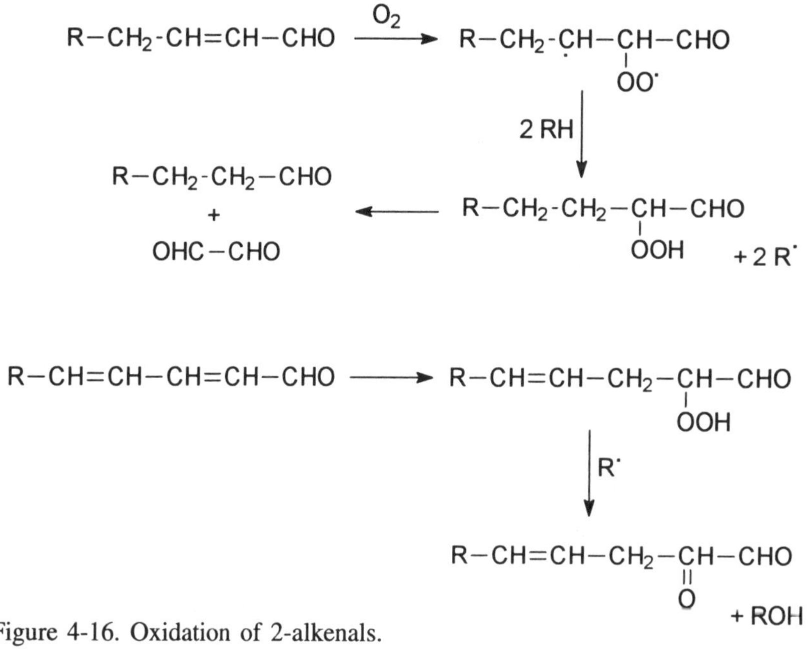

HYDROPEROXIDE DECOMPOSITION

$$R-CH_2-CH=CH-CHO \xrightarrow{O_2} R-CH_2-\overset{\cdot}{C}H-CH-CHO$$
$$\underset{OO\cdot}{|}$$

$$\Big\downarrow 2\,RH$$

$$\begin{array}{c} R-CH_2-CH_2-CHO \\ + \\ OHC-CHO \end{array} \longleftarrow R-CH_2-CH_2-CH-CHO \\ \underset{OOH}{|} \quad + 2\,R\cdot$$

$$R-CH=CH-CH=CH-CHO \longrightarrow R-CH=CH-CH_2-CH-CHO$$
$$\underset{OOH}{|}$$

$$\Big\downarrow R\cdot$$

$$R-CH=CH-CH_2-\underset{\underset{O}{\|}}{C}H-CHO$$
$$+ ROH$$

Figure 4-16. Oxidation of 2-alkenals.

from linolenate hydroperoxides produce more propanal and methyl 9-oxononanoate than the corresponding monomers but less methyl octanoate and much less 2,4-heptadienal and 2,4,7-decatrienal. The major volatile decomposition products expected from different peroxide-linked dimers are shown in Figure 4-17. The formation of methyl 9-oxononanoate as major volatile product can be explained by cleavage between the peroxide link and the olefinic side of the 9- and 16-hydroperoxide of linolenate dimers. The formation of propanal and methyl octanoate can be explained by cleavage on the opposite side of the peroxide link. Although early studies showed that the formation of dimeric materials after processing and deodorization of vegetable oils contribute to poor oxidative stability, little is known regarding the relative contribution of dimers and oligomers on the flavor deterioration of unsaturated fats.

3. Hydroperoxy Epoxides, Epidioxides and Bicyclo-Endoperoxides

A number of volatile epoxy acids and aldehydes are found in autoxidized linoleate and unsaturated fats. *Trans*-2,3-epoxyoctanoic acid is an important volatile compound produced by double bond cleavage of the epoxyhydroperoxide found in autoxidized methyl linoleate. Cleavage at the allylic position of the hydroperoxide forms the corresponding unsaturated epoxy aldehyde (Figure 4-18).

Trans-4,5-epoxy-2-heptenal ($R_1 = C_2H_5$) found in oxidized butterfat is presumed to come from *trans*-15,16-epoxy-12-hydroperoxy-9,13-dienoic acid derived from the 16-hydroperoxide of linolenic acid because 4,5-epoxy-2-heptenal is also found in thermally decomposed methyl linolenate

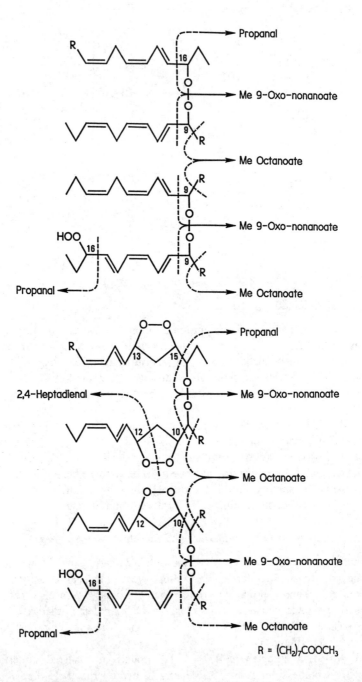

Figure 4-17. Volatile decomposition products expected from peroxide-linked dimers of methyl linolenate (Frankel *et al.* 1988).

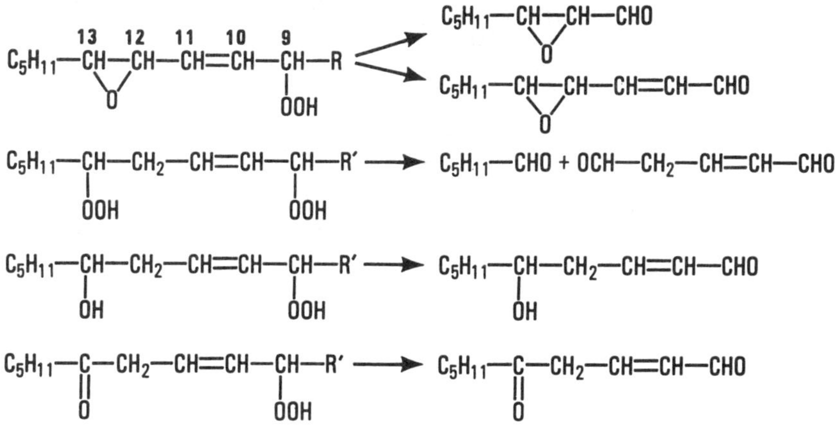

HYDROPEROXIDE DECOMPOSITION

$$C_5H_{11}-\underset{\underset{O}{\diagdown\diagup}}{CH}-\overset{13}{CH}-\overset{12}{CH}=\overset{11}{CH}-\overset{10}{CH}-\overset{9}{R} \quad \underset{\longrightarrow}{\longrightarrow} \quad \begin{array}{l} C_5H_{11}-\underset{\underset{O}{\diagdown\diagup}}{CH}-CH-CHO \\ \\ C_5H_{11}-\underset{\underset{O}{\diagdown\diagup}}{CH}-CH-CH=CH-CHO \end{array}$$

$$C_5H_{11}-\underset{OOH}{CH}-CH_2-CH=CH-\underset{OOH}{CH}-R' \longrightarrow C_5H_{11}-CHO + OCH-CH_2-CH=CH-CHO$$

$$C_5H_{11}-\underset{OH}{CH}-CH_2-CH=CH-\underset{OOH}{CH}-R' \longrightarrow C_5H_{11}-\underset{OH}{CH}-CH_2-CH=CH-CHO$$

$$C_5H_{11}-\underset{O}{\overset{\|}{C}}-CH_2-CH=CH-\underset{OOH}{CH}-R' \longrightarrow C_5H_{11}-\underset{O}{\overset{\|}{C}}-CH_2-CH=CH-CHO$$

Figure 4-18. Formation of unsaturated epoxy aldehydes from the epoxyhydroperoxide of methyl linoleate.

hydroperoxides. 4,5-Epoxy-2-decenal identified in thermally oxidized trilinoleoyl glycerol comes from the corresponding 12,13-epoxy-9-hydroperoxy-10-enoic acid ($R_1 = C_5H_{11}$) from the 13-hydroperoxide of linoleate. Epoxyaldehydes are apparently important flavor precursors because they have a significant sensory impact when tested by the sniffing technique at the exit port of a gas chromatograph (Chapter 5).

The hydroperoxy epidioxides formed from photosensitized oxidized methyl linoleate are important precursors of volatile compounds which are similar to those formed from the corresponding monohydroperoxides. Thus, 13-hydroperoxy-10,12-epidioxy-*trans*-8-enoic acid produces hexanal and methyl 10-oxo-8-decenoate as major volatiles (Figure 4-19). The 9-hydroperoxy-10,12-epidioxy-*trans*-13-enoic acid produces 2-heptenal and methyl 9-oxononanoate. Other minor volatile products include two volatiles common to those formed from the monohydroperoxides, pentane and methyl octanoate, and two that are unique, 2-heptanone and 3-octene-2-one. The hydroperoxy epidioxides formed from autoxidized methyl linolenate produce the volatiles expected from the cleavage reactions of linolenate hydroperoxides, and significant amounts of the unique compound 3,5-octadiene-2-one. This vinyl ketone has a low threshold value or minimum detectable level, and may contribute to the flavor impact of fats containing oxidized linolenate (Chapter 5).

As observed with the monohydroperoxides (Section D), thermal decomposition of the hydroperoxy bicyclo-endoperoxides from methyl linolenate produced more complex mixtures of volatile compounds than acid decomposition with acidic methanol. The thermal decomposition products included methyl 9-oxononanoate, propanal, 2,4-heptadienal, methyl octanoate, methyl 13-oxo-9,11-tridecadienoate and ethane (Figure 4-20). The acid decomposition products, analysed as the di- and tetramethyl acetals, comprised

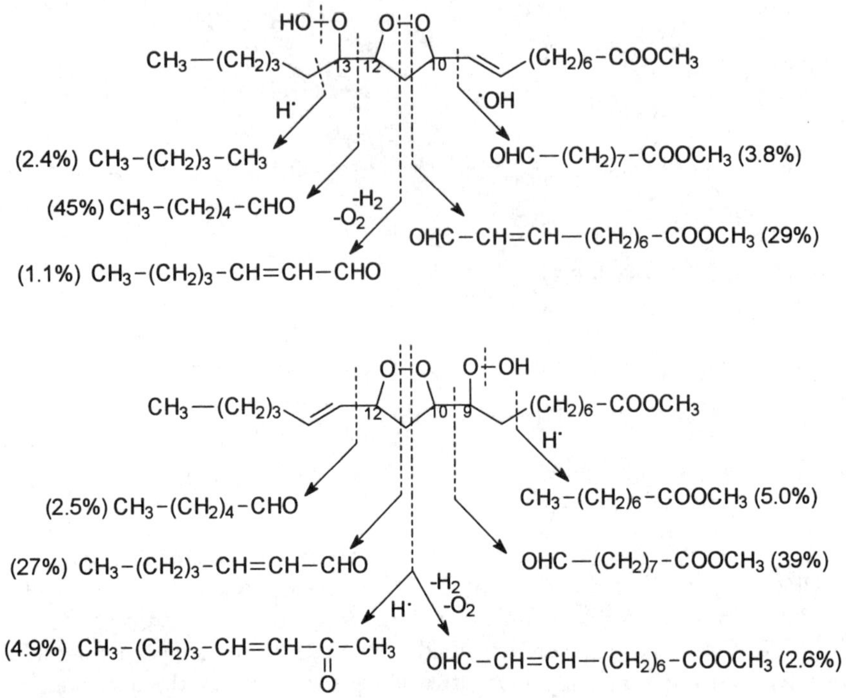

74 LIPID OXIDATION

$$CH_3-(CH_2)_3 \underset{H^\cdot}{\overset{HO\dagger O \quad O\vdots O}{\diagdown}} (CH_2)_6-COOCH_3$$
$$\qquad\qquad\qquad \text{OH}$$

(2.4%) $CH_3-(CH_2)_3-CH_3$ $OHC-(CH_2)_7-COOCH_3$ (3.8%)

(45%) $CH_3-(CH_2)_4-CHO$ $\overset{-H_2}{-O_2}$

                     $OHC-CH=CH-(CH_2)_6-COOCH_3$ (29%)

(1.1%) $CH_3-(CH_2)_3-CH=CH-CHO$

$$CH_3-(CH_2)_3 \diagdown \overset{O\vdots O \quad O\dagger OH}{} (CH_2)_6-COOCH_3$$
$$\qquad\qquad\qquad\qquad\qquad H^\cdot$$

(2.5%) $CH_3-(CH_2)_4-CHO$ $CH_3-(CH_2)_6-COOCH_3$ (5.0%)

(27%) $CH_3-(CH_2)_3-CH=CH-CHO$ $OHC-(CH_2)_7-COOCH_3$ (39%)

                    $H^\cdot \overset{-H_2}{-O_2}$

(4.9%) $CH_3-(CH_2)_3-CH=CH-\underset{\underset{O}{\|}}{C}-CH_3$ $OHC-CH=CH-(CH_2)_6-COOCH_3$ (2.6%)

Figure 4-19. Volatile decomposition products from hydroperoxy epidioxides of methyl linoleate oxidized with singlet oxygen (Frankel *et al.* 1982).

only propanal, methyl 9-oxononaoate, and malonaldehyde. As with the monohydroperoxides, by thermal decomposition the bicyclo-endoperoxides are cleaved on either side of the hydroperoxide group, whereas by acid decomposition they are cleaved only between the hydroperoxide group and the allylic double bond. The bicyclo-endoperoxides undergo cleavage across the ring also to produce malonaldehyde (cf. Figure 4-3), which was only detected under the conditions of acid decomposition (as the tetramethyl acetal derivative). This dialdehyde was too unstable to be detected under the conditions of thermal decomposition.

4. *Triacylglycerols*

The main volatile compounds detected from autoxidized trilinolein monohydroperoxides include pentane, hexanal, 2-heptenal and 2,4-decadienal. The volatiles from trilinolenin monohydroperoxides include propanal, 2,4-heptadienal and 2,4,7-decatrienal. Mixtures of 1:1 trilinolein and trilinolenin autoxidized at low peroxide values (PV 34) show equal contribution of volatiles derived from linoleate and linolenate hydroperoxides.

Soybean oil containing a mixture of oleate, linoleate and linolenate triacylglycerols can produce 14 positional hydroperoxides by autoxidation and

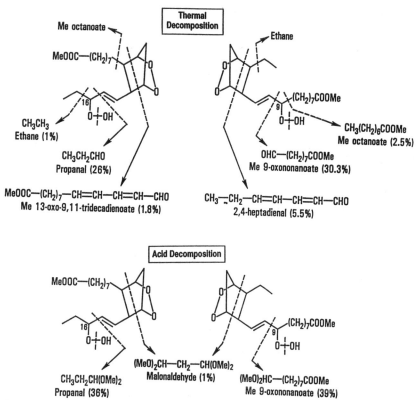

Figure 4-20. Volatile products formed by acid versus thermal decomposition of hydroperoxy bicyclo-endoperoxides of methyl linolenate (Frankel et al. 1984). Aldehydes produced under acid conditions were analysed as the dimethyl acetals, and malonaldehyde as the tetramethyl acetal.

photosensitized oxidation (Chapters 2 and 3). Soybean oil and other vegetable oils, such as safflower oil and corn oil, that do not contain linolenate, produce an unexpectedly high concentration of the 12-hydroperoxide isomer at peroxide values below 50. The 12-hydroperoxide isomer appears to be derived from photosensitized oxidation because its concentration is decreased in the presence of singlet oxygen quenchers such as β-carotene and α-tocopherol (Chapter 3).

A large proportion of the volatiles identified in vegetable oils are derived from the cleavage reactions of the hydroperoxides of oleate, linoleate, and linolenate (Section D). A wide range of hydrocarbons (ethane, propane, pentane and hexane) appear to be formed in soybean oil oxidized to low peroxide values. A number of volatiles identified in vegetable oils that are not expected as primary cleavage products of monohydroperoxides include dialdehydes, ketones, ethyl esters, nonane, decane, undecane, 2-pentylfuran, lactone, benzene, benzaldehyde, and acetophenone. Some of these volatiles may be derived from secondary oxidation products, but the origin of many volatiles still remains obscure. However, studies of volatile decomposition products should be interpreted with caution because the conditions used for isolation and identification may cause artifacts especially when fats are subjected to elevated temperatures.

Soybean oil and other linolenate-containing oils such as rapeseed oil (referred to as canola oil in North America) are notorious for their flavor deterioration noted at unusually low levels of oxidation, sometimes at peroxide values below 1. This flavor defect is known as "reversion," a term derived from their flavor characteristic of crude soybean oil described as "beany" or "grassy". Oils susceptible to this flavor defect contain oxidatively derived dimers and oligomers produced during deodorization by thermal decomposition of hydroperoxides present initially and formed during processing (Section C). These high-molecular weight compounds, also known as oxidative dimers, are easily decomposed into volatile compounds with high impact on flavor. The volatile oxidation products containing an n-3 double bond derived from linolenate have an unusually low threshold values, thus contributing to the flavor deterioration detected at low peroxide values (Chapter 5).

BIBLIOGRAPHY

Evans,C.D., Frankel,E.N., Cooney,P.M. and Moser,H.A. Effect of autoxidation prior to deodorization on oxidative and flavor stability of soybean oil. *J. Am. Oil Chem. Soc.* **42**, 764-770 (1960).
Evans,C.D., McConnell,D.C., Frankel,E.N. and Cowan,J.C. Chromatographic studies on oxidative and thermal fatty acid dimers. *J. Am. Oil Chem. Soc.* **42**, 764-770 (1965).
Frankel,E.N., Evans,C.D. and Cowan,J.C. Thermal dimerization of fatty ester hydroperoxides. *J. Am. Oil Chem. Soc.* **37**, 418-424 (1960).
Frankel,E.N., Evans,C.D., Moser,H.A., McConnell,D.G. and Cowan,J.C. Analysis of lipids and oxidation products by partition chromatography. Dimeric and polymeric products. *J. Am. Oil Chem. Soc.* **38**, 130-134 (1961).
Frankel,E.N. Volatile lipid oxidation products. *Prog. Lipid Res.* **22**, 1-33 (1982).
Frankel,E.N., Neff,W.E. and Selke,E. Analysis of autoxidized fats by gas chromatography-mass spectrometry: VII. Volatile thermal decomposition products of pure hydroperoxides from autoxidized and photosensitized oxidized methyl oleate, linoleate, and linolenate. *Lipids* **16**, 279-285 (1981).
Frankel,E.N., Neff,W.E., Selke,E. and Weisleder,D. Photosensitized oxidation of methyl linoleate. Secondary and volatile thermal decomposition products. *Lipids* **17**, 11-18 (1982).
Frankel,E.N., Neff,W.E. and Selke,E. Analysis of autoxidized fats by gas chromatography-mass spectrometry. VIII. Volatile thermal decomposition products of hydroperoxy cyclic peroxides. *Lipids* **18**, 353-357 (1983).
Frankel,E.N. and Neff,W.E. Formation of malonaldehyde from lipid oxidation products. *Biochim. Biophys. Acta* **754**, 264-270 (1983).
Frankel,E.N., Neff,W.E. and Selke,E. Analysis of autoxidized fats by gas chromatography-mass spectrometry. IX. Homolytic vs heterolytic cleavage. *Lipids* **19**, 790-800 (1984).
Frankel,E.N. Chemistry of autoxidation. Mechanism, products and flavor significance. In *Flavor Chemistry of Fats and Oils*, pp. 1-37 (1985) (edited by D. B. Min and T. H. Smouse), American Oil Chemists' Society, Champaign, Illinois.
Frankel,E.N. Secondary products of lipid oxidation. *Chemistry and Physics of Lipids.* **44**, 73-85 (1987).
Frankel,E.N., Neff,W.E., Selke,E. and Brooks,D.D. Thermal and metal-catalyzed decomposition of methyl linolenate hydroperoxides. *Lipids* **22**, 322-327 (1987).

Frankel,E.N., Neff,W.E., Selke,E. and Brooks,D.D. Analysis of autoxidized fats by gas chromatography-mass spectrometry. X. Volatile thermal decomposition products of methyl linolenate dimers. *Lipids* **23**, 295-298 (1988).

Frankel,E.N., Neff,W.E. and Miyashita,K. Autoxidation of polyunsaturated triacylglycerols. II. Trilinolenoylglycerol. *Lipids* **25**, 40-47 (1990).

Frankel,E.N., Selke,E., Neff,W.E. and Miyashita,K. Autoxidation of polyunsaturated triacylglycerols. IV. Volatile decomposition products from triacylglycerols containing linoleate and linolenate. *Lipids* **27**, 442-446 (1992).

Grosch,W. Reactions of hydroperoxides - Products of low molecular weight, in *Autoxidation of Unsaturated Lipids*, pp. 95-139 (1987) (edited by H.W.-S. Chan), Academic Press, London.

Gardner,H.W. Reactions of hydroperoxides - Products of high molecular weight, in *Autoxidation of Unsaturated Lipids*, pp. 51-93 (1987) (edited by H.W.-S. Chan), Academic Press, London.

Hopia,A.I., Lampi,A-M., Piironen,V.I., Hyvonen,L.E.T. and Koivistoinen,P.E. Application of high-performance size-exclusion chromatography to study the autoxidation of triacylglycerols. *J. Am. Oil Chem. Soc.* **70**, 779-784 (1993).

Hopia,A. Analysis of high molecular weight autoxidation products using high performance size exclusion chromatography: I. Changes during autoxidation. *Wiss. u.- Technol.* **26**, 563-567 (1993).

Keppler,J.G. Twenty-five years of flavor research in a food industry. *J. Am. Oil Chem. Soc.* **54**, 474-477 (1977).

Lillard,D.A. and Day,E.A. Degradation of monocarbonyls from autoxidizing lipids. *J. Am. Oil Chem. Soc.* **41**, 549-552 (1964).

Neff,W.E., Frankel,E.N. and Weisleder,D. Photosensitized oxidation of methyl linoleate monohydroperoxides: Hydroperoxy cyclic peroxides, dihydroperoxides, keto esters and volatile thermal decomposition products. *Lipids* **18**, 868-876 (1983).

Neff,W.E., Frankel,E.N. and Fujimoto,K. Autoxidative dimerization of methyl linolenate and its monohydroperoxides, hydroperoxy epidioxides and dihydroperoxides. *J. Am. Oil Chem. Soc.* **65**, 616-623 (1988).

Neff,W.E., Frankel,E.N. and Miyashita,K. Autoxidation of polyunsaturated triacylglycerols. I. Trilinoleoylglycerol. *Lipids* **25**, 33-39 (1990).

Selke,E., Frankel,E.N. and Neff,W.E. Thermal decomposition of methyl oleate hydroperoxides and identification of volatile components by gas chromatography-mass spectrometry. *Lipids* **13**, 511-513 (1978).

Schiberle,P. and Grosch,W. Model experiments about the formation of volatile carbonyl compounds. *J. Am. Oil Chem. Soc.* **58**, 602-607 (1981)

Smouse,T.H. and Chang,S.S. A systematic characterization of the reversion flavor of soybean oil. *J. Am. Oil Chem. Soc.* **44**: 509-514 (1967).

CHAPTER 5

METHODS TO DETERMINE EXTENT OF OXIDATION

In previous chapters, lipids were shown to be oxidized into complex mixtures of primary and secondary products that decompose into low-molecular weight compounds. These mixtures vary in nature and composition according to the conditions and extent of oxidation. The quality of food lipids is greatly influenced by the nature of these complex oxidation products and the levels and significance of their concentrations as they relate to the acceptability of the food. A large repertoire of methods is available to determine oxidative deterioration of food lipids. The usefulness of the results can be evaluated according to the chemical and physical properties measured, the specificity and precision of the analyses, and how they relate to real life storage of food products. For routine analyses, many methods have been developed to appraise the extent and nature of oxidative deterioration. The most commonly used methods are evaluated in this chapter according to the usefulness of the results in determining how closely they relate to flavor deterioration under ambient storage and processing conditions. More specialized methods based on HPLC separation of primary and secondary lipid oxidation products and their structural characterization by GC-MS and NMR were discussed in Chapters 2-4. Methods to determine the oxidative stability of unsaturated lipids are covered in Chapter 6.

A. SENSORY METHODS

Sensory evaluation is a specialized discipline using trained panels to measure and analyse the characteristics of food lipids evoked by the senses of taste, smell, sight and mouth feel. Sensory analyses are those most closely associated with the quality of food lipids, but their usefulness is limited because they are costly and require a well-trained taste and odor panel and the proper facilities. However, sensory analyses provide sometimes a useful approach to identify flavor or odor defects in the processing of food lipids that cannot be detected by other more objective chemical or instrumental analyses. For example, certain flavor defects characterized as "grassy" or "fishy" in linolenate-containing oils such as soybean oil occur at such low levels of oxidation that they can only be detected by sensory analyses. The so-called "flavor reversion" of soybean oil is based on the characteristic of this oil of undergoing flavor deterioration at unusually low levels of oxidation that cannot be measured by peroxide value determination.

The main limitation of the sensory technique is the poor reproducibility of the data generated by a panel of human subjects. The approach currently recommended is to use precise chemical and instrumental methods to complement and support the sensory analyses. The two important types of taste panels include:

a) the "analytical panel," using 5 to 20 experienced subjects who are trained to differentiate and rank samples according to a numerical scale, using reference standards; and

b) the "consumer panel," using 50 or more untrained subjects who are asked to express preference or acceptability of finished products.

Liquid vegetable oils can be evaluated for odor and flavor according to a 10 point scale, to rate either intensity (bland to strong), or quality (excellent to bad). A freshly deodorized soybean oil is described as "bland;" a weakly oxidized soybean oil is described as "buttery." As oxidation progresses, this oil is described as "beany" and "grassy." In the more advanced stages of oxidation, the oil becomes "rancid" and "painty." Because of the subjective nature of taste panel testing, different investigators vary widely in their descriptions of flavor defects in various oils. Further differences exist in the descriptive vocabulary used by panels from different laboratories, different regions and different countries.

The types of flavors imparted by lipid oxidation in foods is more difficult to assess because there is wide variation in the sensory impact of different volatile products, and in the vocabulary used by taste or odor panels to describe their effect on quality. The diversity of sensory vocabulary used by different investigators to describe the same flavor defect in an edible oil has led to controversy on the origin of different flavor defects in soybean and other vegetable oils.

The sensitivity of taste or odor panels can be measured by the ability of individuals to detect sensory characteristics. "Threshold values" are measures of the least concentrations of volatile compounds detected in a food matrix (oil or water) by 50% of the panelists. There is a considerable difference in the flavor significance of volatile decomposition products formed in oxidized or "rancid" lipids. Hydrocarbons have the highest threshold values ranging from 90 to 2150 ppm, and the least impact on flavor. Substituted furans with threshold values of 2-27 ppm, vinyl alcohols with threshold values of 0.5-0.3 ppm, and 1-alkenes with threshold values of 0.02-9 ppm, are also relatively insignificant in flavor impact. Volatile compounds of increasing flavor intensity include 2-alkenals with threshold values of 0.04-2.5 ppm, alkanals with threshold values of 0.04-1.0 ppm, *trans,trans*-2,4-alkadienals with threshold values of 0.04-0.3 ppm, isolated alkadienals with threshold values of 0.0003-

METHODS TO DETERMINE EXTENT OF OXIDATION 81

0.1 ppm, isolated cis-alkenals with threshold values of 0.0003-0.1 ppm, *trans*-3, *cis*-4-alkadienals with threshold values of 0.002-0.006 ppm and vinyl ketones with threshold values of 0.00002-0.007 ppm.

The sensory data are calculated by determining the overall mean scores for intensity or quality: total points divided by the number of panelists for each sensory session. In some panel procedures, the scores are discarded if the means differ by more than two units from the average score. This procedure requires at least 10 or 12 final judgments for statistical analyses. The significance of the overall mean scores is calculated statistically by two-way analysis of variance (ANOVA). Sensory scores can also be correlated with the results of other tests of lipid oxidation by regression analyses. If an objective test correlates well with sensory analyses it is usually interpreted as giving similar information regarding the level of oxidation. However, correlation data must be interpreted with care because they can only be used to show trends between two sets of analyses, and cannot be used to obtain cause and effect relationships.

B. PEROXIDE VALUE

This is one of the oldest and most commonly used measurement of the extent of oxidation in oils. The standard iodometric procedures measure, by titration, or colorimetric or electrometric methods, the iodine produced by potassium iodide added as a reducing agent to the oxidized sample dissolved in a chloroform-acetic acid mixture. The liberated iodine is titrated with standard sodium thiosulfate to a starch endpoint. The peroxide value (PV) is expressed as milliequivalents of iodine per kg of lipid (meq/kg), or as millimole of hydroperoxide per kg of lipid (referred to as peroxide). PV as meq/Kg = 2 x PV mmol/kg.

The standard iodometric method for peroxide value requires a sample of 5 g for peroxide values below 10, and about 1 g for higher peroxide values. The sensitivity of this method is about 0.5 meq/kg, and can be improved by determining the iodine starch-end point colorimetrically, or the liberated iodine electrometrically by reduction at a platinum electrode maintained at constant potential.

The results of peroxide value determinations vary according to the procedures used as a result of interference from oxygen in the air, exposure to light, and absorption of iodine by the unsaturated fatty acid in the oils. Although this interference can be reduced by carrying out the procedure in the absence of air and in the dark, these precautions are not generally taken for routine analyses. The method is therefore highly empirical and dependent on the technique of the operator.

The ferric thiocyanate method for peroxide value is based on the oxidation of ferrous to ferric ions, which are determined colorimetrically as ferric thiocyanate. This method is more sensitive and requires a smaller sample (about 0.1 g) than does the iodometric method (Table 5-1). However, the

TABLE 5-1.
Comparison of different methods to determine extent of oxidation.[a]

Method	Lo-OOH [b] Absorbance	Sens [c]	Me linoleate PV 475	Sens [c]	Me linolenate PV 450	Sens [c]
Conj. Diene	7.42	1	0.55	1	0.91	1
Conj. Triene	0.07	<0.1	0.06	0.1	0.29	0.3
Fe thiocyanate	26.4	3.6	5.2	9.4	5.7	6.3
TBA	0.07	<0.1	0.03	<0.1	0.42	0.46
Anisidine value	0.12	<0.1	0.16	0.3	0.75	0.8

[a] Adapted from Tsoukalas and Grosch (1977).
[b] Linoleic acid hydroperoxides in cyclohexane-diethyl ether solution.
[c] Sensitivity = calculated absorbance values divided by normalized diene absorption for a solution diluted 1:10.

values obtained by the ferric thiocyanate method are higher by a factor of 1.5 to 2 relative to those of the iodometric method. The peroxide values obtained by both methods are of only relative significance. The ferric thiocyanate method is commonly applied to dairy products, which undergo oxidative deterioration at relatively low peroxide values. Other colorimetric methods for peroxide values include the determination of the blue starch-iodine complex in the iodometric method, the red color developed with 1,5-diphenylcarbohydrazide, the color developed with ferric ions and xylenol orange, ferrous chloride and 2,6-dichlorophenolindophenol, titanium dichloride, and xylenol orange.

The determination of peroxide value is useful for bulk oils that can be analysed directly. For foods, emulsions or muscle tissues, the lipid is extracted with mixtures of solvents that must be carefully removed by evaporation without decomposition of hydroperoxides. This solvent extraction is not quantitative for samples that are highly oxidized or subjected to drastic conditions of storage or processing. Solvents should not contain antioxidants as stabilizers, which would act as reducing agents and interfere with the peroxide value determination.

The peroxide value of an oil is an empirical measure of oxidation that is useful for samples that are oxidized to relatively low levels (peroxide values of less than 50), and under conditions sufficiently mild so that hydroperoxides are not markedly decomposed. During autoxidation, the peroxide value of an oil reaches a maximum followed by a decrease at more advanced stages varying according to the fatty acid composition of the oil and the conditions of oxidation (Figure 5-1). The peroxide value maximum occurs at earlier stages of oxidation in the more polyunsaturated oils, such as soybean and rapeseed oils containing linolenate, and fish oils containing n-3 polyunsaturated fatty acids, because their hydroperoxides decompose more readily. Hydroperoxides are rapidly decomposed during oxidation at temperatures above 100°C, or in the presence of metals, or when exposed to light. The peroxide value of oilsloxidized under these accelerated conditions are erroneously low

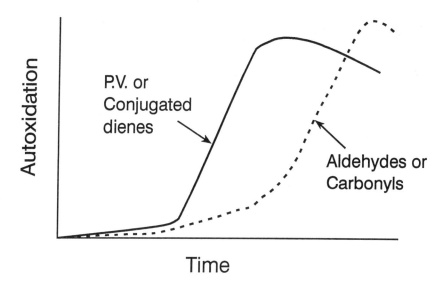

Figure 5-1. Development of hydroperoxides, as measured by peroxide value (P.V.) or conjugated dienes, and decomposition of hydroperoxides, as measured by aldehydes and carbonyls, during the autoxidation of an unsaturated fat.

especially for polyunsaturated oils. Therefore, a knowledge of the history of the oil samples is important to better interpret the significance of peroxide value measurements (see Chapter 6).

C. CONJUGATED DIENES

The conjugated diene hydroperoxides produced from polyunsaturated lipids can be determined quantitatively by their strong absorption maximum at 234 nm (molecular extinction of 25,000 for a solution of 1 mole hydroperoxides/L methanol). A weighed sample is simply diluted in methanol (for esters) or in isooctane (for triacylglycerols) and the absorbance at 234 nm measured spectrophotometrically. Aqueous emulsions can also be measured directly by dissolving in methanol even if the solutions becomes slightly cloudy, provided a suitable blank is used with the same diluted unoxidized sample. The conjugated diene value of lipids can be determined by their UV absorbance at 234 nm and expressed in μmol hydroperoxides/g sample.

The measurement of conjugated dienes is a sensitive method to follow the early stages of lipid oxidation under conditions in which hydroperoxides undergo little or no decomposition. The method requires 50 to 100 mg of sample, is simple and does not depend on chemical or color reactions. As with peroxide value determinations, the conjugated diene determination will reach a maximum during the progress of oxidation and decrease when the rate of decomposition of hydroperoxides exceeds the rate of their formation (Figure 5-1). At more advanced stages of oxidation, the hydroperoxides decompose into secondary and polymeric products that also absorb at 234 nm, but with lower extinction than the monohydroperoxides, and at 270 nm due to conjugated trienes. Although the hydroperoxides of oxidized methyl linolenate have conjugated diene structures that absorb mainly at 234 nm, they produce secondary products that absorb at 270 nm for conjugated trienes and carbonyl compounds that contribute to other tests (carbonyl, TBA and anisidine values) discussed below (Table 5-1).

The conjugated diene method is not suitable for oils that have been heated under conditions that decompose hydroperoxides. Interference will occur from small amounts of conjugated diene and triene products found in hydrogenated fats, and small amounts of conjugated trienes in linolenate-containing oils (soybean and rapeseed oils) produced by deodorization at elevated temperatures (above 250°C). The conjugated diene method is also not suitable for oils containing low levels of polyunsaturated fatty acids such as olive oil, high-oleic sunflower oil and high-oleic safflower oil.

D. CARBONYL COMPOUNDS

Total carbonyl content can be determined in oxidized lipids by the reaction with 2,4-dinitrophenylhydrazine, and the colored hydrazone (2,4-DNPH) derivatives are measured spectrophotometrically at 430-460 nm, and expressed as nmol hexanal/kg sample. The classical method for total carbonyl value using trichloroacetic acid catalyst causes further decomposition of hydroperoxides. This interference can be eliminated by reducing the hydroperoxides with stannous chloride, or by separating the hydrazone derivatives by vacuum distillation. During autoxidation of unsaturated lipids, the carbonyl value increases slowly followed by a rapid rate after the hydroperoxides reach a maximum corresponding to their decomposition into aldehydes and carbonyl products (Figure 5-1). Because the carbonyl value is related to decomposition carbonyl compounds which contribute to flavor deterioration, it provides information relevant to sensory evaluation.

The **anisidine test** measures high-molecular-weight saturated and unsaturated carbonyl compounds in triacylglycerols. The anisidine value is defined as the absorbance of a solution resulting from the reaction of 1g fat in 100 ml of isooctane solvent and reagent (0.25% anisidine in glacial acetic acid). The anisidine test is often attributed to 2-alkenals but no direct evidence is published for it. Oxidized methyl linolenate produces a significantly higher anisidine value than oxidized methyl linoleate because of the greater tendency of linolenate hydroperoxides to decompose into carbonyl compounds (Table 5-1). The test provides useful information on non-volatile carbonyl compounds formed in oils during processing. This type of deterioration can be serious in poorly processed oils containing linolenate (soybean and rapeseed) and n-3 polyunsaturated fatty acids (fish). Several high molecular-weight decomposition products discussed previously, including oxidative dimers of

METHODS TO DETERMINE EXTENT OF OXIDATION 85

triacylglycerols, aldehydo-glycerides or core aldehydes (Chapter 4.C and 4.D), contribute to the anisidine test in poorly processed fats. Like the chromatographic determination of dimers, the anisidine test on freshly processed oils can be used as a rough predictor of their future storage stability. The so-called **Totox value** (anisidine value + 2 PV) is used as a measure of the precursor non-volatile carbonyls present in a processed oil plus any further oxidation products developed after storage. A good quality oil should have a Totox value of less than 4.

Other methods for determining carbonyl compounds involve derivatization followed by column chromatography or by HPLC or by GC. Total carbonyls in lipid extracts are converted to their 2,4-DNPH derivatives by using a column of diatomaceous earth impregnated with 2,4-DNPH, followed by extraction and spectrophotometric measurement at 340 nm. By reaction with methylhydrazine, saturated carbonyls (formaldehyde, acetaldehyde and hexanal) are converted to hydrazones, α,β-unsaturated carbonyls (acrolein and 4-hydroxy-nonenal) to pyrazolines, and β-dicarbonyls (malonaldehyde) to methyl pyrazole derivatives, and determined by GC.

E. 2-THIOBARBITURIC ACID (TBA) VALUE

The TBA test is an old and popular colorimetric method used to measure rancidity in some foods and oxidation products in biological systems. This test is based on the pink color absorbance at 532-535 nm formed between TBA and oxidation products of polyunsaturated lipids. Because this reaction is not specific and produced by a large number of secondary oxidation products, these products are referred to as TBA-reactive substances or TBARS. The colored complex was originally characterized as being due to the condensation of two moles of TBA and one mole of malonaldehyde, a decomposition product of lipid hydroperoxides under the thermal acidic conditions of the test (Figure 5-2). On this basis, the test is standardized by using malonaldehyde generated by acid hydrolysis of 1,1,3,3-tetraethoxypropane, the tetraethoxyacetal of malonaldehyde. The TBA value is defined as the mg of malonaldehyde per kg of sample. However, malonaldehyde is not present in many oxidized lipids and often constitutes a minor secondary oxidation product.

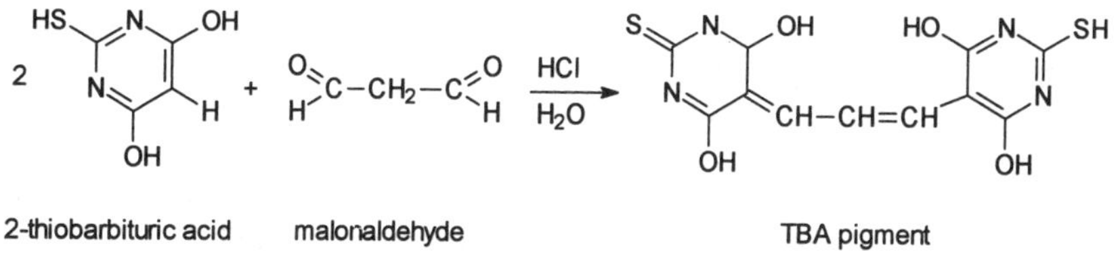

Figure 5-2. Proposed structure of TBA pigment as a colored adduct between TBA and malonaldehyde. From Sinnhuber et al. (1958).

TABLE 5-2.
Determination of malonaldehyde (MDA) by the TBA test and by HPLC of the dansyl hydrazine derivative.[a]

Samples[b]	Peroxide value (meq/Kg)		MDA by TBA (ppm)		MDA by HPLC (ppm)	
	20 hr	50 hr	20 hr	50 hr	20 hr	50 hr
Soybean	11.2	31.4	0.86	4.31	0.28	0.79
Corn	48.0	92.2	0.70	6.09	0.23	0.35
Rapeseed	17.1	28.8	5.00	6.39	0.82	1.93
Olive	8.8	21.6	0.70	1.05	0.27	0.33
Peanut	20.4	37.9	0.80	1.10	0.33	0.35
Methyl oleate	18.5	—	2.58	—	0.44	—
Methyl Linoleate	62.5	—	5.36	—	0.75	—
Methyl Linolenate	799.3	—	147.0	—	2.63	—

[a] From; Hirayama et al. (1983).
[b] Vegetable oils were oxidized at 100°C and fatty esters at 37°C in the dark.

Many factors affect the production of the pink color in the TBA test, including temperature and time of heating, pH, metal ions, and antioxidants. Numerous variations of this test are aimed either at increasing the sensitivity by heating in acids and adding ferric ions, or at reducing the artifactual production of decomposition materials under the conditions of the test, by adding an antioxidant such as BHT or a chelator such as EDTA. Heating in acids with iron causes further oxidation and decomposition of oxidized lipids. Although this decomposition can be reduced by BHT, such addition confounds the application of the TBA test in studies of the effect of natural or added antioxidants. Other variations are aimed at increasing the selectivity of the test by separating TBARS by steam distillation of the volatile components, or by HPLC of the peak corresponding to the TBA-malonaldehyde adduct.

Many precursors of malonaldehyde have been suggested including monocyclic endoperoxides and bicyclic endoperoxides, which are produced as secondary products in polyunsaturated lipids containing 3 or more double bonds (Chapter 4). The TBA test is consequently more sensitive when used with these polyunsaturated lipids and underestimates the oxidation products from lipids containing mainly oleic and linoleic acids.

Pure malonaldehyde has been determined by acid decomposition of various pure fatty acid oxidation products by GC as the tetramethylacetal derivative produced by acetalation under acid methanolic conditions (Table 4-1). The yields of malonaldehyde varied from 0.18-0.26 mol% for the monohydroperoxides of methyl linoleate to 0.22-0.44 mol% for the monohydroperoxides of methyl linolenate, from 2.20-2.34 mol% for the hydroperoxy epidioxide of methyl linolenate to 24.2 mole% for the hydroperoxy epidioxide of methyl linoleate oxidized with singlet oxygen, 4.14 mol% for the hydroperoxy bicyclo-endoperoxides of linolenate, and from 0.81 for the 1,3-dihydroperoxides of linolenate to 4.76 for the 1,3-dihydroperoxides of linolenate. This determination of pure malonaldehyde as the tetramethylacetal derivative by GC showed no correlation with the malonaldehyde determined by the TBA method, which was not specific for most of the pure fatty acid oxidation products examined (Table 4-1, Table 5-1). As expected from the acid cleavage mechanism discussed in Chapter 4, the 1,7- and 1,8-dihydroperoxides of linolenate produced no malonaldehyde by the GC analysis of the tetramethyl acetal derivative. In contrast, the TBA method gave values corresponding to 2.74 and 7.63 mol% malonaldehyde. Several other studies compared the TBA test and methods measuring pure malonaldehyde by HPLC. Determinations of malonaldehyde in oxidized vegetable oils were 2 to 17 times higher by the TBA test than by HPLC of the dansyl hydrazine (Table 5-2). For autoxidized fatty esters, the malonaldehyde values by the TBA test were higher than by HPLC by a factor of 6 for methyl oleate, 7 for methyl linoleate and 56 for methyl linolenate. Malonaldehyde levels determined in muscle tissues by the TBA test were 4 to 5 times higher than HPLC analyses of malonaldehyde. The TBA test evidently measures many other decomposition products than malonaldehyde.

In addition to malonaldehyde, pigments are formed by other decomposition products under the TBA conditions, including alkanals, 2-alkenals, and 2,4-alkadienals absorbing at 450 nm, and by alkenals and 2,4-alkadienals absorbing at 532 nm. Other components present in foods and biological systems, such as browning reaction products, protein and sugar degradation products, amino acids, nucleic acids and nitrite, interfere with the formation of the TBA color complex. Amino acids and carbohydrates decompose in the presence of iron to produce TBARS. Significant interference is observed with the TBA test by the unavoidable presence of metals in food lipids, which also contain natural antioxidants. The literature reports conflicting studies correlating TBA values with sensory evaluations of food lipids. The TBA test is thus notoriously unspecific and unsuitable for complex food materials and biological systems containing non-lipid materials contributing to the color reaction.

F. GAS CHROMATOGRAPHIC METHODS

GC methods are capable of determining volatile oxidation products that are either directly responsible for or serve as markers of flavor development in oxidized lipids. GC analyses for volatile compounds not only correlate with flavor scores by sensory analyses, but also provide sensitive methods to detect low levels of oxidation in various oils and food lipids. Volatile compounds suitable for GC determinations are mainly aldehydes and hydrocarbons. Three commonly used GC methods for volatiles are discussed below.

1. Static Headspace Method

This method includes four steps: i) measuring and sealing a sample into a vial or suitable container, ii) heating to a given temperature until the components of interest diffuse and vaporize into the gas phase and reach or approach equilibrium, iii) injecting an aliquot of the gas phase headspace directly into the gas chromatographic column, either manually with a heated gas-tight syringe, or by using automated systems after pressurizing the heated sample container, and iv) separating the volatile compounds on either a packed or capillary column heated isothermally or by temperature programming.

After equilibration, the concentration determined in the headspace is proportional to the concentration of the volatiles in the sample. Proportionality is based on the equilibrium established when the sample is heated. The method depends on the mass balance of solutes in the sealed system by taking account of their distribution or partition constants, K.

$$K = C_l / C_g$$

where C_l = concentration of solute in liquid phase, and C_g = concentration of solute in gas phase

Quantification is thus based on the distribution of the volatile compounds between the static gas phase and the sample matrix under defined discontinuous gas extraction conditions, keeping the volumes of sample phase and of the gas phase constant. Because this method is very dependent on the nature of the sample matrix, the identity and quantity of each component are determined by comparison with known pure standards added in the same matrix as the sample analysed.

This simple technique permits the quantitative analysis of volatile compounds in various liquid, semi-liquid or solid foods, in biological fluids and tissues, and environmental contaminants in water, air and soils. The method is very sensitive to the equilibrium solute distribution between phases at the temperature selected for the analysis. Equilibration is greatly dependent on the solubility and viscosity of the samples. This method is particularly suited to highly volatile compounds because they have a favorable equilibrium between liquid (or solid) phase and its headspace, producing a higher concentration of volatile compounds in the headspace.

The static headspace method is rapid and suitable for routine consecutive analyses of many samples. This method has the important advantage in not requiring any cleaning between sample injection because only the volatile components are injected into the gas chromatograph and the nonvolatile portion of the sample is retained in the vial. Since this method is applicable to a wide range of liquid, solid and semi-solid samples without manipulation, complex food systems can be analysed directly without manipulations or extractions. This is an important advantage over other methods that are designed for pure oil samples, or require extraction of the lipids from complex food or biological systems. Such extractions can create many artifacts and poor recoveries that may confound the analyses of rancid foods. The main disadvantage of the static headspace method is the difficulty of reaching complete equilibrium with viscous and semi-solid samples and with oxidized polyunsaturated lipid samples that can be easily decomposed during the equilibration heating step. Although equilibration of volatiles can be accelerated by heating, the volatile precursors decompose rapidly at temperatures above 100°C and additional volatiles are then produced. For rapid routine analyses, the samples are heated between 40 and 60°C for 15 to 20 min prior to injection without achieving complete equilibration. Under these non-equilibrium conditions, rigorous standardized control of temperature and time of heating is required to obtain reproducible results.

2. Dynamic Headspace Method

This method also known as "purge-and-trap" is more elaborate than the static headspace method. It includes four steps, i) purging or sweeping a liquid sample with nitrogen or helium in a heated tube or vessel, ii) trapping the vaporized volatiles into a short column containing a porous polymer (Tenax™, Chromosorb™ or Porapaq™) or charcoal with or without cooling, iii) desorbing the volatiles from the trap at elevated temperatures and transferring, by backflushing with carrier gas, into the capillary inlet of the gas chromatograph, and iv) separating the volatile compounds by GC. Commercial instruments are available that allow collection of volatiles from many samples into a series of traps, which are subsequently desorbed sequentially in an automatic temperature-controlled system prior to injection into a crysoscopically cooled capillary GC column for separation of volatiles. Quantification of the volatile compounds is based on the recovery of a suitable internal standard subjected to the same conditions as the sample. The recovery of volatile compounds by this method is based on their relative adsorption and desorption from the trap used under defined continuous gas extraction conditions. The "breakthrough volume" of the porous sorbent used in the trap is defined as the volume of gas required to move a particular compound through the trap column. This property of the sorbent determines the type and amount of volatile compound retained by the trap.

The sampling temperature used in the dynamic headspace method had a significant effect on the volatile profiles of soybean oil (Figure 5-3). At 60°C the volatiles derived from linolenate (2-/3-hexenal and 2,4-heptadienal) contributed 16%, and the volatile derived from linoleate (pentane, hexanal, 2-heptenal and 2,4-decadienal) contributed 50-54% of the total peak area. At 180°C the proportion of linoleate volatiles increased to 74-76% and the linolenate volatiles to 19%.

The dynamic headspace method has the advantage of using lower temperatures than the static headspace method, and permitting enhancement of trace components in complex mixtures of a wide range of volatile compounds. However, this method is slow and not suitable for routine analyses of many samples. Direct purging of liquid oil samples must be avoided to prevent fogging of the volatile compounds that results in serious contamination of the

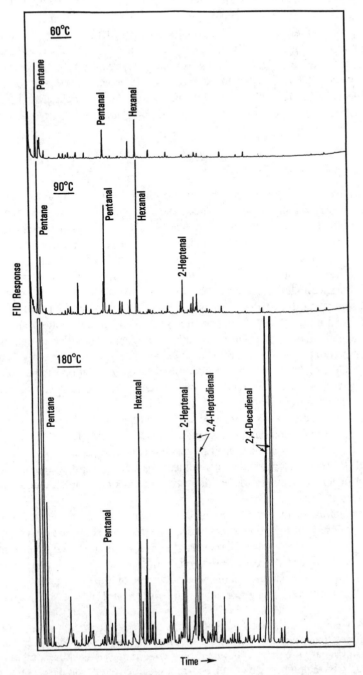

Figure 5-3. Effect of temperature on the dynamic headspace analyses of volatile compounds in oxidized soybean oil (P.V. 9.5) (Selke *et al.*, 1987). Courtesy of the *Journal of the American Oil Chemists' Society*.

transfer lines. The dynamic headspace method requires heated transfer lines and valves and multi-step manipulation of volatiles that must be controlled rigorously. The transfer lines and valves must be cleaned between analyses. Several blank runs are usually necessary to avoid sample contamination causing memory effects in the gas chromatograms. This method thus requires appreciable instrument downtime and maintenance.

3. Direct Injection Method

The sample is transferred into the GC instrument either through an injector insert packed with glass wool or column packing, or through a temperature-controlled external unit. The volatiles generated at the elevated temperature of the injector port are trapped at the head of a packed or capillary GC column either maintained at ambient temperature or cooled cryoscopically. The column is then subjected to temperature programming as required for separation. The direct injection method was first developed to analyse only hydrocarbons by using alumina columns, which completely adsorb the carbonyl and other volatile oxygenated polar compounds in oxidized oils. By this approach, ethane and pentane can be exclusively separated and analysed in oils and foods. The determination of pentane has correlated well with sensory evaluations of oxidized vegetable oils. The direct injection method has been applied successfully for the determination of pentane and ethane in the breath of animals as a non-invasive method to follow *in vivo* oxidation (Chapter 12). This method provided a useful approach to the study of hydroperoxide precursors of volatile decomposition products (Chapter 4).

The direct injection method operates by the same steps as the dynamic headspace method, including *purging* at the injector port, *trapping* at the head of the GC column and *desorbing* by heating the column. However, desorption of the volatiles is slower depending on the temperature program used for separation of the volatiles. For capillary GC columns, better separation of volatile compounds can be achieved by cryoscopic cooling of the head of the column or the entire oven of the gas chromatograph with liquid nitrogen or dry ice before temperature programming.

The direct injection method has similar advantages to the dynamic headspace method in detecting both low-molecular-weight and high-molecular weight volatile compounds. Although the direct injection method permits some enrichment of the volatiles, it is not as efficient as the dynamic headspace method. The direct injection method is much simpler than the dynamic headspace method, more rapid and inexpensive because it does not require modification of the gas chromatograph. The direct injection method is therefore suitable for routine analyses of many samples. However, it requires elevated temperatures which cause further thermal decomposition of hydroperoxides and other lipid oxidation precursors. The sample inserts must be thoroughly cleaned or replaced between injections. It is necessary to run blanks between analyses to avoid or minimize contamination between injected samples.

TABLE 5-3.
Flavor significance of volatile compounds in oxidized soybean oil (peroxide value: 9.5)[a]

Major volatiles	Threshold values	Relative %			Weighted %[b]			Relative order		
		SHS	DHS	DI	SHS	DHS	DI	SHS	DHS	DI
t,t-2,4-decadienal	0.10	0.3	40.5	46.9	0.03	4.1	4.7	7	2	2
t,c-2,4-decadienal	0.02	1.0	21.5	23.8	0.5	10.8	11.9	3	1	1
t,t-2,4-heptadienal	0.04	2.0	13.3	6.5	0.5	3.3	1.6	3	3	3
t-2-heptenal	0.20	8.3	6.7	3.1	0.4	0.33	0.16	5	7	7
t,c-2,4-heptadienal	0.10	2.5	5.4	3.1	0.25	0.54	0.31	6	6	6
hexanal	0.08	24.7	5.4	6.9	3.1	0.68	0.86	2	5	5
pentane	340	38.6	3.7	4.8	1.1[c]	0.11[c]	0.14[c]	10	10	11
t-2-pentenal	1.00	1.2	1.4	1.9	0.01	0.01	0.02	8	8	9
1-octen-3-ol	0.01	0.3	1.1	1.4	0.3	1.1	1.4	4	4	4
2-pentylfuran	2.00	0.5	1.0	1.2	2.5[c]	6.0[c]	6.0[c]	9	9	10
propanal	0.06	20.6	0.0	0.5	3.4	0.0	0.08	1	0	8

Source: Frankel (1985).
[a] Abbreviations: SHS = static headspace, DHS = dynamic headspace, DI = direct injection, t = *trans*, t,c = *trans,cis*, t,t = *trans,trans*
[b] Calculated on the basis of 1-octen-3-ol which has the lowest threshold value
[c] × 10^{-3}

The three GC methods described above were compared in the analysis of volatile compounds in a sample of oxidized soybean oil (Table 5-3). Each method produced significantly different GC profiles (Figure 5-4). The direct injection and dynamic headspace techniques produced much higher proportions of 2,4-decadienal and 2,4-heptadienal than the static headspace method. The static headspace method showed more low-molecular-weight and larger proportions of 2-heptenal, hexanal, pentane and propanal.

As discussed in Section A, the volatile lipid oxidation products vary widely in their sensory impact. It is therefore important to estimate the relative flavor significance of different volatile compounds in addition to their relative concentration in an oil. The flavor significance of the volatiles listed in Table 5-3 were estimated by calculating weighted percentages on the basis of 1-octen-3-ol, which has the lowest threshold value of 0.01. The volatile compositions showed wide differences in flavor significance according to the GC method employed. The direct injection method and the dynamic headspace methods showed the same seven most flavor significant compounds decreasing in the order: *trans,cis*-2,4-decadienal, *trans,trans*-2,4-decadienal, *trans,trans*-2,4-heptadienal, 1-octen-3-ol, hexanal, *trans,cis*-2,4-heptadienal, and 2-heptenal. This similarity in trend indicates that the direct injection and dynamic headspace methods operate on the same principle of purging, trapping and desorbing steps. However, by the static headspace method, which operates by equilibration, the seven most flavor significant and volatile compounds are propanal, followed by hexanal, *trans,trans*- 2,4-heptadienal, 1-octen-3-ol, 2-

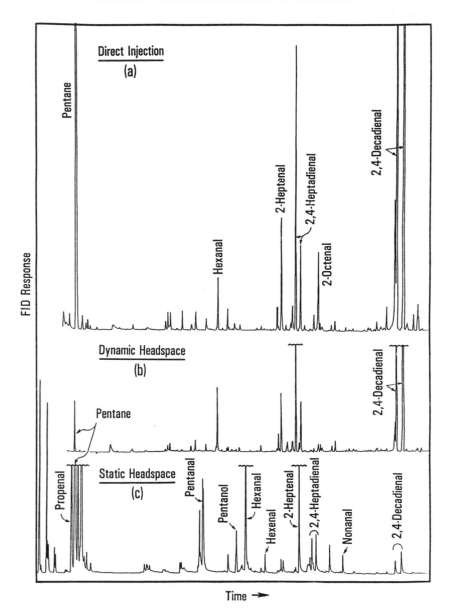

Figure 5-4. Comparison of three gas chromatographic methods for the analyses of volatile compounds in oxidized soybean oil (P.V. 9.5, heated to 180°C) (Snyder et al., 1988). Courtesy of the *Journal of the American Oil Chemists Society*.

heptenal, *trans,cis*-2,4-heptadienal and *trans,trans*-2,4-decadienal. Results of volatile GC analyses by the three methods described above correlated very well with sensory analysis of vegetable oils oxidized at different levels.

G. FLUORESCENCE METHODS

Schiff base compounds formed by the interaction of oxidation products with proteins, phospholipids and nucleic acids produce chromophores showing characteristic fluorescence spectra. The Schiff base formed between malonaldehyde and amino acids is attributed to the conjugated structure: -NH=CH-CH=CH-NH-. Lipid-soluble fluorescence chromophores are produced from oxidized phospholipids and from oxidized fatty acid esters in the presence of phospholipids. These chromophores have fluorescence emission maxima at 435-440 nm and excitation maxima at 365 nm. The Schiff base of malonaldehyde and phospholipids has a higher wavelength maximum for emission (475 nm) and excitation (400 nm). The interaction between oxidized arachidonic acid and dipalmitoyl phosphatidylethanolamine produce similar fluorescence spectra (maximum excitation at 360-90 nm and maximum emission at 430-460 nm). The products from oxidized arachidonic acid and DNA have characteristic fluorescence spectra, with excitation maximum at 315 nm and emission maximum at 325 nm. Similar fluorescence spectra, with excitation maximum at 320 nm and emission maximum at 420 nm, are obtained from the interaction of either lipid hydroperoxides and secondary oxidation products with DNA in the presence of metals and reducing agents, or different aldehydes, ketones and dimeric compounds from oxidized linolenate. Therefore, the Schiff base produced from various oxidized lipids and phospholipids and DNA may be considered to be due to a mixture of closely related chromophores.

Several fluorescence techniques have been used to investigate oxidation products in animal tissues, in frozen fish muscle during storage, in freeze-dried meats, and in oxidatively damaged soybeans. Mixtures of tissues are extracted with chloroform-methanol (2:1, by volume), homogenized, mixed with water and centrifuged. The chloroform layer is analysed for fluorescence using quinine sulfate to calibrate fluorescence intensity and wavelength. Fluorescence has also been measured directly in the chloroform-methanol extract of ground samples of oxidatively damaged soybeans. Fluorescence spectra for oxidized meats show maximum excitation at 350 nm and emission at 440 nm, and for damaged beans maximum excitation at 364 nm and emission at 437 nm.

Fluorescence techniques are very sensitive. On a molar basis the amount of malonaldehyde detected by fluorescence is 10 to 100 times more sensitive than the colorimetric TBA assay. However, this method is not specific as it measures complex mixtures resulting from the interactions of oxidized lipids, unsaturated aldehydes and malonaldehyde with proteins, peptides, amino acids, phospholipids, DNA and nucleic acids. These interactions involve oxidized or polymerized species of proteins or amino acids also. The fluorescence method provides a non-specific but sensitive measure of oxidative deterioration in complex foods such as meats and fish which are often difficult to analyse for carbonyl compounds that are complexed and covalently bonded with proteins.

METHODS TO DETERMINE EXTENT OF OXIDATION

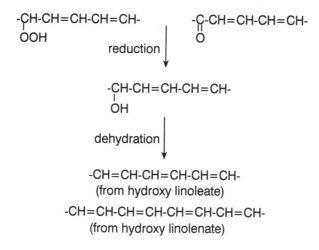

Figure 5-5. Production of conjugatable oxidation products by reduction and dehydration sequence.

H. OCTANOATE METHOD

This method determines the residual ester product remaining attached to triacylglycerol after thermal cleavage of oxidized triacylglycerols. We have seen in Chapter 4 that decomposition of hydroperoxides produce aldehydes and esters by cleavage on either side of the hydroperoxide group (Figure 4-7). Methyl octanoate was an important cleavage product of hydroperoxides of fatty acid esters. The corresponding hydroperoxides of triacylglycerols produce methyl octanoate after alkali transmethylation. In order to reflect the total oxidation of an oil, the oxidized sample is heated under vacuum at 180°C to decompose all oxidation compounds and remove interfering volatile decomposition products. After thermal decomposition, methyl octanoate can be determined quantitatively by GC, as a measure of oxidation of triacylglycerols after the initial hydroperoxides are decomposed during processing. The octanoate assay thus provides a measure of the oxidative state of crude oils and oils at different stages of processing prior to deodorization. The results provide the same information as the anisidine method for aldehyde esters and aldehydo glycerides. Although the octanoate method is more specific than the colorimetric anisidine test, it may not be as sensitive.

I. CONJUGATABLE OXIDATION PRODUCTS

More conjugated chromophores are produced by converting PUFA hydroperoxides and oxodienes in oxidized lipids into the corresponding allylic alcohols by chemical reduction with sodium borohydride, followed by acid dehydration with ethanolic sulfuric acid (Figure 5-5). This reaction sequence

TABLE 5-4.
Analyses of conjugable oxidation products (COP) in different food lipids.[a]

Food lipids [b]	Storage at 25°C, time	P.V.	Oxodiene	COP
Margarine	0 time	nil	nil	0.5
Margarine	120 days	31.4	0.3	11.9
Pastry mix	0 time	nil	0.3	0.5
Pastry mix	147 days	2.0	0.5	1.9
Salad cream	0 time	1.2	0.2	2.3
Salad cream	110 days	51.0	0.8	23.0
Roasted peanuts	0 time	nil	0.2	4.5
Roasted peanuts	112 days	17.3	0.3	10.3
Soup powder	0 time	2.4	0.2	1.0
Soup powder	113 days	2.3	0.3	1.0

[a] Adapted from Parr and Swoboda (1976)
[b] Lipid extracted with a 1:2:0.8 (v/v) mixture of chloroform, methanol and water.

has been used to characterize hydroperoxides of oleate, linoleate, and linolenate.

Accordingly, conjugated trienes absorbing at 268 nm are produced by dehydration of hydroxy linoleate, and conjugated tetraenes absorbing at 301 nm are produced by dehydration of hydroxy linolenate and hydroxy esters from PUFA containing more than three double bonds. The chromophores produced are measured by absorbance in the ultraviolet. The sum of the absorbance increases at 268 and 301 for a 1% (w/v) lipid solution after the dehydration step, is defined as "conjugatable oxidation product value". The concentration of oxodienes present in the original mixture is estimated by determining the decrease in absorbance at 275 nm observed after reduction. This procedure is useful as a quality control measure to distinguish between types of unsaturated fatty acids and unsaturated carbonyls in oxidized food lipids (Table 5-4).

BIBLIOGRAPHY

Barthel,G. and Grosch,W. Peroxide value determination-Comparison of some methods. *J. Am. Oil Chem. Soc.* **51**, 540-544 (1974).

Csallany,A.S., Guan,M.D., Manwaring,J.D. and Addis,P.B. Free malonaldehyde determination in tissues by high-performance liquid chromatography. *Anal. Biochem.* **142**, 277-283 (1980).

Dahle,L.K., Hill,E.G. and Holman,R.T. The thiobarbituric acid reaction and the autoxidations of polyunsaturated fatty acid methyl esters. *Arch. Biochem. Biophys.* **98**, 253-261 (1962).

Dennis,K.J. and Shibamoto,T. Gas chromatographic analysis of reactive carbonyl compounds formed from lipids upon UV-irradiation. *Lipids* **25**, 460-464 (1990).

Esterbauer,H. and Cheeseman,K.H. Determination of aldehydic lipid peroxidation products: malonaldehyde and 4-hydroxynonenal. *Methods in Enzymology* **186B**, 407-421 (1990).

Fioriti,J.A., Kanuk,M.J. and Sims,R.J. Chemical and organoleptic properties of oxidized fats. *J. Am. Oil Chem. Soc.* **51**, 219-223 (1974).

Fishwick,M.J. and Swoboda,P.A.T. Measurement of oxidation of polyunsaturated fatty acids by spectrophotometric assay of conjugated derivatives. *J. Sci. Food Agric.* **28**, 387-393 (1977).

Frankel,E.N. Analytical methods used in the study of autoxidation processes, in *Autoxidation in Food and Biological Systems*, pp. 141-170 (1980) (edited by M.G. Simic and M. Karel), Plenum Press, New York.
Frankel,E.N. Chemistry of autoxidation: Mechanism, products and flavor significance, in *Flavor Chemistry of Fats and Oils*, pp. 1-37 (1985) (edited by D.B. Min and T.H. Smouse), American Oil Chemists' Society, Champaign, Illinois.
Frankel,E.N. In search of better methods to evaluate natural antioxidants and oxidative stability in food lipids. *Trends Food Sci. Technol.* **4**, 220-225 (1993).
Frankel,E.N. and Neff,W.E. Formation of malonaldehyde from lipid oxidation products. *Biochim. Biophys. Acta* **754**, 264-270 (1983).
Frankel,E.N., Nash,A.M. and Snyder,J.M. A methodology study to evaluate quality of soybeans stored at different moisture levels. *J. Am. Oil Chem. Soc.* **64**, 987-992 (1987).
Frankel,E.N., Neff,W.E., Brooks,D.D. and Fujimoto,K. Fluorescence formation from the interaction of DNA with lipid oxidation degradation products. *Biochim. Biophys. Acta* **919**, 239-244 (1987).
Fritsch,C.W. and Gale,J.A. Hexanal as a measure of rancidity in low fat foods. *J. Am. Oil Chem. Soc.* **54**, 225-228 (1977).
Fujimoto,K., Neff,W.E. and Frankel,E.N. The reaction of DNA with lipid oxidation products, metals and reducing agents. *Biochim. Biophys. Acta* **795**, 100-107 (1984).
Gaddis,A.M., Ellis,R., Currie,G.T. and Thornton,F.E. Carbonyls in oxidizing fat. X. Quantitative differences in individual aldehydes isolated from autoxidized lard by mild methods of extraction. *J. Am. Oil Chem. Soc.* **43**, 242-244 (1966).
Gray,J.I. Measurement of lipid oxidation. *J. Am. Oil Chem. Soc.* **55**, 539-546 (1978).
Gray,J.I. Simple chemical and physical methods for measuring flavor quality of fats and oils, in *Flavor Chemistry of Fats and Oils*, pp. 223-239 (1985) (edited by D.B. Min and T.H. Smouse), American Oil Chemists' Society, Champaign, IL.
Henick,A.S., Benca, M.F. and Mitchell, J.H., Estimating carbonyl compounds in rancid fats and foods. *J. Am. Oil Chem. Soc.* **42**, 839-841 (1965).
Hirayama,T., Yamada,N., Nohara,M. and Fukui,S. The high-performance liquid chromatographic determination of total malonaldehyde in vegetable oils with danzyl hydrazine. *J. Sci. Food Agric.* **35**, 338-344 (1984).
Hirayama,N., Yamada,N., Nohara,M. and Fukui,S. High-performance liquid chromatographic determination of malonaldehyde in vegetable oils. *J. Assoc. Off. Anal. Chem.* **66**, 304-308 (1983).
Holm,U., Ekbom,K. and Wode,G. Determination of the extent of oxidation of fats. *J. Am. Oil Chem. Soc.* **34**, 606-609 (1957).
Hoyland,D.V. and Taylor,A. J. A review of the methodology of the 2-thiobarbituric acid test. *Food Chem.* **40**, 271-291 (1991).
Janero,D.R. Malonaldehyde and thiobarbituric acid-reactivity as diagnostic indices of lipid peroxidation and tissue injury. *Free Radical Biol. Med.* **9**, 515-540 (1990).
Kamarei,A.R. and Karel,M. Assessment of autoxidation in freeze-dried meats by a fluorescence assay. *J. Food Sci.* **49**, 1517-1520, 1524 (1984).
Karel,M. Lipid oxidation, secondary reactions, and water activity in foods, in *Autoxidation in Foods and Biological Systems*, pp. 191-206 (1980) (edited by M.G. Simic and M. Karel), Plenum Press, New York.
List,G.R., Evans,C.D., Kwolek,W.F., Warner,K. and Boundy,B.K. Oxidation and quality of soybean oil: A preliminary study of the Anisidine test. *J. Am. Oil Chem. Soc.* **51**, 17-21 (1974).
Marcuse,R. and Johansson,L. Studies on the TBA test for rancidity grading: II. TBA reactivity of different aldehyde classes. *J. Am. Oil Chem. Soc.* **50**, 387-391 (1973).
O'Mahony,M. *Sensory Evaluation of Food. Statistical Methods and Procedures*, (1986) Marcel Dekker, New York.
Parr,L.J. and Swoboda,P.A.T. The assay of conjugable oxidation products applied to lipid deterioration in stored foods. *J. Food Technol.* **11**, 1-12 (1976).
Patton,S. Malonaldehyde, lipid oxidation, and the thiobarbituric acid test. *J. Am. Oil Chem. Soc.* **51**, 114 (1974).
Peers,K.E. and Swoboda,P.A.T. Octanoate: An assay for oxidative deterioration in oils and fats. *J. Sci. Food Agric.* **30**, 876-880 (1979).
Selke,E. and Frankel,E.N. Dynamic headspace capillary gas chromatographic analysis of soybean volatiles. *J. Am. Oil Chem. Soc.* **64**, 749-753 (1987).

Sinnhuber,R.O., Yu,T.C. and Yu,T.C. Characterization of the red pigment formed in the 2-thiobarbituric acid determination of oxidative rancidity. *Food Res.* **23**, 626-634 (1958).
Snyder,J.M., Frankel,E.N. and Selke,E. Capillary gas chromatographic analyses of headspace volatiles from vegetable oils. *J. Am. Oil Chem. Soc.* **62**, 1675-1679 (1985).
Snyder,J.M., Frankel,E.N., Selke,E. and Warner,K. Comparison of gas chromatographic methods for volatile lipid oxidation compounds in soybean oil. *J. Am. Oil Chem. Soc.* **65**, 1617-1620 (1988).
Tappel,A.L. Measurement of and protection from in vivo lipid peroxidation, in *Free Radicals in Biology*, Volume IV, pp. 1-47 (1980) (edited by W.A. Pryor), Academic Press, New York.
Tsoukalas,B. and Grosch,W. Analysis of fat deterioration. Comparison of some photometric tests. *J. Am. Oil Chem. Soc.* **54**, 490-493 (1977).
Waltking,A.E. and Goetz,A.G. Instrumental determination of flavor stability and its correlation with sensory flavor responses. *CRC Critical Rev. Food Sci. Nutr.* **19**, 99-132 (1983)
Ward,D.D. The TBA assay and lipid oxidation: an overview of the relevant literature. *Milchwissenschaft* **40**, 583-588 (1985).
Warner,K. and Eskins,N.A.M. (editors) *Methods to Assess Quality and Stability of Oils and Fat-Containing Foods*, (1995), American Oil Chemists' Society Press, Champaign, Illinois.
Warner,K., Frankel,E.N. and Mounts,T.L. Flavor and oxidative stability of soybean, sunflower and low-erucic acid rapeseed oils. *J. Am. Oil Chem. Soc.* **66**, 558-564 (1989).
Warner,K. and Nelsen,T. AOCS collaborative study on sensory and volatile compound analyses of vegetable oils. *J. Am. Oil Chem. Soc.* **73**, 157-166 (1996).

CHAPTER 6
STABILITY METHODS

A. STABILITY AS PARAMETER OF QUALITY CONTROL

Stability can be defined as the resistance of a lipid to oxidation and to the resulting deterioration due to the generation of flavor and odor causing oxidative rancidity and decreasing food quality. Tests for *oxidative stability* attempt to accelerate the normal oxidation process to yield results that can be translated into quality parameters for different food lipids. Stability methods thus constitute an important quality control tool to predict the shelf-life of foods. Stability methods are also important to evaluate the effect of changes in food formulations, processing parameters and other factors on lipid oxidation. Stability methods are particularly valuable in research and evaluation of antioxidants and their effects on protection of foods against lipid oxidation (Chapter 8).

The literature on the evaluation of oxidative stability of food lipids is extensive but the results of different investigators are difficult to interpret because of variations in test conditions and end-point of lipid oxidation. Conditions used by many workers for accelerated oxidation have been too drastic. The use of stability tests requiring elevated temperatures of oxidation (100°C and higher) is particularly questionable, because samples develop excessive levels of rancidity, which are not relevant to normal storage conditions. Further complications arise in multi-component foods in which interactions between lipids, carbohydrates and proteins at elevated temperatures produce materials acting as catalysts or inhibitors. As new oil phases are formed, the solubility and partition of different reactants between different phases change with temperature. Metals and exposure to light may cause decomposition of hydroperoxides and produce misleadingly low peroxide values even though rancidity increases and sensory quality decreases.

The oxidative stability of food lipids has been evaluated by a variety of methods using a wide range of conditions and techniques to measure oxidation. The choice of appropriate methods and oxidation conditions is critical in the interpretation of stability data. Thus, interpretation of data pertaining to lipid oxidations should take into account the limitation of the methodology used. The oxidative stability of unsaturated lipids can be estimated by determining the extent of oxidation produced under various defined and standardized conditions. The susceptibility of the samples to oxidation can be measured reliably if the samples are initially fresh (with a peroxide value close to 0) and the conditions of oxidation are sufficiently mild and relevant to normal storage conditions. Excessive deteriorations beyond normal levels are to be avoided for many reasons discussed below. The results of stability measurements are useful in comparing the quality of raw materials in various food applications, in attempting to predict future quality performance of fats used as food ingredients, and in comparing the performance of antioxidants and stabilizers (Chapters 8 and 9).

For the judicious choice of stability methods we need to address the following questions:

a) what is happening to the unsaturated lipid or food under a range of realistic storage conditions?
b) what oxidative stress should be used? Effect of temperature, metals, light;
c) what end-point is relevant to the sample in question? Acceptable level of quality deterioration.
d) what lipid oxidation methods should be used? How significant are the results with respect to quality deterioration? Different products should be measured, including primary and secondary decomposition products.
e) how is stability influenced by the interfacial relationship between food, oxidants and antioxidants?

Methods to measure lipid oxidation have generally been unspecific and not sufficiently sensitive to measure flavor and low levels of oxidative deterioration of food lipids. Several specific complementary methods are needed to determine the contribution of lipid oxidation products formed at different stages of the multi-step process of oxidative deterioration in foods. Since polyunsaturated hydroperoxides decompose more readily at higher temperatures to form aldehydes causing rancidity, it is important to measure both aldehydes and hydroperoxides to monitor lipid oxidation. For the reliable prediction of shelf-life of foods containing polyunsaturated lipids, it is essential to use more than one specific method to determine oxidation and to store samples at several temperatures, preferably between 40 and 60°C.

B. DEVELOPMENT OF RANCIDITY IN FOODS DURING STORAGE

Lipid oxidation in food products develops slowly initially, and then accelerates at later stages during storage (Figure 6-1). The *induction period* (IP), or shelf-life at room temperature, is defined as the time to reach a constant percent oxidation of the fat or the end of shelf-life. The induction period is measured as the time required to reach an end-point of oxidation corresponding to a level of detectable rancidity. Alternatively, the induction time is taken as the time required for a sudden change in rate of oxidation, by estimating the intersect or cross-point between the initial and final rates of oxidation (IP_A and IP_B). The end-points or acceptability limits for vegetable

STABILITY METHODS

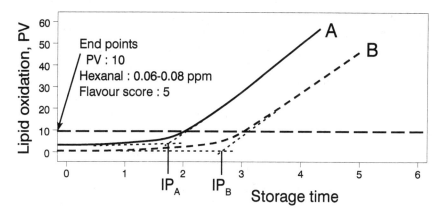

Figure 6-1. Development of rancidity in foods during storage. Storage time are given in arbitrary values. PV = peroxide value; IP = induction period.

oils are shown in Figure 6-1 as a broken line: peroxide value of 10 for a polyunsaturated oil, or hexanal content of 0.06-0.08 ppm, or flavor score of 5. This end-point is reached after 2 days/weeks/months (depending on conditions) in sample A, compared to 3 days/weeks/months in sample B. If we measure peroxide value or hexanal content, we refer to this evaluation as *oxidative stability*, if we measure a flavor score, then we refer to it as *flavor stability*. The difference in stability between samples A and B could be due to their different initial levels of oxidation at the start of the storage period. From the crude stage to different stages of processing, some oxidation can take place, which will affect the stability of the food product. This oxidation may influence the subsequent stability during storage.

The distinction between oxidative stability and flavor stability is particularly important in the evaluation of linolenate-containing oils, such as soybean and rapeseed oils. These oils undergo flavor deterioration sooner and at an earlier stage of oxidation than linoleate-containing oils, such as cottonseed and corn oils. The difference between these oils can be attributed to the lower threshold values of volatiles produced from oxidation of linolenate compared to volatiles produced from oxidation of linoleate (Chapter 4).

C. STABILITY OF DIFFERENT SYSTEMS

We may consider three types of food systems:

Bulk oil system : Oil is the continuous phase and air bubbles or air-surface is the dispersed phase. Hydrophilic antioxidants and prooxidants may be oriented in the air-oil interface, and lipophilic antioxidants and prooxidants remain in solution in the oil phase.

Oil-in-water emulsion: Water is the continuous phase and oil is the dispersed phase. This dispersed phase is stabilized by an emulsifier. Lipophilic antioxidants and prooxidants are generally sufficiently surface active to be oriented in the oil-water interface, whereas hydrophilic antioxidants and prooxidants move to the water phase and become diluted.

Liposomes: Phospholipids spontaneously form bilayers called liposomes when dispersed in water. Hydrophilic antioxidants and prooxidants would be oriented at the polar end of the phospholipid bilayers, while the lipophilic antioxidants and prooxidants would be located between the long chain of the lecithin and the water phase. Multiphase food systems will be discussed further in Chapter 9.

D. ACCELERATED PARAMETERS

To estimate the oxidative stability or susceptibility of a fat to oxidation, the sample is subjected to an accelerated oxidation test under standardized conditions and a suitable end-point is chosen to determine appropriate levels of oxidative deterioration (Figure 6-1). Several parameters such as temperature (60-140°C), metal catalysts (5-100 ppm), oxygen pressure (3-165 psi), or variable shaking to increase reactant contact, are manipulated to accelerate oxidation and development of rancidity in oils and emulsions. The oxidation level used for an end-point varies widely according to the time desired to obtain stability data. For practical purposes, predictions of oxidative stability in foods and oils based on measurements of induction period should be related to actual product shelf-life, and the conditions used should be as close as possible to those under which the food is stored. To translate the induction period obtained under accelerated conditions to the actual shelf-life of a product, it is necessary to use an arbitrary factor based on prior experience with the desired product.

E. EFFECT OF TEMPERATURE

Temperature is the most important factor to be considered in evaluating the oxidative stability of unsaturated fats, because the mechanism of oxidation changes with temperature, and different hydroperoxides of linoleate and linolenate, acting as precursors of volatile flavors, decompose at different temperatures. Because the rate of oxidation is exponentially related to temperature, the shelf-life of a food lipid decreases logarithmically with increasing temperature (Section J). At ambient temperatures, the rate of lipid oxidation is independent of the oxygen pressure (Chapter 1). However, at elevated temperatures, the rate of lipid oxidation becomes dependent on oxygen pressures because the solubility of oxygen decreases and becomes limiting during the progress of the reaction. In stability tests, it is important therefore to use several storage temperatures and in a range as low as practical.

The activation energy (or the additional energy needed by reactants to form products) of lipid oxidation is much higher in the presence of antioxidants than in their absence, because antioxidants lower the rates of oxidation by increasing the overall energy of activation. An Arrhenius plot of log induction period versus the temperature shows that the effectiveness of antioxidants

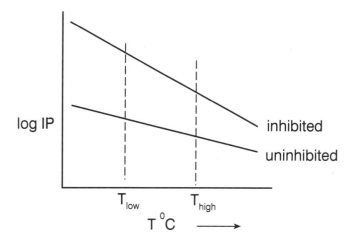

Figure 6-2. Arrhenius plots for inhibited and uninhibited lipid oxidation. IP = induction period; T = temperature. Adapted from Ragnarsson and Labuza (1977).

increases as temperature decreases (Figure 6-2). The overall rate constant for autoxidation is proportional to the reciprocal of the induction period. The temperature coefficients (or how much faster a sample is oxidized by raising the temperature) are different for a fat according to the relative concentration and effectiveness of natural or added antioxidants. Thus, the temperature coefficient is higher for an animal fat which has very low levels of antioxidants, than a vegetable oil containing natural antioxidants. Consequently, any prediction for lower temperatures requires testing at several different temperatures.

The accelerated temperature used for stability tests is greatly dependent on the relative fatty acid composition of the oils tested. In general, the more polyunsaturated the oils the lower temperatures should be used to test their oxidative stability. The significance of the induction period depends on the degree of unsaturation of the oils tested. With a highly polyunsaturated oil, the level of oxidation detected at the end-point at high temperatures is too high and beyond the point in which rancid flavors are detected. The effect of heating foods containing phospholipids and carbohydrates is also confounded by the formation of browning reaction products that may have reducing properties and increase the oxidative stability of lipids (see Chapter 10).

F. STABILITY OF OXIDATION PRODUCTS

The large number of precursors of volatile decomposition products affecting the flavor of oils has been discussed in Chapter 4. Only qualitative information is available on the relative oxidative stability of hydroperoxides, aldehydes and secondary oxidation products. As observed with the unsaturated fatty ester precursors, the stability of hydroperoxides and unsaturated aldehydes decreases with higher unsaturation. Different hydroperoxides of unsaturated lipids, acting as precursors of volatile flavor compounds, decompose at different temperatures. Hydroperoxides of linolenate decompose more readily and at lower temperatures than hydroperoxides of linoleate and oleate. Similarly, the alkadienals, are less stable than alkenals, which in turn are less stable than alkanals. For secondary products, dimers are less stable than dihydroperoxides, which are less stable than cyclic peroxides.

Common vegetable oils containing mixtures of oleate, linoleate and linolenate produce 14 positional isomers of hydroperoxides by autoxidation and photosensitized oxidation (Chapters 3 and 4). More secondary products, dimers and oligomers are produced from linolenate-containing oils than from linoleate-containing oils at lower levels and temperatures of oxidation. Fish oils containing long-chain n-3 polyunsaturated fatty acids produce volatile decomposition products at lower temperatures (below 40°C) than vegetable oils containing linoleate and linolenate (above 60°C).

G. STANDARD ACCELERATED STABILITY TESTS

To estimate oxidative stability or susceptibility of a fat to oxidation, the sample is subjected to an accelerated oxidation test under standardized conditions and a suitable end-point is chosen to determine signs of oxidative deterioration.

1. Storage at Ambient Conditions

Although ambient conditions are ideal, this test is too slow to be of practical value. Results at ambient conditions are very difficult to reproduce because environmental variables are difficult to control over prolonged periods of storage.

a. *Light oxidation.* By exposure to light, the stability of vegetable oils can be screened rapidly. This test is especially useful for soybean oil and other vegetable oils that are very sensitive to light oxidation. However, the mechanism of photosensitized oxidation is different from that of autoxidation (Chapter 3), and different flavor precursors are formed with different volatile breakdown products which may have different flavor significance.

b. *Metal accelerated oxidation.* The addition of metal catalysts such as iron or copper provides a rapid screening test for stability, which emphasizes the breakdown products more than the flavor precursors. Because lipid oxidation is generally catalysed by trace metals, they become the dominating catalyst at moderately elevated temperatures such as 40°C, and adding metals may be considered as a meaningful way of accelerating oxidation. Hemin and metallic catalysts are useful for estimating the oxidative stability of food emulsions.

2. Storage at Elevated Temperatures

a. *Weight-gain method.* This method consists of measuring the increase in weight of an oil sample due to oxygen absorption. The test is not sensitive and the end-point is questionable because it requires too high a level of oxidation, that is beyond the point where flavor deterioration occurs, especially with vegetable oils containing linolenic acid and fish oils containing n-3 polyunsaturated fatty acids.

b. *Schaal oven.* This test, also referred to as *oven test*, involves heating the sample at 50 to 60°C until it becomes rancid by smell or taste, or reaches a suitable end-point based on peroxide value, conjugated diene or carbonyl value. This test uses relatively mild temperatures and has the fewest problems. The results of this test correlate best with actual shelf-life because the end-point represents a lower degree of oxidation. The determinations of peroxide values, conjugated diene or carbonyl values of oils are more meaningful with oils heated at 60°C or below. At 100°C and higher temperatures, the results of these determinations become questionable and difficult to interpret because the hydroperoxides are thermally decomposed and side reactions such as polymerization and cleavage become significant and breakdown products are formed in significant amounts.

c. *Active Oxygen Method (AOM), and Rancimat.* Air is bubbled through a sample of oil in special test tubes heated at 98-100°C and the progress of oxidation is followed by peroxide value determination in the AOM test, and by conductivity measurements in the Rancimat test. In the automated Rancimat test, the effluent gases are led into a tube containing distilled water and the changes in conductivity measured are due to the formation of volatile acids by thermal oxidation.

Both AOM and Rancimat tests are not reliable because they are run at elevated temperatures at which the mechanism of lipid oxidation changes. At these high temperatures, oxidation becomes more dependent on oxygen concentration, the solubility of oxygen decreases, and the concentration of available oxygen changes drastically because of the rapid rates of oxidation. The dependence on oxygen increases with degree of oxidation of the oils. Lipid oxidation proceeds erratically at elevated temperatures by increasing the activity of contaminating metals and by promoting the prooxidant activity of tocopherols in vegetable oils (Chapter 8). At elevated temperatures, polymerization and cyclization become important, and these side-reactions are not significant at normal storage temperatures. Volatile acids measured by the automated Rancimat method are only produced in significant amounts at elevated temperatures and, therefore, are not relevant to normal storage conditions. In spite of these serious pitfalls, the AOM and the Rancimat methods are commonly used to compare the relative stability of very similar samples. However, the results of these high-temperature stability tests can be misleading in the evaluation of the effectiveness of antioxidants (Chapter 8). Antioxidants are markedly influenced by the effect of oxidation temperature, and widely diverging results are frequently obtained when antioxidants are compared at temperatures below 60°C with temperatures above 60°C.

H. ALTERNATIVE STABILITY METHODS

1. Gas Analysis

This method measures oxygen in the headspace of an oil stored in a suitable container under variable conditions by gas chromatography using a thermal conductivity detector. This is a more sensitive and sophisticated method than oxygen uptake by the classical manometric methods. The information provided is limited, however, because it is not related to breakdown products causing flavor deterioration.

2. Analyses of Volatiles by Gas Chromatography

These tests are the most useful and objective for estimating oxidative and flavor stability, because the volatile products measured correlate well with rancidity if they are formed under relatively mild conditions of oxidation (60°C or below). These methods are discussed in detail in Chapter 5, Section F.

3. Hydroperoxides

These primary products of oxidation can be determined by gas chromatography, after preparing suitable derivatives, and by HPLC using either an ultraviolet or refractive index detector. Although hydroperoxides are the most important precursors of flavor compounds, their importance depends on their relative stability and the composition of the volatiles they produce under different conditions. The yields of hydroperoxides decrease as their rates of decomposition increase, in the order: oleate, linoleate, linolenate, higher polyunsaturated fatty acids (containing more than three double bonds).

4. Carbonyl Compounds

These secondary products of oxidation can also be determined by gas chromatography or HPLC, either directly or after preparing suitable derivatives. The HPLC method is less sensitive than the direct gas chromatography methods. HPLC can be carried out under mild conditions and is subject to less artifacts than gas chromatographic methods, which require higher temperatures. However, HPLC requires the use of organic solvents which cause losses of volatile carbonyl compounds. HPLC is also unsuitable for the determination of volatile hydrocarbons formed by decomposition of hydroperoxides.

TABLE 6-1.
Ranking of lipid oxidation methods.

Methods	Sensitivity	Precision	Information
Sensory	High	Low	High
Volatiles by GC	High	Low	High
Ultraviolet absorption	High	High	Low
Carbonyls	Low	High	Low
Anisidine value	Low	High	Low
Peroxide value	Low	High	Low
Oxygen uptake	Low	High	Low
TBA	Low	High	Low
Carotene bleaching	High	Low	Low
Volatile acids (Rancimat)	Low	Low	Low

5. Thermal Analyses

These techniques measure the effect of temperature on physical or chemical changes, such as specific heat and sample weight, in a lipid material. *Differential scanning calorimetry* measures the heat absorbed or gained (ΔH) by the sample to obtain the temperatures and heat of the exothermic oxidation. This technique involves heating the sample and reference compounds to continuously increasing elevated temperature (120-150°C), with either purged oxygen or oxygen pressure (500 psig), for the rapid estimation of oxidative stability. The end-point is measured as the time where a rapid exothermic reaction occurred.

Thermogravimetric analysis measures the effect of temperature on change in weight resulting from oxygen absorption during oxidation. The technique can also be carried out as an accelerated stability test either isothermally or by linear temperature programming from ambient to elevated temperatures. When run isothermally, the end-point is the time where a rapid weight gain is observed. When run by temperature programming, the end-point is the temperature where the rapid weight gain is detected. Although these methods correlate well with the standard AOM test, they have the same problems in requiring elevated temperatures where oxidation is too high and beyond the level of detection of rancidity flavors.

I. LIPID OXIDATION METHODS

To evaluate oxidative stability, different methods are used to measure lipid oxidation after the sample is oxidized under standardized conditions to a suitable end-point. In Table 6-1, different lipid oxidation tests are ranked in decreasing order of usefulness in predicting the stability or shelf-life of a food product. Methods for sensory evaluations, conjugated diene, gas chromatography of volatiles, peroxide values and thiobarbituric acid-reacting substances were discussed in Chapter 5.

1. Sensory Methods

These methods provide the most useful information because it is related to consumer acceptance of the food product based on odor and taste. This technique is very sensitive and provides information on flavor stability; unfortunately it is expensive and dependent on taste panels that must be properly trained and on scoring, which is not reproducible from laboratory to laboratory.

2. Analysis of Volatiles by Gas Chromatography

This approach is most closely related to flavor evaluation and suitable for comparison with results of sensory panel tests. Although it is more precise than sensory methods, the results vary with different unsaturated oils, and with different additives such as antioxidants and metal inactivators. This method provides useful data on the origin of volatiles and flavor precursors. While total and individual volatiles can be correlated with flavor scores, their flavor significance and impact on flavor stability are not clearly established and are difficult to evaluate.

3. Ultraviolet Absorption Measurements

The determination of conjugated dienes by ultraviolet spectrophotometry is related to the contents of polyunsaturated hydroperoxides, which act as flavor precursors. These measurements are sensitive and reproducible, but the information is only useful to determine precursors of volatiles formed from polyunsaturated oils.

4. Carbonyls and Anisidine Values

These tests provide similar information as gas chromatographic analyses for volatiles, but they include also non-volatile precursors. The results are less sensitive but more precise than gas chromatographic analyses. Although the information may be related to flavor scores, the proportion of flavor significant carbonyls is difficult to assess.

5. Peroxide Values

This measurement is not sensitive but its precision is relatively high. The results are related to flavor precursors such as hydroperoxides but not to the flavor compounds. This method is empirical and the results are misleading in samples that have been thermally abused or subjected to light oxidation. The peroxide value of samples heated under the activated oxygen method or AOM conditions has no relation to the actual level of oxidation in the same samples stored under ambient conditions.

STABILITY METHODS 109

6. Oxygen Absorption Methods

These methods measure the amount of oxygen absorbed during lipid oxidation. Because they measure a reactant and not a product of oxidation, they are of limited sensitivity and require high levels of oxidation as end-point for induction periods. The information provided is not closely related to flavor stability.

7. Thiobarbituric Acid (TBA) Method

Although less sensitive this method is relatively precise. However, the test is not specific, and the results can be misleading because many primary and secondary oxidation products form TBA-reactive substances. This test is more sensitive with polyunsaturated fats containing three or more double bonds and not sensitive for the oxidation products of oleic and linoleic acids. The results overestimate the amount of lipid oxidation because extraneous materials in foods such as browning reaction products, protein and sugar degradation products, interfere with the TBA test (see Chapter 5).

8. Carotene Bleaching

This spectrophotometric method measures carotene bleaching occurring by co-oxidation of linoleic acid. It is simple and sensitive, but not specific and subject to interference from oxidizing and reducing agents present in crude extracts. Linoleic acid is not an appropriate substrate for food lipids consisting mainly of triacylglycerols because it forms micelles in aqueous systems. Micelles have markedly different colloidal properties and oxidative stabilities than emulsified triacylglycerols; antioxidants and oxidants behave differently in micelles than in triacylglycerol systems (see Chapter 9).

9. Volatile Acids by Rancimat

This automated instrumental test measures the conductivity of low-molecular weight fatty acids (mainly formic acid) produced during autoxidation of fats at elevated temperatures. (Section G.2.c.). The results obtained by this method require relatively high levels of oxidation (peroxide values of 50-100); the end-points are unreliable because high temperatures are necessary to obtain detectable amounts of organic acids, which are produced beyond the normal level for rancidity.

J. SHELF-LIFE PREDICTION

We have seen earlier that since the rate of oxidation is exponentially related to temperature, the shelf-life of a food lipid decreases logarithmically with increasing temperature (Section E). Assuming the Arrhenius model, the true shelf-life of a food product at ambient temperature can be thus approximated by extrapolating measurements of rancidity obtained at higher temperatures projected to a lower temperature (Figure 6-3). A well recognized temperature acceleration factor, known as Q_{10}, is based on the increase in oxidation rate produced by a 10°C increase in temperature (T). The shelf-life is defined as the reciprocal of the accelerator factor, Q_{10}, as follows:

Q_{10} = quality measure at T + 10 / quality measure at T
Q_{10} = shelf-life at T / shelf life at T + 10

For a constant Q_{10} of 2, the shelf life of a product is doubled for every 10°C decrease in storage temperature, e.g. if the shelf-life is 2 weeks at 50°C, a shelf-life of 4 weeks can be predicted at 40°C (Table 6-2). The Q_{10} of oxidation is related to the activation energy which can be determined from an Arrhenius plot (Section E) of ln k versus the reciprocal of absolute temperature. The activation energy of oxidation and Q_{10} can be calculated from the slope of the Arrhenius equation. Lipid oxidation in foods has an activation energy between 15 and 25 kcal/mole, depending of course on the presence of prooxidant metal catalysts, which lower the activation energy, or the presence of antioxidants, which increase it. Unusual effects may occur from competitive reactions occurring at lower temperatures, such as during freezing; the concentration of catalysts in the unfrozen liquid may accelerate the oxidation. On the other hand, if oxygen availability is a limiting factor in a food, raising

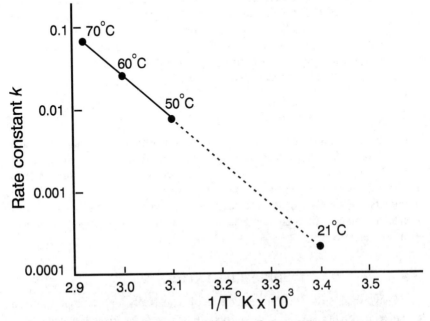

Figure 6-3. Application of Arrhenius plot to approximate shelf-life of a food. Adapted from Labuza, 1984).

the temperature decreases the solubility of oxygen in fat or water by about 25% for every 10°C increase, and may result in a decrease in quality deterioration.

The shelf-life of a food at ambient conditions has been estimated by plotting the log of induction periods (or other suitable end-point of stability) versus accelerated temperature and extrapolating to room temperature. However, this approach can lead to errors if the stability test is based on an end-point that is beyond the rancidity detected by sensory methods. Extrapolation of the stability results obtained by the Rancimat test to ambient storage leads to either over-prediction or under-prediction of the actual shelf-life depending on the fatty acid composition of the oils (Figure 6-4). This discrepancy in prediction can be attributed to the different end-point between ambient conditions (peroxide value of 10 for rancidity of oils), and the much higher oxidation levels (peroxide value of 50 or higher under conditions of the automated AOM test) occurring at the elevated temperatures of the Rancimat test. This test run between 100 and 130°C estimates the shelf-life of a much more oxidized oil than other oven tests carried out between 40 and 60°C.

Shelf-life testing is a relatively expensive undertaking that for most practical purposes must be applied to specific food products. Results of such tests are sparse in the literature. Many factors affect the prediction of shelf-life of foods. Phase changes due to melting of fats at elevated temperatures do not reflect normal changes at lower temperatures. The effect of storage temperature

TABLE 6-2.
Effect of storage temperature on shelf-life[a.]

Temp. °C	Shelf life (weeks)		
	$Q_{10} = 2$	$Q_{10} = 2.5$	$Q_{10} = 3$
60	1	0.8	0.67
50	2	2	2
40	4	5	6
30	8	12.5	18
20	16	31.25	54

[a] Adapted from Labuza and Riboh (1982).

on the moisture content of foods and their water activity (discussed in Chapter 10) may be larger than predicted by the Arrhenius model. Among other factors influencing the shelf-life of foods we may include differential partitioning of catalysts and antioxidants between different phases, changes in moisture, pH and protein denaturation at high temperatures.

K. RECOMMENDED STABILITY TESTING PROTOCOL

The desired objective is to use sensitive methods to detect end-points at low levels of oxidation relevant to flavor deterioration. To evaluate the oxidative or

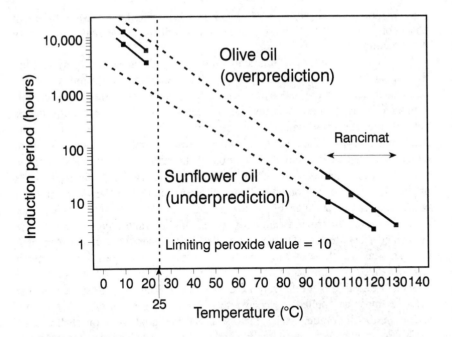

Figure 6-4. Shelf-life plots based on induction periods of olive oil and sunflower oil extrapolated to long time storage at 25°. Adapted from Kaya et al. (1993).

flavor stability of food lipids, it is critical to establish how the experimental data are influenced by the methodology used and the nature of the sample materials. No short cut approach can be used to food stability tests because it is influenced by complex phenomena. The stability of each system has to be researched under various test conditions. The following factors must be considered:

a. Substrates can include triacylglycerols or phospholipids, in bulk, or emulsions or liposomes. To eliminate the confounding effects of natural antioxidants, evaluations should be made with polyunsaturated edible oils stripped of (treated to remove) natural tocopherols. Various carbon bleaching or silicic acid chromatographic methods are available to remove natural tocopherols. These methods should remove the tocopherols efficiently without affecting the original fatty acid composition of the oil.

b. Different storage temperatures are required for testing to obtain reliable comparison of oxidative stabilities of food lipids, preferably between 40 and 60°C, with efficient shaking depending on the rate of oxidation. Each test should be calibrated for each individual fat and accelerated conditions must be kept as close as practical to the actual storage conditions.

c. More than one method should be used to determine the degree of oxidation at suitable time intervals by measuring different types of products, including initial products of lipid oxidation (e.g. hydroperoxides) and their decomposition products (e.g. hexanal). Hydroperoxides formed initially can be estimated by peroxide values or conjugated dienes. Decomposition products of oxidation can be measured by analyses of carbonyl compounds or volatiles by gas chromatographic methods. Initial products provide information on precursors of flavor compounds, and decomposition products are directly related to the flavor compounds affecting quality. Flavor stability is measured by sensory evaluations which provide the most appropriate criterion of consumer quality.

BIBLIOGRAPHY

Buzas,I., Kurucz,E. and Hollo,J. Study of the thermooxidative behavior of edible oils by thermal analysis. *J. Am. Oil Chem. Soc.* **56**, 685-688 (1979).
Cross,C.K. Oil stability. A DSC alternative method for the active oxygen method. *J. Am. Oil Chem. Soc.* **47**, 229-230 (1970).
deMan,J. M. and deMan,L. Automated AOM test for fat stability. *J. Am. Oil Chem. Soc.* **61**, 534-536 (1984).
Frankel,E.N. Soybean oil and flavor stability, in Handbook of Soy Oil Processing and Utilization, pp. 229-244, (1980) (edited by D.R. Erickson, E.H. Pryde, O.L. Brekke, T.L. Mounts, and R.A. Falb), American Soybean Association, St. Louis, and American Oil Chemists' Society, Champaign.
Frankel,E.N. In search of better methods to evaluate natural antioxidants and oxidative stability in food lipids. *Trends in Food Science & Technology* **4**, 220-225 (1993).
Frankel,E.N. Formation of headspace volatiles by thermal decomposition of oxidized fish oils vs. oxidized vegetable oils. *J. Am. Oil Chem. Soc.* **70**, 767-772 (1993).
Frankel,E.N. and Huang,S-W. Improving the oxidative stability of polyunsaturated vegetable oils by blending with high-oleic sunflower oil. *J. Am. Oil Chem. Soc.* **71**, 255-259 (1994).
Frankel,E N., Huang,S-W., Kanner,J. and German,J.B. Interfacial phenomena in the evaluation of antioxidants: bulk oils versus emulsions. *J. Agr. Food Chem.* **42**, 1054-1059 (1994).
Gray,J.I. Measurement of lipid oxidation: a review. *J. Am. Oil Chem. Soc.* **55**, 539-546 (1978).
Hassel,R.L. Thermal analysis: An alternative method of measuring oil stability. *J. Am. Oil Chem. Soc.* **53**, 179-181 (1976).
Karel,M. Chemical effects in food stored at room temperature. *J. Chem. Ed.* **61**, 335-339 (1984).
Kaya,A., Tekin,A.R. and Öner,M.D. Oxidative stability of sunflower and olive oils: comparison between a modified active oxygen method and long term storage. *Lebensm.-Wiss. U.-Technol.* **26**, 464-468 (1993).
Kwolek,W.F. and Bookwalter,G.N. Predicting storage stability from time-temperature data. *Food Technol.* **25**, 51-63 (1971).
Labuza,T.P. Kinetics of lipid oxidation in foods. *CRC Critical Revs. Food Technol.* **2**, 355-405 (1971).
Labuza,T.P. *Shelf-life Dating of Foods*. (1982) Food and Nutrition Press, Inc., West Port, Conn.
Labuza,T.P. Application of chemical kinetics to deterioration of foods. *J. Chem. Ed.* **61**, 348-358 (1984).
Labuza,T.P. and Riboh,D. Theory and application of Arrhenius kinetics to the prediction of nutrient losses in foods. *Food Technol.* **32**, 66-74 (1982).
Labuza,T.P. and Schmidl,M.K . Accelerated shelf-life testing of foods. *Food Technol.* **39**, 57-64,134 (1985).
Läubli,M.W. and Bruttel,P.A. Determination of the oxidative stability of fats and oils: comparison between the active oxygen method and the Rancimat method. *J. Am. Oil Chem. Soc.* **63**, 772-795 (1986).
Nieschlag,H.G., Hagemann,J.W., Rothfus,J.A. and Smith,D.L. Rapid gravimetric estimation of oil stability. *Analyt. Chem.* **46**, 2215-2217 (1974).
Ragnarsson,J.O. and Labuza,T.P. Accelerated shelf-life testing for oxidative rancidity in foods - a review. *Food Chem.* **2**, 291-308 (1977).
Rossell,J.B. Measurement of rancidity, in *Rancidity in Foods*, pp. 21-45 (1983) (Edited by J.C. Allen, and R.J. Hamilton), Applied Science, London.
Snyder,J.M., Frankel,E.N. and Warner,K. Headspace volatile analysis to evaluate oxidative and thermal stability of soybean oil. Effect of hydrogenation and additives. *J. Am. Oil Chem. Soc.* **63**, 1055-1058 (1986).

Warner,K., Frankel,E.N. and Mounts,T.L. Flavor and oxidative stability of soybean, sunflower and low-erucic acid rapeseed oils. *J. Am. Oil Chem. Soc.* **66**, 558-564 (1989).

Warner,K. and Frankel,E.N. Flavor stability of soybean oil based on induction periods for the formation of volatile compounds by gas chromatography. *J. Am. Oil Chem. Soc.* **62**, 100-103 (1985).

Waters,W.A. The kinetics and mechanism of metal-catalyzed autoxidation. *J. Am. Oil Chem. Soc.* **48**, 427-433 (1971).

CHAPTER 7

CONTROL OF OXIDATION

The development of rancidity in edible oils is a serious problem in some sectors of the food industry because of increasing use of polyunsaturated vegetable and fish oils, discontinuing of the use of synthetic antioxidants, and fortification of some foods with iron. Important nutritional and economic benefits can be expected by improved strategies to control lipid oxidation in foods, including improvement in safety, by protecting unstable polyunsaturated fatty acid components and vitamins, and reduction of food product loss by deterioration during processing and storage.

Effective control methods against lipid oxidation include:

a) inactivating prooxidant metals
b) processing by reducing or eliminating metal contamination and light exposure, and minimizing the loss of natural tocopherols
c) partially hydrogenating polyunsaturated fats
d) blending polyunsaturated fats with more stable monounsaturated fats
e) packaging to minimize exposure to air and moisture
f) using antioxidants (see Chapter 8)
g) genetic modification of fatty acid composition

Significant progress has recently been achieved by the genetic modification of soybeans, rapeseeds and sunflowers to produce edible oils of improved oxidative stability by changing their fatty acid compositions. However, the commercialization of these new crops will require further improvements in their agronomic and economic properties in various climatic environments and countries.

A. METAL INACTIVATION

The use of citric acid was one of the earliest measure discovered to significantly stabilize soybean oil against oxidation. The effectiveness of citric acid was attributed to its metal chelation properties. Many other metal chelators were developed to inactivate prooxidant metals in oils, including tartaric, phytic, phosphoric and carboxymethylmercaptosuccinic acids. By forming stable coordination complexes with prooxidant metals, these chelating compounds inhibit effectively both the metal-catalysed initiation of free radical oxidation, and the decomposition of hydroperoxides producing volatile compounds that lower the flavor and oxidative stability of oils (Chapters 1 and 4).

Citric acid and phosphoric acid are the most commonly used commercially because they are economical. Although citric acid is less oil soluble than phosphoric acid, it is very effective and most widely used during processing of vegetable oils. Isopropyl citrate, stearyl citrate and monoglyceride citrate have been used to increase the solubility of citric acid in oils. However, in properly processed oils free citric acid is generally effective at levels below 50 ppm. The effectiveness of citric acid in vegetable oils is also increased by a synergistic action with tocopherols that are naturally occurring in vegetable oils in varying concentrations (see Table 8.6). Because traces of copper can completely inactivate the effects of natural and added synthetic antioxidants, metal inactivators such as citric acid act as "preventive antioxidants" by reinforcing the effect of and stabilizing tocopherols and added synthetic antioxidants. This well-established phenomenon is referred to as "synergism" and will be discussed in Chapter 8.

Phosphatides and purified lecithin are also useful metal inactivators but they must be used at low concentrations (a few ppm) to minimize brown color formation. The use of phosphatides and phosphoric acid is also limited to low concentrations because above 0.02-0.05% these materials impart a "mellony" or "cucumber" off-flavor.

Crude extracted oils contain varying amounts of iron and copper, in either ionized or complex form (such as heme complexes), and these are very detrimental to the oxidative and flavor stability of the processed oils. Crude soybean oil may contain as much as 3 to 5 ppm of iron and 0.1 to 0.2 ppm of copper. Alkali refining removes 95 to 98% of the initial iron and copper found in crude oils. However, the flavor stability of soybean oil can be significantly reduced with as little as 0.1 ppm of iron and 0.01 ppm of copper. The judicious use of metal inactivators is therefore essential to the production of reasonably stable vegetable oils. Although citric acid is active as a metal inactivator in oils subjected to normal storage conditions, it is decomposed and becomes ineffective at the elevated temperatures of cooking and frying.

Citric acid and other metal inactivators are more effective in soybean oil after they are heated, apparently because metals form prooxidant complexes with preformed hydroperoxides that must be thermally destroyed before the chelating agents become effective.

$$LOOH + Me \rightleftharpoons [Me-OOL] + citric\ acid \xrightarrow{210°C} [citric\ acid-Me] + LOOH$$

$$LOOH \xrightarrow{210°C} decomposition\ products$$

By this mechanism the chelating agent replaces the hydroperoxides after its thermal decomposition to form a more stable metal coordination complex that can no longer catalyse lipid oxidation. If added before deodorization, citric acid is thermally decomposed into a mixture of organic acids (aconitic, itaconic and citraconic acids) that are less effective as metal chelators. For this reason citric acid is generally added to vegetable oils on the cooling cycle of deodorization generally below 100°C. Because citric acid is added during processing, it is considered a processing aid in the USA and does not require to be declared on food labels.

Water-soluble metal chelators, such as ethylenediamine tetraacetic acid (EDTA) and its calcium disodium salt, phosphates and ascorbic acid, are effective in improving the oxidative stability of aqueous food emulsion systems (*e.g.* salad dressings, mayonnaise and margarine). The EDTA salts are also used to inactivate metal contaminants in salt, spices and other food ingredients. The salt purified in the presence of EDTA to remove heavy metals can be added to foods without a labeling requirement. Ascorbic acid has multiple functions in aqueous foods, and its effectiveness as a metal chelator may be compromised because it is known to have prooxidant effects in the presence of significant amounts of metals (Chapter 8).

B. PROCESSING

Any oxidation during processing is injurious to the flavor and oxidative stability of processed oils, either directly by the formation and accumulation of oxidation products, or indirectly by the loss of protective naturally occurring tocopherol antioxidants. Optimum oxidative stability can be achieved therefore by minimizing exposure of edible oils to air, light and oxidation during processing. The effect of antioxidants in relation to processing will be discussed in Chapter 8. Good processing measures required to minimize oxidation include careful control of refining temperature, vacuum bleaching, nitrogen blanketing between processing steps, and minimizing light exposure (Figure 7-1). During processing, polyunsaturated vegetable oils are more susceptible to oxidation after removal of most of the phospholipids during the water-degumming step. It is important therefore to minimize oxidation during the subsequent alkali refining, bleaching and deodorization steps. Bleaching lowers the oxidative stability of partially processed oil by removal of the phospholipids and by exposure to the large surface area of the adsorption clays at elevated temperatures. Deaeration of the bleaching earth and vacuum bleaching are thus desirable. After bleaching, soybean oil is particularly susceptible to oxidation and the oil must be carefully protected by nitrogen blanketing or by minimizing storage in air before deodorization. The oxygen content and peroxide value of refined and bleached soybean oil are significantly decreased by storage in nitrogen compared to air (Table 7-1).

Deodorization at elevated temperatures (200-275°C) under vacuum is the last and most critical step in vegetable oils processing (Figure 7-1). The quality of vegetable oils can be appreciably improved by deaerating the oil at low temperature to reduce dissolved oxygen (to less than 0.1% vol.), avoiding any

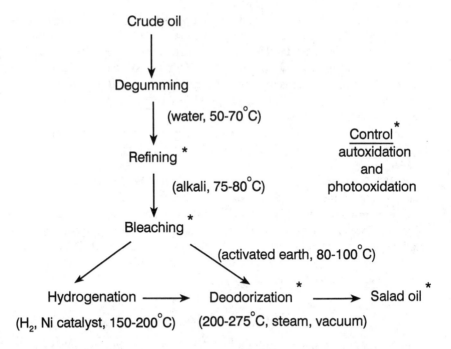

Figure 7-1. Key steps in processing of soybean that require control of autoxidation and photooxidation.

contact of hot oil with air, and cooling the deodorized oil to below 40°C before exposure to air or cooling under nitrogen atmosphere. Storage of the deodorized oils or packaging under nitrogen is of course most desirable to minimize oxidation.

Any hydroperoxides and other secondary oxidation products formed in oils during processing are polymerized during and after deodorization. These materials are found at levels of 1 to 2% and include 70% dimers, trimers and higher molecular weight compounds, aldehydo-glycerides, and esters, carbonyls and alcohols. These high-molecular weight compounds are readily

TABLE 7-1.

Effect of storing refined and bleached soybean oil in tanks with air or nitrogen atmospheres.[a]

Characteristics	Nitrogen	Air
Oil temperature: range	16-34°C	18-31°C
average	23°C	24°C
Oxygen in headspace	1.4%	21%
Peroxide value: initial	1.0	1.0
final	1.5	5.0

[a] From: Going (1968).

TABLE 7-2.
Effect of oxidation prior to deodorization on dimer and secondary products in soybean oil.[a]

Peroxide value [b]	Dimer, %	Secondary products, %
8.1	1.8	0.0
8.8	1.6	0.0
43.6	2.4	0.6
94.6	3.8	1.4
859.0	11.9	12.5

[a] From: Frankel et al. (1961)
[b] Oils oxidized at 60°C before deodorization, and analysed for dimers chromatographically after deodorization.

decomposed during storage to produce volatile compounds contributing to the flavor deterioration of oils (Chapter 4).

The formation and decomposition of oxidative polymers during soybean oil processing has been referred to as "hidden oxidation" because it cannot be measured by the conventional peroxide value determination. However, these oxidative polymers can be determined by silicic acid chromatography or estimated by the anisidine test (Chapter 5). Such high-molecular weight materials are detrimental to the oxidative and flavor stability of vegetable oils and particularly fish oils. The peroxide value of soybean oils before deodorization is directly related to the dimer content determined chromatographically after deodorization (Table 7-2). The production of oxidative dimers by oxidation prior to deodorization decreases the stability of oils. Flavor and oxidative stability of soybean oil correlates with the concentration of oxidative dimers determined chromatographically. The oxidative stability of vegetable oils can therefore be markedly improved if any oxidation products present in partially processed oils can either be removed or chemically reduced to stable hydroxides before deodorization. Several patented processes have been developed to improve the oxidative and flavor stability of vegetable oils by reducing lipid oxidation during processing. In some processes, oxidation products are removed chromatographically or by improved bleaching with better adsorbents such as silicic acid. In other processes, degummed oils are treated with alkaline sodium borohydride to reduce the peroxide content of oils and improve the oxidative stability of the processed oils. Fish oils, containing large amounts of oxidizable long-chain n-3 polyunsaturated fatty acids, are sometimes processed by molecular distillation at lower temperatures than conventional deodorization (180°C or below) under high vacuum to minimize polymer formation.

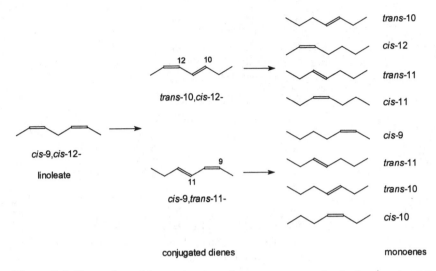

Figure 7-2. Formation of isomeric *cis* and *trans* monoenes by hydrogenation of linoleate.

C. HYDROGENATION

The treatment of polyunsaturated vegetable oils with hydrogen gas in the presence of activated nickel catalysts reduces their linolenate and linoleate components into mixtures of isomeric diunsaturated and monounsaturated oils. According to the temperature, hydrogen pressure and activity of the catalyst, linolenate and linoleate triacylglycerols can be more or less selectively hydrogenated without the formation of saturated products. Partial hydrogenation of linolenic acid in soybean and rapeseed oils, and of n-3 polyunsaturated fatty acids in fish oils provides a practical means of increasing their oxidative and flavor stability. However, hydrogenation does not completely control the flavor deterioration of these oils because this process produces isomeric diunsaturated fatty acids that can oxidize and decompose into volatile compounds contributing to flavor defects designated as "hardening" or "hydrogenation" that are considered objectionable, especially during frying. Furthermore, hydrogenation of polyunsaturated oils forms varying amounts of products with *trans* double bonds according to the conditions used in this process. More recently, hydrogenation is becoming less attractive because *trans* isomers are considered nutritionally questionable.

On catalytic hydrogenation, linoleate produces conjugated dienes (diunsaturated fatty acids), and linolenate produces conjugated diene-trienes (with two conjugated double bonds and one isolated double bond) as intermediates. These conjugated products are more reactive with hydrogen and become rapidly reduced. The conjugated dienes of linoleate produce isomeric monoenes (monounsaturated fatty acids) with *cis* and *trans* double bonds distributed between carbon-9 and carbon-12 (Figure 7-2). The conjugated

CONTROL OF OXIDATION   121

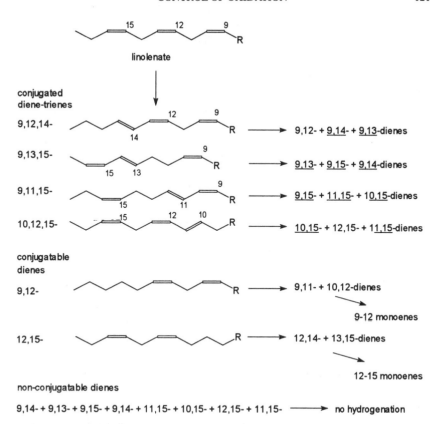

Figure 7-3. Formation of isomeric monoenes, and dienes by hydrogenation of linolenate. The "conjugatable dienes" are hydrogenated into isomeric monoenes. The "nonconjugatable dienes" or "isolinoleate" are not hydrogenated and accumulate in the product.

diene-trienes of linolenate produce two kinds of isomeric dienes (Figure 7-3). The "conjugatable dienes" are methylene interrupted and produce conjugated dienes that can hydrogenate into a mixture of monoenes with double bonds distributed between carbon-9 and carbon-15. The "non-conjugatable dienes" have two isolated double bonds separated by more than one methylene group. These dienes are referred to as "isolinoleate" and accumulate in the partially hydrogenated oils because they cannot be conjugated and do not react with hydrogen. These isomeric dienes behave like two oleate functions in the same molecule and are, therefore, difficult to oxidize. Soybean oil is thus stabilized by hydrogenating linolenate to produce isolinoleate. However, the oxidation of isolinoleate produces hydroperoxides that are readily decomposed into undesirable volatile flavor compounds.

The oxidation of methyl 9,15-octadecadienoate (Chapter 2. C), used as a model for isolinoleate, produced a mixture of 8-, 9-, 10- and 11-hydroperoxides by oxygen attack at the 9-double bond, and a mixture of 14-,

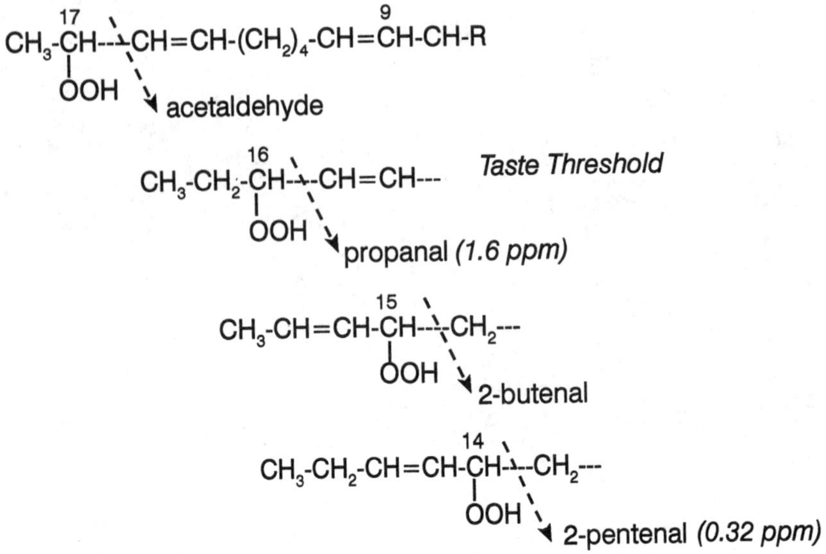

$$CH_3\text{-}\overset{17}{CH}\text{-}\!\!\!\!\diagdown\!CH=CH\text{-}(CH_2)_4\text{-}CH=\overset{9}{CH}\text{-}CH\text{-}R$$
$$\underset{OOH}{|} \searrow \text{acetaldehyde}$$

$$CH_3\text{-}CH_2\text{-}\overset{16}{\underset{|}{CH}}\!\!\!\!\diagdown\!CH=CH\text{---} \quad \textit{Taste Threshold}$$
$$OOH \searrow \text{propanal } (1.6 \text{ ppm})$$

$$CH_3\text{-}CH=CH\text{-}\overset{15}{\underset{|}{CH}}\!\!\!\!\diagdown\!CH_2\text{---}$$
$$OOH \searrow \text{2-butenal}$$

$$CH_3\text{-}CH_2\text{-}CH=CH\text{-}\overset{14}{\underset{|}{CH}}\!\!\!\!\diagdown\!CH_2\text{---}$$
$$OOH \searrow \text{2-pentenal } (0.32 \text{ ppm})$$

Figure 7-4. Volatile decomposition products from 14-, 15-, 16- and 17-hydroperoxides formed from the autoxidation of methyl *cis*-9, *cis*-15-octadecadienoate.

15-, 16- and 17-hydroperoxides resulting from oxidation of the 15-double bond (Figure 2-9). Thermal decomposition of the external 14-, 15-, 16- and 17-hydroperoxides would produce a mixture of acetaldehyde, propanal, 2-butenal and 2-pentenal, which is similar to that produced from the hydroperoxides of linolenate (Figure 7-4). On the other hand, decomposition of the internal mixture of 8-, 9-, 10- and 11-hydroperoxides produces a mixture of unsaturated aldehydes, including 2,8-undecadienal, 2,7-decadienal, 6-nonenal and 5-octenal (Figure 7-5). Because these unsaturated aldehydes have particularly low threshold values, ranging from 0.00032 to 0.32 ppm, they are very potent flavor compounds that contribute to the undesirable "hardening" or "hydrogenation" off-flavors of partially hydrogenated polyunsaturated vegetable and fish oils. Hydrogenation provides therefore a practical but incomplete measure to control oxidatively derived flavor problems of polyunsaturated oils.

For most consumer applications that do not require elevated temperatures, polyunsaturated vegetable oils, such as soybean, rapeseed and sunflower oils, can be used without hydrogenation when they are properly processed. These oils can also be used for either cooking or for single or pan frying foods (see Chapter 11). However, the flavor of polyunsaturated oils will deteriorate, especially if the oils contain linolenic acid, when they are used repeatedly for cooking and deep-fat frying. These oils produce undesirable fishy odors after prolonged and repeated heating. Partially hydrogenated linolenate-containing oils are commonly used commercially for continuous or intermittent frying operations even when mixed with fresh make up oils. In current industrial

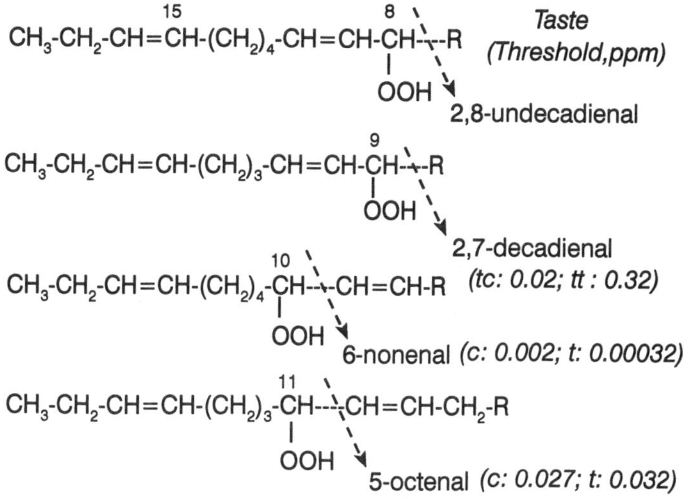

$$CH_3\text{-}CH_2\text{-}CH=CH\text{-}(CH_2)_4\text{-}CH=CH\text{-}\underset{\underset{OOH}{|}}{\overset{8}{CH}}\text{-}R$$

Taste (Threshold,ppm)

2,8-undecadienal

$$CH_3\text{-}CH_2\text{-}CH=CH\text{-}(CH_2)_3\text{-}CH=CH\text{-}\underset{\underset{OOH}{|}}{\overset{9}{CH}}\text{-}R$$

2,7-decadienal (tc: 0.02; tt: 0.32)

$$CH_3\text{-}CH_2\text{-}CH=CH\text{-}(CH_2)_4\text{-}\underset{\underset{OOH}{|}}{\overset{10}{CH}}\text{-}CH=CH\text{-}R$$

6-nonenal (c: 0.002; t: 0.00032)

$$CH_3\text{-}CH_2\text{-}CH=CH\text{-}(CH_2)_3\text{-}\underset{\underset{OOH}{|}}{\overset{11}{CH}}\text{-}CH=CH\text{-}CH_2\text{-}R$$

5-octenal (c: 0.027; t: 0.032)

Figure 7-5. Volatile decomposition products from 8-, 9-, 10- and 11-hydroperoxides formed from the autoxidation of methyl cis-9,cis-15-octadecadienoate.

practices, these polyunsaturated oils are lightly hydrogenated to a linolenate content of 2 to 4%, and blended with more stable oils to minimize the *trans* isomers and the "hydrogenation" flavor that may be objectionable in some consumer markets.

D. BLENDING WITH MONOUNSATURATED OILS

The flavor and oxidative stability of linolenate-containing oils can be improved substantially by blending them with different levels of high-oleic oils to lower the linolenate content. High oleic sunflower and safflower oils are relatively stable because they have a fatty acid composition similar to that of olive oil with an oleate content ranging from 65 to 80%. The mixing of different proportions of high-oleic sunflower oil with soybean or rapeseed oils provides a simple method to prepare more stable edible oils with a wide range of desired fatty acid compositions (Table 7-3). The relative oxidizability of oils and mixtures can be calculated on the basis of their fatty acid compositions by assuming that oleate does not oxidize significantly (in the presence of polyunsaturated fatty acids) and linolenate oxidizes twice as fast as linoleate (based on data in Table 1-1). As expected, blending polyunsaturated oils with high-oleic sunflower oil lowers their calculated oxidizability significantly. Mixtures of rapeseed oil and high oleic-sunflower oil containing 1 and 2% linolenate can be prepared with the same or better oxidative stability than that of partially hydrogenated rapeseed oil containing 1% linolenate. Catalytic hydrogenation of linolenate-containing vegetable oils can be avoided by blending with high-oleic sunflower oil or other stable oils. Oxidatively stable

TABLE 7-3.
Fatty acid composition of mixtures of soybean, rapeseed and corn oils with high-oleic safflower oil (Relative %).[a]

Oils and mixtures	Saturates	C18:1	C18:2	C18:3	Oxidizability [b]
Hi-Ol Sun (HOSO) [b]	8.9	77.4	12.5	0.1	0.13
Soybean (SBO)	14.6	24.7	53.8	6.9	0.75
SBO/HOSO, 27:73	10.5	63.5	24.0	2.0	0.30
SBO/HOSO, 41:59	11.4	55.6	30.0	3.0	0.39
SBO/HOSO, 63:37	12.7	43.7	39.1	4.5	0.53
Rapeseed oil (RO) [c]	5.7	61.3	21.9	9.5	0.50
RO/HOSO, 7:93	8.3	78.7	12.0	1.0	0.15
RO/HOSO, 17:83	8	76.8	13.2	2.0	0.19
RO/HOSO, 27:73	7.8	75.0	14.2	3.0	0.23
Corn oil (CO)	12.9	26.3	59.8	1.0	0.63
CO/HOSO, 18:82	9.2	70.4	20.0	0.4	0.21
CO/HOSO, 59:41	11.1	48.2	40.0	0.7	0.42

[a] From: Frankel and Huang (1994)
[b] Oxidizability = [(% C18:2 x 1) + (% C18:3 x 2)] ÷ 100
[c] Includes erucic acid (22:0) for the balance to 100%

all cis oils can, therefore, be produced by blending high-oleic sunflower oil with polyunsaturated oils as an alternative technology to hydrogenation.

Mixing different proportions of high-oleic sunflower or safflower oils with soybean, rapeseed and corn oils provides a simple method to prepare more stable edible oils with a wide range of desired fatty acid compositions. A number of vegetable oil blends are now available on the U.S. market that are produced from mixtures of soybean, rapeseed, sunflower, corn and olive oils. These blended oils are designed for improved nutritional value by lowering their saturated fatty acid and raising their oleic acid contents. However, blends of soybean and rapeseed oils containing significant amounts of linolenic acid cause flavor deterioration and poor frying performance when used for repeated and prolonged frying, and reduce oxidative stability. Although the detrimental effects of oxidation products of linolenic acid to the stability and ultimately to the safety of the products are well known, the amounts of linolenic acid needed in the diet for its nutritionally benefits are not adequately established (Chapter 12).

E. GENETIC MODIFICATION

For many years many varieties of soybeans have been developed to improve their oxidative stability and frying performance by lowering the content of linolenic acid. Conventional breeding did not succeed in producing variety lines that maintained this characteristic in succeeding generations. Recent breeding programs were more successful and mutant soybean varieties low in linolenic acid were produced by recurrent selection and mutation breeding. New varieties of soybean oils containing linolenic acid ranging from 3 to 4.8% show higher oxidative and flavor stability than the control soybean oil

TABLE 7-4.
Fatty acid composition and stability of new varieties of soybean oils.[a]

Oils	C 16:0	C 18:0	C 18:1	C 18:2	C 18:3	Flavor score (PV) [b]
1985 SBO	10.5	3.6	23.8	54.4	7.7	5.5 (5.8)
HWSBO	9.2	3.9	47.8	36.1	3.0	n.r.
1983 A5	10.8	4.5	45.0	36.4	3.3	6.6 (3.0)
1985 C1640	10.6	3.6	24.6	56.4	4.2	6.6 (4.9)
1982 PI	13.1	3.6	16.6	61.9	4.8	7.0 (5.4)

[a] From: Mounts et al. (1988). Abbreviations: SBO = standard soybean oil, HWSBO = hydrogenated-winterized soybean oil, n.r. = not reported, PV = peroxide value
[b] Flavor evaluations and PV determinations made after storage at 60°C for 8 days.

containing 7.7% linolenic acid (Table 7-4). These new low-linolenate soybean oils show even more significant improvements in frying performance (Chapter 11). Therefore, genetically modified vegetable oils provide improved fatty acid compositions and oxidative stability. However, these new varieties are reportedly not resistant to environmentally changes and produce lower yields than the conventional crops.

New high-oleic soybeans have been developed recently by introducing cloned mutant genes into soybeans to create transgenic lines. The oils from these transgenic soybeans contain more than 80% oleic acid, less than 4% stearic acid and less than 2% linolenic acid, and would be expected to be oxidatively stable. Such transgenic soybeans are not yet commercially available.

F. PACKAGING

Ideal containers for oils, shortenings and other lipid-rich foods should be impermeable to air and moisture, and should be opaque to light to prevent further oxidation during prolonged storage under ambient conditions. Liquid oils were traditionally packaged in clear glass containers; brown bottles were sometimes used to protect unstable oils from light oxidation. Clear glass containers were, however, later perceived to be preferred by the consumer. Glass bottles are now generally replaced by plastic containers. Polyvinyl chloride (monomer free) is preferred because it is less permeable to oxygen and superior to polyethylene, which is permeable to oxygen, and to acrylonitrile-butadiene-polystyrene, which is more permeable to moisture. Although polymers excluding light or colored or opaque plastic material are preferable to protect the oils from light oxidation, clear plastic containers are generally used because of consumer preference. Tin or aluminum cans and paper cartons are also used for shortenings and margarine which may be wrapped in aluminum foil-laminated paper. Environmental factors affecting lipid oxidation may be controlled by novel packaging techniques, e.g. by incorporation of oxygen scavengers, or desiccants, or volatile antioxidants in packaging materials.

Highly oxidizable oils such as fish oils and oil-soluble vitamins can be protected by a process known as *microencapsulation* which coats the oils with a thin matrix of protein (gelatin, or gluten, or casein, or albumin), carbohydrate (starch, or dextran, or sucrose), cellulose (carboxymethylcellulose, or cellulose derivatives) and lecithin. Microencapsulation provides protection against oxidation and imparts oxidative stability. The use of carboxymethylcellulose and cyclodextrins as coatings is claimed to provide better protection of oils by improved oxygen barrier properties. For special applications as nutritional supplements, fish oils enriched in n-3 PUFA are microencapsulated, in the presence of antioxidants, into a powder product that is relatively stable for storage at ambient temperatures.

BIBLIOGRAPHY

Brekke,O.L., Mounts,T.L. and Pryde,E.H. Summary and Recommendations, in *Handbook of Soy Oil Processing and Utilization*, pp. 551-560 (1980) (edited by D.R. Erickson, E.H. Pryde, O.L. Brekke, T.L. Mounts and R.A. Falb), American Soybean Association, St. Louis, MO, and American Oil Chemists' Society, Champaign, IL.

Cooney,P.M., Evans,C.D., Schwab,A.W. and Cowan,J.C. Influence of heat on oxidative stability and on effectiveness of metal-inactivating agents in vegetable oils. *J. Am. Oil Chem. Soc.* **35**, 152-156 (1958).

Crossley,A., Davies,A.C. and Pierce,J.H. Keeping properties of edible oils. III. Refining by treatment with alumina. *J. Am. Oil Chem. Soc.* **39**, 150-165 (1962).

Evans,C.D., Schwab,A.W., Moser,H.A., Hawley,J.E. and Melvin,E.H. The flavor problem of soybean oil. VII. Effect of trace metals. *J. Am. Oil Chem. Soc.* **28**, 68-73 (1951).

Evans,C.D., Cooney,P.M., Moser,H.A., Hawley,J.E. and Melvin,E.H. The flavor problem of soybean oil. X. Effects of processing on metallic content of soybean oil. *J. Am. Oil Chem. Soc.* **29**, 61-65 (1952).

Evans,C.D., Frankel,E.N., Cooney,P.M. and Moser,H.A. Effect of autoxidation prior to deodorization on oxidative and flavor stability of soybean oil. *J. Am. Oil Chem. Soc.* **37**, 452-456 (1960).

Frankel,E.N. Soybean oil and flavor stability, in *Handbook of Soy Oil Processing and Utilization*, pp.229-244 (1980) (edited by D.R. Erickson, E.H. Pryde, O.L. Brekke, T.L. Mounts and R.A. Falb), American Soybean Association, St. Louis, MO, and American Oil Chemists' Society, Champaign, IL.

Frankel,E.N., Evans,C.D. and Cowan,J.C. Thermal dimerization of fatty ester hydroperoxides. *J. Am. Oil Chem. Soc.* **37**, 418-424 (1960).

Frankel,E.N., Evans,C.D., Moser,H.A., McConnell,D. G. and Cowan,J.C. Analyses of lipids and oxidation products by partition chromatography. Dimeric and polymeric products. *J. Am. Oil Chem. Soc.* **38**, 130-134 (1961).

Frankel,E.N. and Huang,S.W. Improving the oxidative stability of polyunsaturated vegetable oils by blending with high-oleic sunflower oil. *J. Am. Oil Chem. Soc.* **71**, 255-259 (1994).

Going, L.H. Oxidative deterioration of partially processed soybean oil. *J. Am. Oil Chem. Soc.* **45**, 632-634 (1968).

Helebra,S.F., Mikulski,R.A. and Cook,M.M. Purification of natural oils with alkali metal borohydrides. *European Patent* No. 116,408 (1984).

Jackson,L.S. and Lee,K. Microencapsulation and the food industry. Food Sci. Technol. *Lebensmittel-Wissenschaft u. Technol.*, **24**, 289-297 (1991).

Karel,M. Chemical effects in foods stored at room temperature. *J. Chem. Ed.* **61**, 335-339 (1984).

Keppler,J.G., Schols,J.A., Feenstra, W.H. and Meijboom, P.W. Components of the hardening flavor present in hardened linseed oil and soybean oil. *J. Am. Oil Chem. Soc.* **42**, 246-249 (1965).

Keppler,J.G., Horikx,M.M., Meijboom,P.W. and Feenstra,W.H. Iso-linoleic acids responsible for the formation of the hardening factor. *J. Am. Oil Chem. Soc.* **44**, 543-544 (1967).

Knowlton,S., Ellis,S.K.B. and Kelly,E.F. Oxidative stability of high-oleic soybean oil. *Paper presented at the American Oil Chemists' Society meeting*, May 11-14, 1997, Seattle, WA.

Mounts,T.L., Warner,K., List,G.R., Kleiman, R., Fehr,W.R., Hammond, e.g. and Wilcox, J.R. Effect of altered fatty acid composition on soybean stability. *J. Am. Oil Chem. Soc.* **65**, 624-628 (1988).

Mounts,T.L., Warner,K. And List,G.R. Performance evaluation of hexane-extracted oils from genetically modified soybeans. *J. Am. Oil Chem. Soc.* **71**, 157-161 (1994).

CHAPTER 8

ANTIOXIDANTS

The inhibition of free radical autoxidation by antioxidants is of considerable practical importance in preserving polyunsaturated lipids from oxidative deterioration. Synthetic antioxidants such as butylated hydroxyanisole (BHA), butylated hydroxytoluene (BHT), propyl gallate (PG) and tert-butylhydroquinone (TBHQ) (Figure 8-1) have been commonly used in the U.S.A. and several other countries to inhibit lipid oxidation and to retard the development of rancidity in foods. Although these synthetic antioxidants are efficient and relatively cheap, special attention has been given to the use of natural antioxidants because of a worldwide trend to avoid or minimize the use of synthetic food additives. The presence of important natural antioxidants in plant foods is attracting further interest because of their clear benefits as anticarcinogenic agents and as inhibitors of biologically harmful oxidation reactions in the body (Chapter 12). The evidence is accumulating that diets rich in plant antioxidants derived from fruits and vegetables are associated with lower risks of coronary heart disease and cancer.

A. MECHANISM OF ANTIOXIDATION

1. Chain-Breaking Antioxidants (AH)

These types of antioxidants inhibit or retard lipid oxidation by interfering with either chain propagation or initiation by readily donating hydrogen atoms to lipid peroxyl radicals (Chapter 1).

$$LH \longrightarrow L^{\cdot} \tag{1}$$

$$L^{\cdot} + O_2 \longrightarrow LOO^{\cdot} \tag{2}$$

$$LOO^{\cdot} + LH \longrightarrow ROOH + L^{\cdot} \tag{3}$$

$$LOO^{\cdot} + AH \rightleftharpoons LOOH + A^{\cdot} \tag{4/-4}$$

$$L^{\cdot} + AH \longrightarrow LH + A^{\cdot} \tag{5}$$

$$A^\cdot + LOO^\cdot \longrightarrow \text{Non-radical products} \quad (6)$$

$$A^\cdot + A^\cdot \longrightarrow \text{Non-radical products} \quad (7)$$

Phenolic compounds with bulky alkyl substituents such as BHA, BHT, TBHQ and tocopherols (Figure 8-1) are effective chain-breaking antioxidants because (a) they produce stable and relatively unreactive antioxidant radicals $A^\cdot$, by reactions (4) and (5), that are too unreactive to propagate the chain, and (b) they are able to compete with the lipid substrate (LH), present in much higher concentration in reaction (3), for the chain-carrying peroxyl radicals LOO$^\cdot$. In the presence of an antioxidant, the rate of hydroperoxide formation is related to the ratio of lipid (LH) concentration to that of the antioxidant (AH) concentration.

$$d[LOOH] / dt = k_3 \, R_1 \, [LH] / (k_4 \, [AH])$$

where k_3 = rate constant for propagation reaction (3)
R_1 = rate of initiation (1)
and k_4 = rate of inhibition reaction (4).

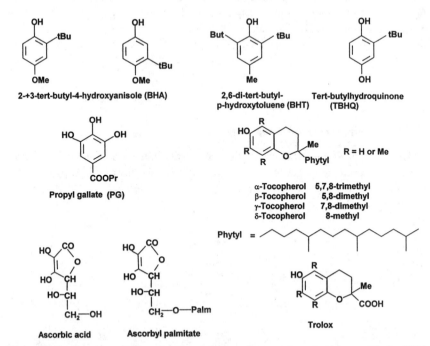

Figure 8-1. Structures of major synthetic and natural antioxidants. Trolox™ is a trade name of Hoffmann La Roche Inc. for 6-hydroxy-2,5,7,8-tetramethyl-chroman-2-carboxylic acid.

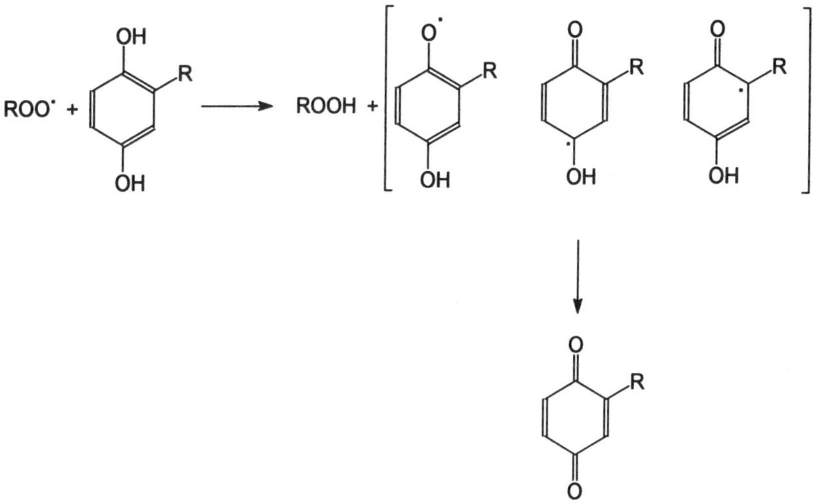

Figure 8-2. Stabilization of an alkylated hydroquinone by electron delocalization.

At elevated temperatures, the oxygen concentration is limiting, inhibition reaction (5) becomes more important, and the rate of hydroperoxide formation is then related to the ratio of oxygen concentration to that of the antioxidant concentration.

$$d[LOOH] / dt = k_3 R_1 [O_2] / (k_5 [AH])$$

where k_5 = rate of inhibition reaction (5).

The antioxidant radicals are stabilized by electron delocalization of their phenoxyl structures (Figure 8-2). The antioxidant effectiveness is directly related to the resonance stabilization (see Introduction) of the phenoxy radicals. Phenolic compounds with electron-donating substituents at the *ortho* and *para* positions are more effective than those with substituents at the meta position. *Ortho*- and *para*-dihydric phenols have high antioxidant activity because of their ability to donate hydrogen atoms and to form semiquinone structures that are readily delocalized (Figure 8-2).

(a) *Prooxidant activity.* Under certain conditions phenolic antioxidants are less effective because of their tendency to act as chain-carriers and become prooxidants. An important factor affecting the effectiveness of an antioxidant is the tendency of the radical it produces to undergo transfer with the lipid substrate. The effectiveness of antioxidants can thus be decreased by direct oxidation by reaction (8), and by a hydrogen chain transfer reaction (9) of the antioxidant radical with the lipid substrate to reinitiate the chains.

$$AH + O_2 \longrightarrow A^{\cdot} + HO_2^{\cdot} \qquad (8)$$

$$A^{\cdot} + LH \longrightarrow AH + L^{\cdot} \qquad (9)$$

Reactions (8) and (9) become important with less substituted phenols, at high concentrations of antioxidants and at elevated temperatures (see Chapter 11).

At high concentrations, phenolic compounds such as α-tocopherol can become prooxidant and chain-carriers by regenerating peroxyl radicals by reversing reaction (4), *i.e.* reaction (-4). Antioxidants also lose their efficiency at elevated temperatures because of homolytic decomposition of hydroperoxides formed by reversing reaction (4) and by producing a chain-carrying peroxyl radical. However, the homolytic decomposition of hydroperoxides producing alkoxyl radicals by reaction (10) is more important than reaction (-4), because it requires less energy to cleave an LO-OH bond (~44 kcal/mol) than an LOO-H bond (90 kcal/mol). Cleavage by reaction (10) is also facilitated by metal catalysts and by heat (Chapter 4).

$$\text{LOOH} \longrightarrow \text{LO}^{\cdot} + \text{OH}^{\cdot} \tag{10}$$

The alkoxyl radicals produced by cleavage reaction (10) can undergo further reactions, including: (11) β-cleavage to produce aldehydes and other decomposition product (short-chain hydrocarbons, and shorter chain fatty acids) that contribute to flavor deterioration of unsaturated lipids (Chapter 4); (12) cyclization to produce epoxy derivatives, and formation of ketone and hydroxy derivatives that do not contribute to flavor deterioration of unsaturated lipids (Chapter 4); and (13) hydrogen abstraction from the lipid substrate to form alcohols and a chain-carrying lipid radical.

$$\text{LO}^{\cdot} \longrightarrow \text{Aldehydes \& other decomposition products} \tag{11}$$

$$\text{LO}^{\cdot} \longrightarrow \text{Secondary products (epoxy, keto, hydroxy)} \tag{12}$$

$$\text{LO}^{\cdot} + \text{LH} \longrightarrow \text{LOH} + \text{L}^{\cdot} \tag{13}$$

(b) *Effect on hydroperoxide decomposition.* Antioxidants can inhibit the decomposition reaction (11) by different pathways, including: (14) hydrogen donation to form stable hydroxy compounds; or (15) by trapping alkoxyl radicals with antioxidant radicals.

$$\text{LO}^{\cdot} + \text{AH} \longrightarrow \text{LOH} + \text{A}^{\cdot} \tag{14}$$

$$\text{LO}^{\cdot} + \text{A}^{\cdot} \longrightarrow \text{LOA} \tag{15}$$

The overall rate of aldehyde formation (11) is directly related to the concentration of hydroperoxides and inversely related to the rate of formation of stable products that do not form aldehydes.

$$d[\text{aldehydes}]/dt = k_{10} k_{11} [\text{LOOH}] / k_{12} + k_{14} [\text{AH}] + k_{15} [\text{A}^{\cdot}]$$

Reaction (14) producing stable hydroxy fatty acid derivatives is important because it inhibits the decomposition of hydroperoxides into aldehydes and

TABLE 8-1.
Effect of concentration of α- and γ-tocopherols on inhibition (in percentage) of hydroperoxide and hexanal formation in bulk corn oil and oil-in-water emulsions oxidized at 60°C.

	bulk oils			oil-in-water emulsions			
	Hydroperoxides		hexanal	hydroperoxides		hexanal	
corn oil	day 3	day 7	day 7	day 2	day 4	day 2	day 4
+ α–tocopherol							
100 ppm	9	88	88	6	70	9	30
250 ppm	-26	85	99	-5	81	24	63
500 ppm	-39	82	99	-24	75	25	72
750 ppm	-85	79	100	-101	57	36	73
1000 ppm	-97	77	99	-100	69	37	85
+ γ-tocopherol							
100 ppm	26	73	56	-5	66	-2	47
250 ppm	27	83	85	26	86	23	69
500 ppm	24	82	99	45	88	20	82
750 ppm	23	80	99	29	86	24	84
1000 ppm	10	79	99	7	82	31	86

From: Huang et al. (1994). Negative values represent prooxidant activity.

other volatile products that decrease the oxidative and flavor stability of polyunsaturated food lipids.

We have seen earlier that in the presence of an antioxidant the rate of hydroperoxide formation is related to the ratio of lipid concentration to that of the antioxidant concentration, which is dependent on the rate of inhibition reaction (4). The effectiveness of an antioxidant is generally based on the balance between the inhibition rate (k_{AH}) of reaction (4) and the transfer reactions (-4), (9) and (13). Therefore, the effect of antioxidants on hydroperoxide decomposition reactions (10) and (11) is an important property that needs to be evaluated. However, most studies of antioxidant actions measure initial events of lipid oxidation based on oxygen absorption, hydroperoxide formation, and peroxide values (Chapters 5 and 6). Very few studies measure the effect of antioxidants on decomposition products of hydroperoxides, such as aldehydes and carbonyl compounds. Yet, these volatile decomposition aldehydes are more relevant to the development of rancidity and to the ultimate quality and stability of food lipids.

For example, α-tocopherol exhibited prooxidant activity with stripped corn oil (that has been treated to remove natural tocopherols) and with corn oil-in-water emulsions at concentrations of 250-1000 ppm, based on conjugated diene formation, but it had antioxidant activity between 100 and 1000 ppm, based on hexanal formation (Table 8-1). γ-Tocopherol was also more effective in inhibiting hexanal than hydroperoxide formation in bulk corn oil but not in the corn oil emulsion. Contrasting results are thus obtained according to the methods used to measure products at different stages of lipid oxidation and depending on the lipid system used in the test. Assuming that hexanal determinations, measuring decomposition of hydroperoxides, are more closely related to flavor deterioration than conjugated diene measurements of hydroperoxide formation, the effects of antioxidants in inhibiting hydroperoxide decomposition may better reflect their activity in reducing flavor deterioration due to lipid oxidation.

(c) *Effectiveness*. An effective antioxidant can stop the reaction of two chain-carrying peroxyl radicals by reactions (4) and (6), and thus break two kinetic chains (reactions producing new radicals) per molecule. The stoichiometry of the antioxidant reaction can be estimated by controlling the initiation reaction with a standard artificial azo initiator such as α,α-azobis-isobutyronitrile (Chapter 1). Under conditions in which free radicals are produced only by decomposition of the added initiator and not from the products of oxidation, any added antioxidant (AH) that can cause an induction period can be considered to be removing propagating free radicals. From the length of the induction period and the known rate of production of initiating free radicals, the number of free radicals reacting with each antioxidant molecule may be determined.

n = the number of radicals that react with a single molecule of antioxidant. An n value of 2 is found for a number of phenolic antioxidants, including α-tocopherol, that corresponds to the two peroxyl radicals trapped by the antioxidant in reactions (4) and (6).

The limitations and pitfalls of using artificial azo initiators were discussed in Chapter 1. Unfortunately, the use of such artificial initiators and radical

$$n\text{LOO}^\cdot + \text{AH} \longrightarrow \text{Stable products} \qquad (16)$$

trapping compounds to evaluate antioxidants is common. The so-called "radical trap" methods (also known as "antioxidant status or capacity") are dependent on the use of either water-soluble or lipid-soluble radical trapping agents. These methods measure the relative reactivity of antioxidants toward artificial radicals, but they provide no quantitative information on what lipid oxidation products are inhibited by the antioxidants. Antioxidants show a marked variation in activity with different lipid systems and under different oxidizing conditions. The radical trap methods may not be useful without knowing what specific lipid substrate is protected and what lipid oxidation products are inhibited.

To learn about the real effects of antioxidants, it is therefore important to obtain specific chemical information about what products of lipid oxidation are inhibited. Several specific assays are needed to elucidate how lipid oxidation products act in the complex multi-step mechanism of lipid oxidative deterioration of foods. The results of several complementary methods are required to determine lipid oxidation products formed at different stages of the free radical chain. Since antioxidants show different activities toward hydroperoxide formation and decomposition, it is important that more than one method be used to monitor lipid oxidation.

2. Other Types of Antioxidants

Several types of compounds can inhibit lipid oxidation by mechanisms that do not involve deactivation of free radical chains. They sometimes have multiple effects and their mechanisms of action are therefore often difficult to interpret.

(a) *Preventive antioxidants.* Metal inactivators are the most important compounds of this type; they deactivate metal ions, which promote the initiation and decomposition of hydroperoxides, and thus retard the formation of deleterious aldehydes. They function either by coordinating the metals and changing their potential by suppressing the redox reactions producing peroxyl and alkoxyl radicals (Chapter 1), or by blocking complex formation with hydroperoxides and preventing their decomposition. Care must be taken, however, with some coordination compounds that have a prooxidant effect by changing the redox properties of metals and promoting the oxidative catalytic properties of metals.

Common inactivating chelating compounds include citric acid, phosphoric acid, ethylenediamine tetraacetic acid (EDTA), and 8-hydroxy-quinoline (Chapter 7). Citric acid and other metal inactivators are more effective in soybean oil after heating, apparently to decompose metal-hydroperoxide complexes (Chapter 7); for this reason, citric acid is added to vegetable oils after heating during the cooling stage of deodorization. Hemin metal complexes behave generally as prooxidant in lipid oxidation. Ascorbic acid may behave either as an antioxidant or prooxidant depending on its relative concentration, the amounts of metals present and the lipid system (Section D.2). By reducing iron or copper to a lower valence state, ascorbic acid becomes a potent prooxidant catalyst of lipid oxidation in foods and biological systems.

(b) *Hydroperoxide destroyers.* These compounds inhibit oxidation by induced decomposition of hydroperoxides by forming either stable alcohols or inactive products by non-radical processes, *e.g.* by reduction or hydrogen donation. α-Tocopherol and its water-soluble carboxylic acid analog called Trolox™ (Figure 8-1), and other phenolic compounds are weak hydroperoxide destroyers by reducing linoleate hydroperoxides into stable hydroxylinoleate in low yields. We have seen above that α-tocopherol is more efficient in inhibiting hexanal formation than hydroperoxide formation during oxidation of polyunsaturated vegetable oils (Table 8-1). Other examples of effective hydroperoxide destroyers include reducing agents, sulfur and selenium compounds, phosphites, and phosphines, which cannot be used in foods because of their toxicity and production of undesirable flavors, however.

TABLE 8-2.
Synergistic effects of tocopherols and citric acid on the oxidative stability of carbon-treated soybean oil at 60°C.[a]

Samples [b]	Tocopherol ppm	Induction period hr	Initial rate PV / hr	Synergism[c] %
Control	1500	20	0.14	—
Control + CA	1500	104	0.023	285
C-treated 1 x + CA	750	235	0.018	226
C-treated 2 x + CA	468	244	0.018	328
C-treated 3 x + CA	214	160	0.032	566
C-treated 4 x + CA	0	7	0.31	—

[a] From: Frankel et al. (1959a).
[b] CA = citric acid, 0.01%
[b] C-treated 1 x = carbon-treated once to remove tocopherols
[b] C-treated 2 x = carbon-treated twice to remove tocopherols
[b] C-treated 3 x = carbon-treated three times to remove tocopherols
[b] C-treated 4 x = carbon-treated four times to remove tocopherols
[c] See text for calculation.

(c) *Ultraviolet light deactivators.* Hydroperoxides and traces of carbonyl compounds acting as photo-initiators of lipid oxidation can be deactivated by certain compounds that can absorb irradiation without formation of radicals. Examples of compounds useful in non-food applications include, pigments like carbon black used in polymers, phenyl salicylate, and aromatic ketones such as *ortho*-hydroxybenzophenone.

(d) *Synergists.* In multi-components systems, these antioxidant compounds can reinforce each other by cooperative effects known as *synergism*. Significant synergism is generally observed when chain-breaking antioxidants are used together with preventive antioxidants or peroxide destroyers because they suppress both initiation and propagation reactions of free radicals. A strong synergistic or combined effect is observed between natural tocopherols in soybean oil in the presence of added citric acid (Table 8-2). The percent synergism is calculated on the basis of the induction periods (IP) observed, as follows:

$$\% \text{ synergism} = \frac{\left(\begin{array}{c}\text{IP tocopherol + citric acid}\\\text{combined}\end{array}\right) - \left(\begin{array}{cc}\text{IP tocopherol} + & \text{IP citric acid}\\\text{alone} & \text{alone}\end{array}\right)}{\left(\begin{array}{cc}\text{IP tocopherol} + & \text{IP citric acid}\\\text{alone} & \text{alone}\end{array}\right)} \times 100$$

The calculated synergism between tocopherols and citric acid in soybean oil was 285%, and increased at the lower concentrations of tocopherol (200-500 ppm) in the oil treated with carbon black (Table 8-2). This synergistic effect between a chain-breaking antioxidant and citric acid is attributed to metal chelation. Another well-recognized synergism between a free radical acceptor such as α-tocopherol and a reducing agent such as ascorbic acid is explained by the regeneration and recycling of the tocopheroxyl radical intermediate to the parent phenol, α-tocopherol (Section C.1).

B. ANTIOXIDANT EVALUATIONS

The effectiveness of antioxidants is very dependent on complex phenomena determined by the relative physical states of the lipid substrates, the conditions of oxidation, the methods used to follow oxidation and the stages of oxidation. The methods used to evaluate antioxidants are complicated by many factors (Chapter 6): (i) The use of different lipid substrates has a significant impact on the activity of various antioxidants according to their hydrophilic or lipophilic nature; (ii) Solubility and partition properties affect the activity of antioxidants in heterogeneous systems where they are distributed differently between aqueous and lipid phases; (iii) Different results can be obtained at different temperatures because the mechanism of oxidation and hydroperoxide decomposition changes with temperature; (iv) Different methods used to follow oxidation can give varying results according to the different effects of antioxidants on formation of hydroperoxides and their decomposition (Table 8-1). Relative antioxidant efficiencies vary markedly from one oxidizing lipid substrate to another. In the same lipid substrate the relative activities of antioxidants often depend on the antioxidant concentrations.

The activity of natural antioxidants is greatly affected by complex interfacial phenomena in emulsions and multi-component foods. The methodology to evaluate natural antioxidants must be carefully interpreted depending on whether oxidation is carried out in bulk oils or in emulsions, and what analytical method is used to determine extent and end-point of oxidation.

Each antioxidant evaluation should be carried out under various conditions of oxidation, using several methods to measure different products of oxidation related to real food quality. There cannot be a short cut approach to determining the activity of antioxidants. Various testing protocols should consider the following parameters:

(i) *Substrates*: triacylglycerols or phospholipids, in bulk, emulsions, or liposomes systems, should be used because they are most important in foods and biological systems. Free fatty acids should be avoided because they influence the relative activities of different antioxidants, and because they form micelles that behave differently from triacylglycerols.

(ii) *Conditions*: different temperatures, metal catalysts, surface exposure (*e.g.* agitation).

(iii) *Analyses*: initial products (hydroperoxides, peroxide value, conjugated dienes) and secondary decomposition products (carbonyls, volatile compounds).

(iv) *Concentrations*: antioxidants should be compared at the same molar concentration of active components and compared to appropriate reference compounds. With crude extracts, compositional data are needed to compare samples.

TABLE 8-3.
Effect of temperature on antioxidant activity in soybean oil [a]

Antioxidants (0.02%)	Antioxidant activity days at 45°C [b]	Antioxidant activity hours at 98°C [b]
Control	7	5
Ascorbic acid	12	43
Ascorbyl palmitate	19	13
BHA	9	9
BHT	10	11
PG	15	14
TBHQ	26	26

[a] From: Cort et al. (1982)
[b] To reach a peroxide value of 70. Oxidation in thin layer at 45°C; oxidation under conditions of the active oxygen method (AOM) at 98°C. BHA, butylated hydroxyanisole; BHT, butylated hydroxytoluene; PG, propyl gallate; TBHQ, tertbutyl hydroquinone (Figure 8-1).

Because of the complexity of real foods, accelerated test systems are difficult to standardize for antioxidants. Each test should be calibrated for each lipid or food. Accelerated oxidation conditions should be as close as practical to the storage conditions under which the food is to be protected (Chapter 7). Ultimately, antioxidants should be evaluated on the food itself.

C. SYNTHETIC ANTIOXIDANTS

For food applications, synthetic antioxidants must be sufficiently active to be used at low concentrations (below 0.02%) and cannot be toxic. To be effective in foods, they must partition favorably between the oil-air interfaces in bulk oil systems, and between oil-water interfaces in emulsion systems (Chapter 9). Food antioxidants must also be stable to processing and cooking conditions. The term "carry through effect" refers to the ability of antioxidants to withstand the thermal treatments of frying or baking to be absorbed by the food and have sufficient activity to stabilize the fried or baked food.

The most important synthetic antioxidants for food applications worldwide include BHA, BHT, PG and ascorbyl palmitate (Figure 8-1). TBHQ can be used in the USA but not in Europe. BHA is more soluble in fats than BHT and both are considered to be more effective in less unsaturated animal fats than polyunsaturated vegetable oils. The trihydric PG is less oil-soluble than the monohydric BHA and BHT. The dihydric TBHQ is more effective in soybean oil than BHA, BHT or PG, and more stable to heat processing. Commercial synthetic antioxidants are commonly formulated with propylene glycol, glyceryl monooleate, mono- and diglycerides, or vegetable oils as carriers to increase their solubility or dispersibility in foods.

Several synergistic mixtures are used commercially including BHA and BHT, BHA and PG, BHA, BHT and TBHQ, with various levels of citric acid.

PG reinforces the antioxidant activity of BHA but not that of BHT. Ascorbyl palmitate (Figure 8-1) is a weak antioxidant when used alone, but it shows synergistic activity in mixtures with natural tocopherols. A patent claims that a combination between ascorbyl palmitate, tocopherols and lecithin is particularly effective for the stabilization of polyunsaturated fish oils (Löliger and Saucy, 1989).

The reported and claimed activities of different antioxidants and formulations are difficult to evaluate because of the different testing conditions used. For example, when tested at 45°C, ascorbic acid was less effective in stabilizing soybean oil than ascorbyl palmitate, which was in turn more effective than BHA and BHT, but not as active as PG and TBHQ. TBHQ was the most active of the antioxidants tested (Table 8-3). However, when tested at 98°C under conditions of the active oxygen method (AOM), the relative antioxidant activities of ascorbic acid and ascorbyl palmitate were reversed and ascorbic acid became more active than PG and TBHQ. These results must be interpreted with caution. The peroxide value used to follow oxidation is not reliable when oils are heated at 98°C because a large part of the hydroperoxides are decomposed at this temperature (Chapter 6). The peroxide value of 70 used as an end-point is also questionable since flavor deterioration occurs in soybean oil at peroxide values below 10.

Phospholipids are generally useful as synergists in reinforcing the antioxidant activity of phenolic compounds. However, the effects of phospholipids are complicated by their multiple functions; they are effective metal chelators, and good emulsifying agents. As emulsifiers, the antioxidant effects of phospholipids may be explained by their ability to improve the affinity of the tocopherols and other phenolic antioxidants toward the lipid substrate. Another confounding effect of phospholipids is their tendency to produce browning materials when heated at the elevated temperatures used for the AOM and Rancimat tests (Chapter 6). Browning materials act as reducing agents, which are effective food antioxidants (Chapter 10).

D. NATURAL ANTIOXIDANTS

Much interest has developed in naturally occurring antioxidants because of the adverse attention received by the synthetic antioxidants, and because of the worldwide trend to avoid or minimize the use of artificial food additives. In the last 15 to 20 years, special attention has been given to the use of natural antioxidants because of the possible, but not well established, hazardous effects of synthetic antioxidants. The action of food processors to remove antioxidants from their formulations has also been motivated by economic concerns to save on testing costs.

1. Tocopherols

Tocopherols are the most important natural antioxidants found in vegetable oil-derived foods. They occur as different homologs varying in the extent of methylation of the chroman ring (α, β, γ and δ) containing a saturated phytyl

TABLE 8-4.
Tocopherols in vegetable oils (ppm or mg/kg) [a]

Oils	α	β	γ	δ
corn	112	50	602	19
cottonseed	389	—	387	—
olive	119	—	7	—
palm [b]	256	—	316	70
peanut	130	—	214	21
rapeseed	210	1	42	—
safflower	342	—	71	—
soybean	75	15	797	266
sunflower	487	—	51	8
menhaden	75	—	—	—
lard	12	—	—	—
tallow	27	—	—	—

[a] From: Gunstone et al. (1994).
[b] Includes 149 ppm of tocotrienols.

(trimethyltridecyl) side chain (Figure 8-1). Vegetable oils are good sources of tocopherols and their concentrations and homologous compositions vary widely (Table 8-4). Soybean oil is rich in γ-tocopherol whereas sunflower oil is rich in α-tocopherol. Tocotrienols contain a tri-unsaturated side chain and occur in significant amounts (about 150 ppm) only in palm oil, but their relative antioxidant activities are not well established. Animal fats and fish oils contain much lower levels of tocopherols than vegetable oils.

α-Tocopherol behaves like a chain-breaking electron donor antioxidant by competing with the substrate for the chain-carrying peroxyl radicals, normally present in the highest concentration in the lipid system. α-Tocopherol reacts rapidly with peroxyl radicals by reaction (4) and donates a hydrogen atom to produce lipid hydroperoxides and a tocopherol radical that is stabilized by resonance. This tocopherol radical does not propagate the chain but forms non-radical products, including stable peroxides, which can be reduced to a tocoquinone adduct and to tocopherol dimers (Figure 8-3). α-Tocopherol is also oxidized by reaction (8) to form α-tocopheroquinone, which acts as an electron acceptor antioxidant by competing with oxygen for alkyl radicals (RQ˙). This competitive reaction would only become important at low oxygen pressure (e.g. elevated temperatures), because the alkyl radicals react extremely rapidly with oxygen under atmospheric conditions.

The well-recognized synergism between α-tocopherol and ascorbic acid, is explained by the regeneration and recycling of the tocopherol radical intermediate to the parent α-tocopherol. During the oxidation of methyl linoleate in the presence of α-tocopherol and ascorbic acid, the induction period was prolonged and the propagation stage was delayed by the sparing effect of ascorbic acid, which is oxidized first followed by the oxidation of α-tocopherol (Figure 8-4). In a liposome system in which lecithin forms lamellar bilayers separated by water layers (see Chapter 9), ascorbic acid regenerates

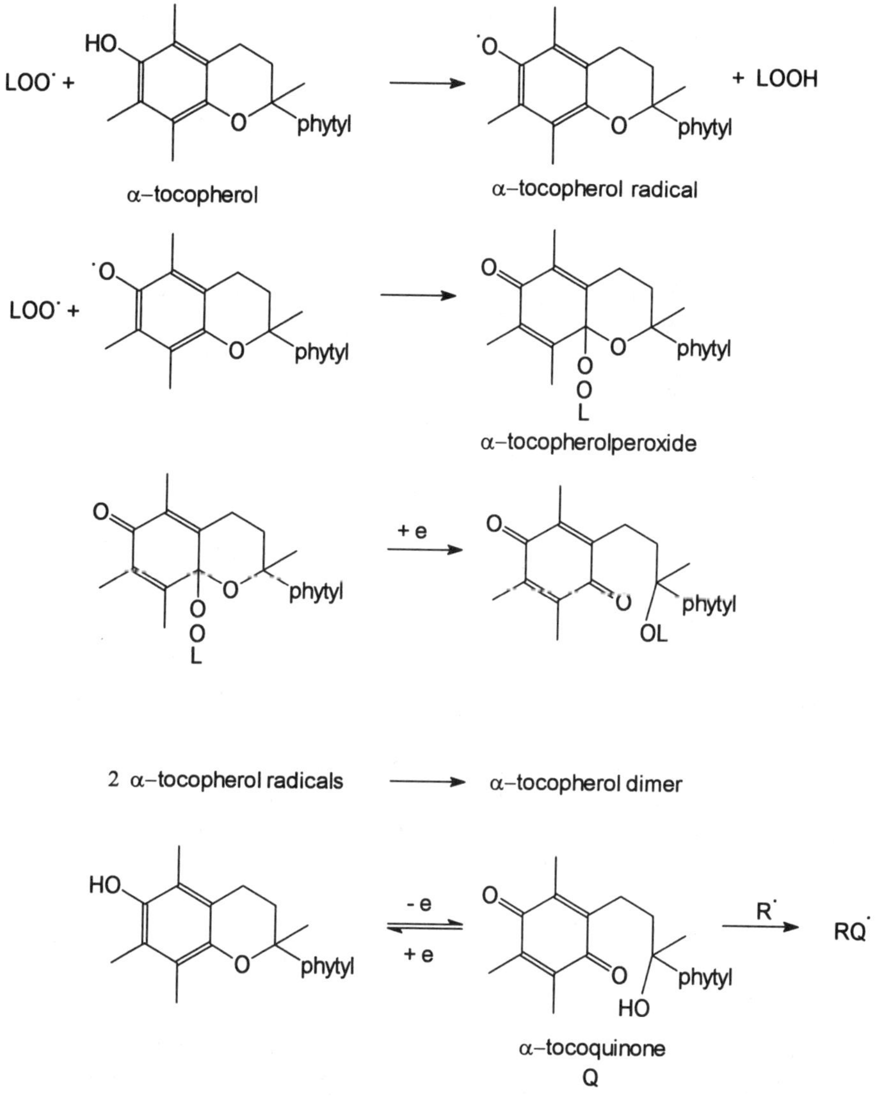

Figure 8-3. Mechanism of α-tocopherol action.

α-tocopherol only with a water-soluble initiator, but not with a lipid-soluble initiator. This observation suggests that ascorbic acid can probably only react with the tocopherol radicals at the surface of the liposome.

In natural food systems, the interaction of tocopherols and ascorbic acid occurs favorably because initiation of lipid oxidation occurs also by metal catalysts present in the water phase. In biological systems, synergism takes place in a cascade of reactions where ascorbic acid is also regenerated at the expense of more oxidizable substrates (Chapter 12).

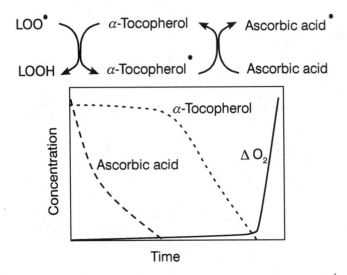

Figure 8-4. Disappearance of α-tocopherol and ascorbic acid during oxidation of methyl linoleate. From Niki et al. (1980).

Tocopherols can interrupt lipid autoxidation by interfering with either the chain propagation or the decomposition processes. In addition to its free radical scavenging activity, α-tocopherol is highly reactive toward singlet oxygen and protects food lipids against photosensitized oxidation (Chapter 3). Although about 20 to 40% of the tocopherols are lost during processing of vegetable oils, the levels of tocopherols left are usually adequate for protection against oxidation. In many studies in the literature, α-tocopherol has been shown to have a prooxidant effect at high concentrations on the basis of methods measuring hydroperoxide formation during the early stages of lipid oxidation. Although the mechanism for the antioxidant effects of α-tocopherol in intercepting peroxyl radicals is well accepted, the prooxidant effects observed at high concentrations are not well understood.

In polyunsaturated vegetable oils, tocopherol mixtures are generally regarded as poor antioxidants. With soybean oil containing 1500 ppm mixed tocopherols, the removal of a portion of the natural tocopherols with carbon-black treatment markedly increased the oxidative stability (Table 8-5). The oxidative stability of soybean oil was optimum at tocopherol concentrations between 400 to 600 ppm. This observation is the basis of the commercial preparation of tocopherol concentrates, which are recovered from the deodorization distillates during soybean oil processing. In soybean oil treated with carbon-black to remove the natural tocopherols almost completely, added α-tocopherol was a less effective antioxidant than either ascorbic acid or citric acid (Table 8-6). The mixture of α-tocopherol and citric acid gave the same induction period as citric acid alone. The effect of metal chelation by citric acid was, therefore, more important than the antioxidant effect of α-tocopherol in soybean oil. Although the activity of BHT alone was about the

TABLE 8-5.
Effect of tocopherol concentration on oxidative stability of soybean oil at 60°C.[a]

Samples[b]	Tocopherol ppm	Oxygen absorption Induction period, hr	Peroxide value Induction period, hr
Control	1500	39	30
C-treated 1 x	1050	84	60
C-treated 2 x	544	88	78
C-treated 3 x	196	6	40
C-treated 4 x	45	—	18

[a] From: Frankel et al. (1959a).
[b] C-treated 1 x = carbon-treated once to remove tocopherols
[b] C-treated 2 x = carbon-treated twice to remove tocopherols
[b] C-treated 3 x = carbon-treated three times to remove tocopherols
[b] C-treated 4 x = carbon-treated four times to remove tocopherols

same as that of α-tocopherol, the mixture of BHT and citric acid was as effective as the mixture of α-tocopherol and citric acid. The addition of 0.01% citric acid decreased the loss of tocopherol in soybean oil and significantly increased its oxidative stability (Table 8-7). The addition of 0.3 ppm of iron caused the rapid loss of tocopherols and a significant decrease in oxidative stability of soybean oil. The addition of 0.01% citric acid was sufficient to almost completely eliminate the prooxidant effect of iron.

The relative antioxidant activity of different tocopherol homologs has received much attention in the literature. On the basis of induction period

TABLE 8-6.
Effect of antioxidants and metal inactivators on oxidative stability of soybean oil at 60°C.[a]

Samples[c]	Peroxide value Induction period, hr
Control A (1500 ppm tocopherol)	26
C-treated (45 ppm tocopherol)[b]	17
+ α-tocopherol	21
+ BHA	18
+ BHT	23
+ propyl gallate	26
+ citric acid (CA)	50
+ ascorbic acid	43
+ α-tocopherol + 0.01% CA	53
+ BHA + 0.01% CA	56
+ BHT + 0.01% CA	53

[a] From: Frankel et al. (1959a).
[b] C-treated = carbon treated to remove tocopherol
Antioxidants concentrations = 0.57 mmol/kg

TABLE 8-7.
Effect of citric acid (CA) and iron on oxidative stability and tocopherol oxidation in soybean oil at 60°C.[a,b]

Samples	Induction period hr	Initial rate PV / hr	Tocopherol loss, ppm / hr
Control	16	0.1	3.5
+ 0.01% CA	105	0.02	1.4
+ 0.3 ppm Fe^{3+}	4	0.41	43
+ 0.01% CA + 0.3 ppm Fe^{3+}	100	0.02	1.5

[a] Adapted from: Frankel et al. (1959b).
[b] Tocopherol content of soybean oil : 1530 ppm; induction period based on peroxide value (PV).

measurements, γ-and δ-tocopherols were relatively more effective than β- and α-tocopherols (Table 8-8). On the basis of initial rates of oxidation of styrene in chlorobenzene solution in the presence of azo-bis(isobutyronitrile) as initiator, the relative activities of tocopherol homologues increased in the same order as their vitamin E activities: $\alpha > \beta > \gamma > \delta$. The temperature of oxidation had a marked effect on the relative activities of tocopherol homologs in linseed esters high in linolenate but not in cottonseed esters in which linoleate was the main unsaturated fatty acid. The discrepancies concerning the relative antioxidant activities of tocopherol homologs can be attributed to the wide differences in unsaturated substrates tested, the level of oxidation used in the tests, and the method used to analyse oxidation.

Tocopherols can behave as antioxidants or prooxidants depending on the test system, the concentrations, the stages of oxidation tested and the methods used to follow oxidation. Initially, α-tocopherol inhibited hydroperoxide formation at 100 ppm, but promoted hydroperoxide formation at 750 and 1000 ppm (Figure 8-5, 3-5 days). At later stages of oxidation (6-7 days), α-tocopherol inhibited hydroperoxide formation at all concentrations tested. After 7 days, α-tocopherol

TABLE 8-8.
Relative order of antioxidant activities of tocopherols.

References	Conditions	α	β	γ	δ
Lea (1960)	Me linoleate, 50°C	4	3	1	2
	Lard esters, 90°C	4	3	2	1
Lea and Ward (1959)	Cottonseed esters, 37°C	4	3	1	2
	Cottonseed esters, 60°C	4	3	1	2
	Linseed esters, 37°C	1	3	2	4
	Linseed esters, 60°C	2	3	1	4
Olcott and Van der Veen (1968)	Menhaden oil 37°C	3		1	2
Burton and Ingold (1981)	Styrene, 30°C	1	2	3	4

From: Frankel (1989).

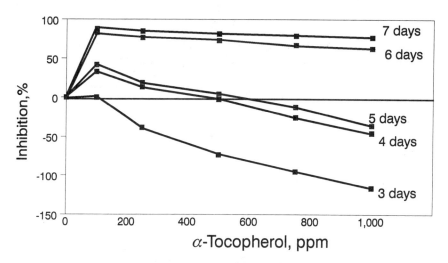

Figure 8-5. Effect of different concentrations of α-tocopherol on the inhibition of hydroperoxide formation at different stages of oxidation of bulk corn oil. From Frankel et al. (1994).

showed its highest inhibition of hydroperoxide formation at 100 ppm. In contrast to hydroperoxide formation, α-tocopherol inhibited hexanal formation at all levels tested (Figure 8-6). Inhibition of hexanal formation increased with concentrations of α-tocopherol and with oxidation time. After 6 and 7 days, hexanal formation was almost completely inhibited at all concentrations of α-tocopherol tested. The same trends were observed in corn oil-in-water

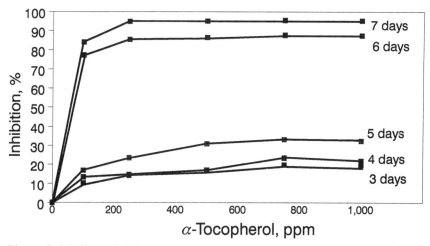

Figure 8-6. Effect of different concentrations of α-tocopherol on the inhibition of hexanal formation at different stages of oxidation of bulk corn oil. From Frankel et al. (1994).

TABLE 8-9.
Percent inhibition of hydroperoxides and hexanal formation by α-tocopherol and Trolox in bulk corn oil and corn oil-in-water emulsion at 60°C.[a]

Antioxidants	Bulk oil		Emulsion	
	hydroperoxides	hexanal	hydroperoxides	hexanal
Corn oil				
+ α-tocopherol				
150 μM	15	35	26	54
300 μM	9	40	88	75
+ Trolox				
150 μM	30	6	15	2
300 μM	84	88	13	-27

[a] From: Huang et al. (1996). Hydroperoxides and hexanal measured after 6 days of oxidation in bulk corn oil, and after 4 days in corn oil emulsion. Negative values represent prooxidant activity.

emulsions. Although α-tocopherol at high concentrations promoted hydroperoxide formation, its ability to inhibit hexanal formation improved at increasing concentrations. The ability of antioxidants in inhibiting hexanal formation may be more relevant to flavor development.

In contrast to α-tocopherol, γ-tocopherol inhibited hydroperoxide and hexanal formation at all levels tested. The inhibition was significantly higher at 250-500 ppm and lower at 100 ppm in both bulk corn oil and in the corresponding oil-in-water emulsions. γ-Tocopherol had no prooxidant activity in either bulk oil or emulsion systems (Table 8-1). In bulk corn oil, δ-tocopherol, which is present in significant amounts in soybean oil (Table 8-4), did not show prooxidant activity at higher concentrations (1000-2000 ppm). At these concentrations, δ-tocopherol was also more effective than α- and γ-tocopherols in inhibiting the decomposition of hydroperoxides. With mixtures of soybean tocopherols, maximum antioxidant activity was observed between 250 and 3000 ppm. The prooxidant activity of tocopherols can be explained by the chain transfer reactions (-4) and (9), which become significant at high concentrations of α-tocopherol. Tocopherol oxidation by reaction (8) is also greater for α-tocopherol than for γ-tocopherol.

The effect of different lipid substrates has a significant impact on the activity of different antioxidants according to their hydrophilic or lipophilic character. For example, α-tocopherol (Figure 8-1) is a lipophilic antioxidant that behaved quite differently in various lipid substrates from its carboxylic acid analog Trolox, which is hydrophilic. In bulk corn oil triacylglycerols, Trolox was a better antioxidant than α-tocopherol, but the opposite trend was observed in the corresponding oil-in-water emulsions (Table 8-9). In emulsified methyl linoleate, Trolox behaved the same way as in corn oil, and was less effective than α-tocopherol. However, in emulsified linoleic acid, Trolox was a better antioxidant than α-tocopherol. Linoleic acid constitutes a unique substrate because of its tendency to form mixed micelles (Figure 9-1). The physical states of lipid systems affect the distribution of antioxidants and influence their activity. α-Tocopherol and Trolox exhibit complex interfacial properties between air-oil and oil-water interfaces that significantly affect their relative activities in different lipid systems (see Chapter 9). In the bulk oil system, the hydrophilic Trolox is apparently more protective by being oriented in the air-oil interface (Figure 9-8). In the emulsion system, the lipophilic α-tocopherol is more protective by being oriented in the oil-water interface. Because of its tendency to form micelles, linoleic acid is not an appropriate lipid to test antioxidants since their behavior in this substrate would be significantly different from that in foods composed mainly of triacylglycerols.

2. Ascorbic Acid

This organic acid is known for its complex multi-functional effects, as an antioxidant, a prooxidant, a metal chelator, a reducing agent or as an oxygen scavenger. A combination of these effects may predominate in many food applications. In aqueous media, ascorbic acid is active as an antioxidant at high concentrations (approximately 1 mM). However, it can act as a prooxidant at lower concentrations (approximately 0.01 mM), especially in the presence of metal ions. Metals such as Fe^{3+} reduced by ascorbic acid are much more catalytically active in their lower valence state (Fe^{2+}) in the homolytic decomposition of hydroperoxides (Chapter 4). Ascorbic acid interacts with Cu^{2+} also to accelerate the decomposition of linoleate hydroperoxides.

In soybean oil treated with carbon black to remove most of the natural tocopherols, ascorbic acid was more effective as an antioxidant than α-tocopherol, BHA, BHT and propyl gallate, but less effective than citric acid during oxidation at 60°C (Table 8-6). However, the relative antioxidant activity of ascorbic acid was markedly influenced by the temperature of oxidation (Table 8-3). Ascorbic acid was less effective at 45°C and more effective at 98°C than ascorbyl palmitate.

Ascorbic acid acts as an efficient reducing agent by regenerating antioxidants and increasing their effectiveness. The mixture of ascorbic acid and α-tocopherol thus shows strong synergism, an effect that has been known for a long time and has been explained by different mechanisms. During oxidation, α-tocopherol produces tocopheroxyl radical intermediates that can be readily reduced by ascorbic acid and regenerate α-tocopherol (Figure 8-4). This redox mechanism has been confirmed by electron spin resonance (ESR), a technique which provides direct evidence for the tocopheroxyl and ascorbyl radicals. Another mechanism involves the metal inactivation properties of ascorbic acid. Tocopherols are inactivated in the presence of metals and their antioxidant activities can be greatly improved in the presence of metal inactivators (citric acid in Table 8-7).

The synergism exhibited by the ternary mixture of α-tocopherol, ascorbic acid and phospholipids has been shown to be due to the stabilization of α-tocopherol, on the basis of ESR studies with methyl linolenate oxidized at 90°C to detect the free radicals of α-tocopherol and ascorbic acid. Evidence was obtained by this technique for the formation of nitroxide radicals (R-N-O˙) in the presence of phosphatidylserine or phosphatidylethanolamine or soybean lecithin and oxidized methyl linolenate. However, as pointed out earlier (Section C), the synergistic activity of this ternary mixture may be derived from antioxidant products formed from the phospholipids at elevated temperatures by the Maillard browning reaction (Chapter 10).

We observed previously that α-tocopherol behaved differently from its carboxylic acid analog, Trolox, in bulk corn oil triacylglycerols compared to the corresponding oil-in-water emulsions. The same behavior is observed when ascorbic acid is compared to ascorbyl palmitate. The hydrophilic ascorbic acid was a more effective antioxidant in bulk corn oil than in emulsified corn oil. In contrast, the lipophilic ascorbyl palmitate was a more effective antioxidant in the corn oil emulsion than in bulk corn oil (Figures 8-7, 8-8). In the bulk oil system, the hydrophilic ascorbic acid is apparently more protective by being oriented in the air-oil interface. In the emulsion system, the lipophilic ascorbyl palmitate is more protective by being oriented in the oil-water interface (Figure 9-8). This interfacial phenomenon is discussed in more details in Chapter 9.

In aqueous solution, ascorbic acid is a good oxygen scavenger. After shaking for 24 hours, an ascorbic acid solution consumed oxygen at a ratio close to the theoretical value of 3.3 mg per cm^3 of headspace. By its oxygen scavenging activity, ascorbic acid is effective in protecting flavor compounds in wine, beer, fruits and vegetables against oxidation. However, anaerobic conditions are necessary for the effectiveness of ascorbic acid because it is readily decomposed in aqueous solution in the presence of air. Ascorbyl palmitate is too insoluble in water to function as an oxygen scavenger.

In summary, the multiple effects of ascorbic acid include: (i) hydrogen donation to antioxidant radicals to regenerate the antioxidant; (ii) metal inactivation to reduce the rate of initiation by metals; (iii) reduction of hydroperoxides to stable alcohols; and (iv) scavenging oxygen in aqueous systems.

3. *Rosemary and Other Spice Extracts*

Rosemary extracts provide a major source of natural antioxidants used commercially at present in foods. The antioxidant activity of commercial rosemary extracts is mainly related to its content of two phenolic diterpenes, carnosic acid and carnosol, which are also found in sage. Carnosic acid has a structure consisting of three six-membered rings including a dihydric phenolic ring, and a free carboxylic acid (Figure 8-9). Carnosol is a derivative of carnosic acid containing a lactone ring. Extracts of rosemary and other herbs

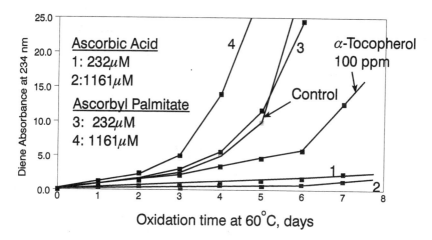

Figure 8-7. Effect of ascorbic acid, ascorbyl palmitate and α-tocopherol on the oxidative stability of bulk corn oil. Adapted from Frankel et al. (1994).

and spices are produced by using organic solvents (e.g. hexane and ethanol) and they are deodorized or bleached. They are available commercially in the form of powders, paste, or liquids. Some formulations contain various proprietary carriers such as propylene glycol, medium chain triacylglycerols or vegetable oils.

In bulk corn oil, the rosemary extract, carnosic acid, rosmarinic acid and α-tocopherol were significantly more active than carnosol (Table 8-10). In contrast, in corn oil-in-water emulsion, carnosol was significantly more active than in bulk oil, whereas the rosemary extract and carnosic acid were less active and rosmarinic acid showed prooxidant activity. The polar hydrophilic

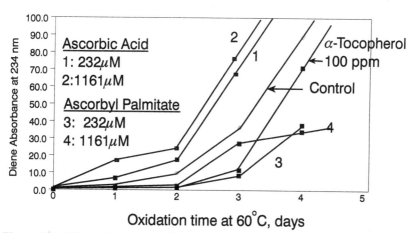

Figure 8-8. Effect of ascorbic acid, ascorbyl palmitate and α-tocopherol on the oxidative stability of corn oil-in-water emulsion. Adapted from Frankel et al. (1994).

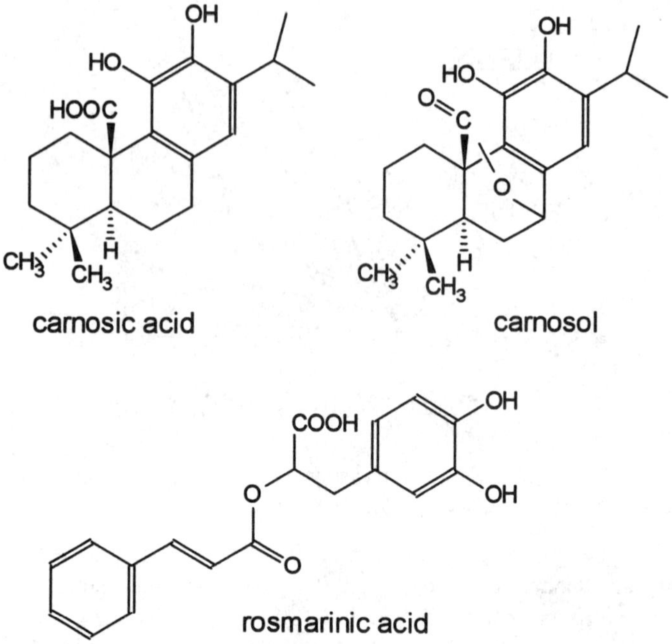

Figure 8-9. Structures of rosemary antioxidants: carnosic acid, carnosol, and rosmarinic acid.

rosemary compounds were less active in emulsion systems because they undergo interfacial partitioning into the water thus becoming less protective than in the bulk oil system (see Chapter 9).

The antioxidant activities of rosemary extracts, carnosol and carnosic acid were also significantly influenced by the oil substrates and the type of system

TABLE 8-10.

Antioxidant activity of a rosemary extract, carnosic acid, carnosol and rosmarinic acid in stripped corn oil and emulsified corn oil at 60°C (% inhibition).[a]

Samples	Hydroperoxides		Hexanal	
	bulk oil 6 days	emulsion 4 days	bulk oil 6 days	emulsion 4 days
rosemary extract [b]				
250 ppm	64	11	-3	35
500 ppm	81	12	98	28
carnosic acid, 50 ppm	70	37	29	83
carnosol, 50 ppm	6	42	-20	84
rosmarinic acid, 50 ppm	68	-11	78	-53

[a] From: Frankel *et al.* (1996a).
[b] Contains 10.3% carnosic acid and 4.5% carnosol.

TABLE 8-11.
Percent inhibition of hydroperoxide formation by rosemary extracts (R-E), carnosol and carnosic acid in bulk oils and oil-in water emulsions.[a]

Samples	Corn oil		Soybean oil		Peanut oil		Fish oil	
	Bulk	Emulsion	Bulk	Emulsion	Bulk	Emulsion	Bulk	Emulsion
R-E:300 ppm	83.9	-81.2	55.0	-211	90.1	-17.7	78.9	25.3
R-E:500 ppm	89.3	-11.9	76.3	-59.6	94.8	-107	88.0	42.1
Carnosol,30 ppm	38.4	-165	32.6	-102	37.3	-135	16.6	-2.5
Carnosol,50 ppm	51.0	-27.5	34.0	-189	65.5	-136	10.3	20.3
Carnosic,30 ppm	74.4	-29.6	5.5	-131	87.6	-107	28.0	36.1
Carnosic,50 ppm	87.7	-14.9	70.5	-97.2	91.7	-104	45.7	32.7

[a] From: Frankel et al. (1996b). Oxidation of vegetable oils were carried out at 60°C for 20 days in bulk and for 12 days in emulsions; oxidation of fish oil were carried out at 40°C for 10 days. Negative values represent prooxidant activity.

TABLE 8-12.
Antioxidant activity of plant extracts (1%) obtained in different carriers in bulk and emulsified corn oil systems.[a]

Plant extracts	MCT[b] carrier		PG[b] carrier	
	Bulk	Emulsion	Bulk	Emulsion
Rosemary	1.7	12	1.7	32
Sage	1.5	4	1.5	19
Thyme	1.1	1.3	1.2	8.4
Oregano	1.0	1.2	1.1	7.0
Cloves	1.0	8.9	1.0	13

[a] From: Löliger et al. (1996). Activity values expressed as "antioxidant index" (AI) = induction period of oil + antioxidant / induction period of control oil.
[b] MCT = medium chain triglycerides, PG = propylene glycol.

tested, bulk oils versus oil-in-water emulsions. The rosemary extracts, carnosol and carnosic acid effectively inhibited hydroperoxide formation in corn oil, soybean oil, peanut oil and fish oil, when tested in bulk (Table 8-11). Test compounds also inhibited hexanal formation in bulk vegetable oils, and propanal and pentenal formation in bulk fish oils. In marked contrast, these test compounds were either inactive or promoted oxidation in the corresponding vegetable oil-in-water emulsions. In fish oil emulsions, however, the rosemary compounds inhibited conjugated diene and pentenal formation, but not propanal.

Interfacial phenomena may explain these differences in activities. Rosemary extracts, carnosol and carnosic acid behaved like other hydrophilic antioxidants such as ascorbic acid and Trolox in being more effective in bulk oil than in oil-in-water emulsion systems. In the bulk oil systems where oil is the main phase, the hydrophilic rosemary antioxidants may be more protective by being oriented in the air-oil interface. In contrast, in the oil-in-water emulsion systems, where water is the main phase, the hydrophilic rosemary antioxidants remain in the water and are less effective in the oil-water interface where oxidation takes place (Chapter 9). The higher antioxidant activities of rosemary antioxidants observed in fish oil emulsions than in vegetable oil emulsions may be explained by their greater affinity toward the more polar oil interface with the water of the fish oil systems.

Plant extracts obtained mechanically with either medium-chain triacylglycerols or propylene glycols exhibited different antioxidant activities and were more active in corn oil emulsion than in bulk corn oil (Table 8-12). In the emulsion system, the rosemary extract in propylene glycol was more active than the extract in medium chain triacylglycerols. The higher activity of the rosemary extract with propylene glycol was attributed to the presence of more polar substances such as rosmarinic acid and flavonoids.

TABLE 8-13.
Antioxidant activities of flavonols (0.01%)
in lard oxidized at 60°C.[a]

Flavones (Figure 8-10)	I.P.[b]
Control	90
3,3',4-trihydroxy-	145
3,7,3',4'-tetrahydroxy-	320
3,5,7,3',4'-pentahydroxy-[c]	410
3,7,8,3',4'-pentahydroxy-	555
3,5,7,3',4',5'-hexahydroxy-[d]	685
3,5,6,7,3',4'-hexahydroxy-	698

[a] From: Mehta and Sheshadri (1959)
[b] I.P. = induction period in hours at 60°C to reach a peroxide value of 25.
[c] Quercetin
[d] Myricetin

Although carnosic acid and carnosol are readily oxidized during oxidation at 60°C and higher temperatures, their antioxidant activities are maintained. The oxidation products formed are apparently active antioxidants at high temperatures and protect oils during frying. They also have carry-over activity by protecting the fried foods. Rosemary extracts are particularly effective in stabilizing dried oats, meat products and roasted nuts. The activities of natural antioxidants are thus very system-dependent and their effectiveness in different real food systems is difficult to predict.

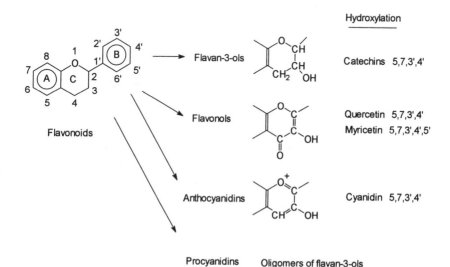

Figure 8-10. Structures of flavonoids.

TABLE 8-14.
Antioxidant activities and phenolic composition (mg/L) of grape extracts in lecithin liposome oxidized at 37°C in the presence of 3 μM copper acetate.[a]

Grape	Type	Total phenols	Inhibition, %		Hydroxy cinnamates (as caffeic acid)	Anthocyanins (as malvin)	Flavonols (as rutin)
			hydro-peroxides	hexanal			
Petite Sirah	RW	1115	100	100	10.4 (0.6)	1685 (98)	19.8 (1.1)
Cabernet Sauvignon	RW	565	71	47	0.8 (0.1)	640 (98)	9.3 (1.4)
Merlot	RW	569	67	46	5.3 (1.1)	437 (93)	26.8 (5.7)
Sauvignon Blanc	WW	264	43	25	6.5 (39)	0	10.0 (61)
Chardonnay	WW	230	27	18	1.4 (25)	0	4.2 (75)
Red Globe	BT	179	26	15	2.7 (11)	19.6 (79)	2.6 (10)
Thompson seedless	WT	198	25	15	3.8 (30)	0	8.9 (70)

[a] From: Yi et al. (1997). Abbreviations: RW, red wine; WW, white wine; BT, blush table; WT, white table. Total phenols expressed as gallic acid equivalents / L. Values in parentheses are the relative percentage of each compound. The extract of Petite Sirah grapes was the only sample containing flavan-3-ols (10.6 mg/L as catechin).

4. Flavonoids and Green Tea Catechins

Flavonoids are among the most active natural antioxidants found in plant foods. Their basic structures have a flavan nucleus (2-phenyl-benzo-γ-pyrane) consisting of two benzene rings (A and B) combined by an oxygen-containing pyran ring C (Figure 8-10). Differences in substitution on ring C define the different classes of flavonoids, which include *flavan*-3-ols, containing a saturated C ring with a methylene at position 4 (*e.g.* catechins with OH on 5,7,3′,4′-positions); *flavonols*, containing a keto group at position 4 in ring C (*e.g.* quercetin with OH on 5,7,3′,4′-positions, myricetin with OH on 5,7,3′,4′,5′-positions generally glycosylated); *anthocyanidins*, having lost the oxygen at carbon-4 of ring C to form fully aromatic flavylium compounds, which exist as cation-forming salts with organic acids (*e.g.* cyanidin with OH on 5,7,3′,4′ and glucoside on the 3-position); and *procyanidins*, which include oligomers of flavan-3-ols. Additional flavonoids found in fruits contain a large number of glycosides, which arise from glycosylation on the 3 and 7 positions.

Flavonoids are recognized for both their antioxidant activity and their metal chelating properties. Antioxidant activity increases with the number of hydroxyl groups in rings A and B (Table 8-13). Metal chelation is contributed by the ortho-diphenol structure in rings A and B and the ketol structure in ring C. Much work has been published on evaluations of natural flavonoids in various lipid systems, but most evaluations were done either at elevated temperatures, under AOM or Rancimat conditions, or were based on the bleaching of β-carotene by co-oxidation with linoleic acid. Unfortunately, from the results of these assays, the usefulness of flavonoids as antioxidants is difficult to predict for foods containing mainly triacylglycerols and stored under ambient or near ambient conditions (Chapter 7).

Extracts of different grapes were tested for their antioxidant activities in a copper-catalysed lecithin liposome system (Table 8-14). Extracts of red wine grapes varieties contained higher concentrations of total phenolic compounds and higher relative antioxidant activities (67-100%) than the white wine and table grapes (25-27%). The relative percent inhibition of conjugated dienes and hexanal correlated with total phenols. Anthocyanins were the most abundant phenolic compounds in extracts of red grapes and flavonols were most abundant in extracts of white grapes.

Green tea extracts consist of a complex mixtures of catechin gallates, including (+)-catechin, (+)-gallocatechin, (-)-epicatechin, (-)-epicatechin gallate, (-)-epigallo-catechin, and epigallocatechin gallate (Figure 8-11). Tea catechins showed different trends in relative antioxidant activity in different lipid systems. In corn oil oxidized at 50°C, epigallocatechin, epigallocatechin gallate and epicatechin gallate were better antioxidants than epicatechin and catechin (Figure 8-12). However, in the corresponding corn oil-in-water emulsions, the tea catechins were prooxidant as were propyl gallate and gallic acid. In contrast, in lecithin liposomes, epigallocatechin gallate was the best antioxidant, followed by epicatechin, epigallocatechin, epicatechin gallate and catechin. Different commercial green teas were active antioxidants in bulk corn

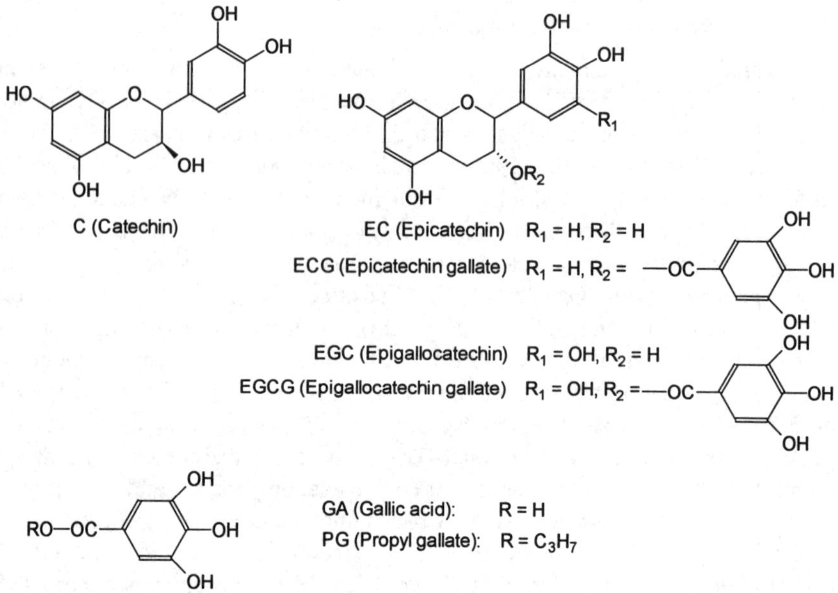

Figure 8-11. Structures of green tea catechins and reference compounds gallic acid and propyl gallate.

oil oxidized at 50°C, but were prooxidant in the corresponding oil-in-water emulsions (Table 8-15). Green teas were also active antioxidants in soybean lecithin liposomes oxidized at 37°C in the presence of a copper catalyst.

The marked variation in activity among green tea samples may be due in part to differences in their relative partition between phases in different lipid systems. The improved antioxidant activity observed for green teas in lecithin liposomes compared to corn oil emulsions was explained by the greater affinity of the polar tea catechin gallates with the polar surface of the lecithin lamellar bilayers (see Chapter 9), thus affording better protection against oxidation. Liposomes may thus be appropriate lipid models to evaluate hydrophilic antioxidants for foods containing phospholipids.

E. ANTIOXIDANT ENZYMES

Enzymes have been evaluated as new types of natural antioxidants in some food applications. Enzymes can be beneficially used to remove oxygen and reactive oxygen species and to reduce lipid hydroperoxides. Glucose oxidase coupled with catalase is the best known commercially available system to remove oxygen from foods. Glucose oxidase utilizes oxygen by catalysing the oxidation of β-D-glucose to produce 2-δ-gluconolactone and hydrogen peroxide. The gluconolactone is spontaneously hydrolysed to D-gluconic acid, and the hydrogen peroxide can be removed by catalase. As legally permitted for food use in the USA, this glucose-oxidase system has been applied in fruit juices, mayonnaise and salad dressings to increase shelf life and prevent off-flavor development. In the presence of superoxide, an active species of oxygen

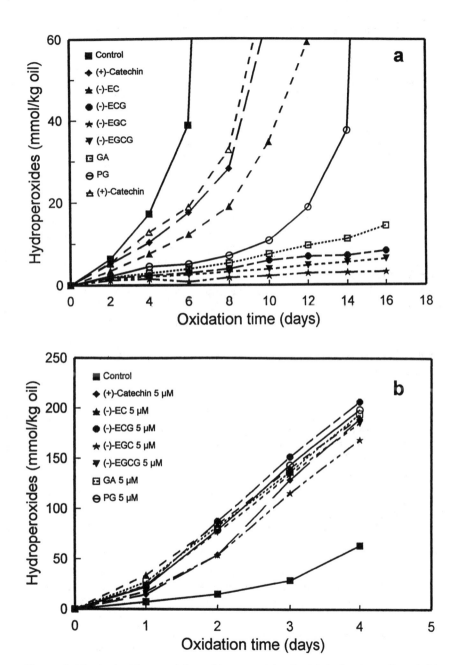

Figure 8-12. Antioxidant activity of tea catechins in (a) bulk corn oil and (b) corn oil-in-water emulsion (From Huang and Frankel (1997). See structures and abbreviations in Figure 8-11. With permission from the American Chemical Society.

TABLE 8-15.
Percent inhibition of hydroperoxides and hexanal by green tea extracts in bulk and emulsified corn oil at 50°C.[a]

Samples[b]	Concn.[c] ppm	Bulk oil		Oil-in-water emulsion	
		Hydroperoxides	Hexanal	Hydroperoxides	Hexanal
G. tea extract I	98	-28	7	-42	-23
G. tea extract II	112	46	29	-45	-46
G. tea extract III	120	31	44	-66	-53
G. tea extract IV	155	52	3	-145	-93
Catechin	125	88	29	—	—

[a] From: Frankel et al. (1997). Negative values represent prooxidant activity.
[b] Samples contained equivalent of approximately 125 ppm of total catechin.
[c] Total catechin concentration. Concentrations of tea extracts used in emulsions were about ten fold less than in the bulk systems.

(see Chapter 12), superoxide dismutase produces oxygen and hydrogen peroxide. When used in conjunction with catalase, to remove hydrogen peroxide, superoxide dismutase had antioxidant activity in a milk fat model system oxidized with iron. However, this system was not effective in food emulsions containing an edible oil. Glutathione transferase was shown to reduce linoleic acid hydroperoxides in the presence of glutathione, but the usefulness of this enzyme was not demonstrated in a food system.

BIBLIOGRAPHY

Aeschbach,R., Baechler,R., Rossi,P., Sandoz,L. and Wille,H.-J. Mechanical extraction of plant antioxidants by means of oils. *Fat Sci. Technol.* **96**, 441-443 (1994).
Aeschbach,R. and Rossi,P. Procédé d'extraction d'antioxydants de matière végétale. *Europeun patent* **A-95**, 102, 439.7 (1995).
Bailey,H.C. The mechanism of anti-oxidant action. *Ind. Chemist* **38**, 215-222 (1962).
Bracco,U., Löliger, J. and Viret, J-L. Production and use of natural antioxidants. *J. Am. Oil Chem. Soc.* **58**, 686-690 (1981).
Burton,G.W. and Ingold,K.U. Autoxidation of biological molecules. I. The antioxidant activity of vitamin E and related chain-breaking phenolic antioxidants in vitro. *J. Am. Chem. Soc.* **103**, 6472-6477 (1981).
Cort,W.M. Antioxidant activity of tocopherols, ascorbyl palmitate and ascorbic acid and their mode of action. *J. Am. Oil Chem. Soc.* **51**, 321-325 (1974).
Cort,W.M. Antioxidant properties of ascorbic acid in foods. *Adv. Chem. Ser.* **200**, 533-550 (1982).
Frankel,E.N. The antioxidant and nutritional effects of tocopherols, ascorbic acid and beta-carotene in relation to processing of edible oils. *Bibliotheca Nutritio et Dieta. Basel, Krager* **43**, 297-312. (1989).
Frankel,E.N. Natural antioxidants in foods and biological systems. Their mechanism of action, applications and implications. *Lipid Technology* **7**, 77-80 (1995).
Frankel,E.N. Antioxidants in lipid foods and their impact on food quality. *Food Chemistry* **57**, 51-55. (1996).
Frankel,E.N., Cooney,P.M., Moser,H.A., Cowan,J.C. and Evans,C.D. Effect of antioxidants and metal inactivators in tocopherol-free soybean oil. *Fette Seifen, Anstrichm.* **10**, 1036-1039 (1959a).
Frankel,E.N., Evans,C.D. and Cooney,P.M. Tocopherol oxidation in natural fats. *J. Agric. Food Chem.* **7**, 438-441 (1959b).

ANTIOXIDANTS 159

Frankel,E.N. and Gardner,HW. Effect of alpha-tocopherol on the volatile thermal decomposition products of methyl linoleate hydroperoxides. *Lipids* **24**, 603-608 (1989).
Frankel,E.N., Huang,S.W. and Aeschbach,R. Antioxidant activity of green teas in different lipid systems. *J. Am. Oil Chem. Soc.* **74**, 1309-1315 (1997).
Frankel,E.N., Huang,S-W., Kanner,J. and German,J.B. Interfacial phenomena in the evaluation of antioxidants: bulk oils *versus* emulsions. *J. Agr. Food Chem.* **42**, 1054-1059 (1994).
Frankel,E.N., Huang,S-W., Aeschbach,R. and Prior,E. Antioxidant activity of a rosemary extract and its constituents, carnosic acid, carnosol and rosmarinic acid in bulk oil and oil-in-water emulsion. *J. Agr. Food Chem.* **44**,131-135 (1996a)
Frankel,E.N., Huang,S-W., Prior,E., and Aeschbach,R. Evaluation of antioxidant activity of rosemary extracts, carnosol and carnosic acid in bulk vegetable oils and fish oil and their emulsions. *J. Sci. Food Agric.* **72**, 201-208 (1996b).
Gunstone,F.D., Harwood,J.L. and Padley,F.B. *The Lipid Handbook*, Second edition, p. 131 (1994), Chapman and Hall, London.
Hopia,A., Huang,S-W. and Frankel,E.N. Effect of α-tocopherol and Trolox on the decomposition of methyl linoleate hydroperoxides. *Lipids* **3**, 357-365 (1996).
Hopia,A.I., Huang,S-H., Schwarz,K., German,J. B. and Frankel,E.N. Effect of different lipid systems on antioxidant activity of rosemary constituents carnosol and carnosic acid with and without α-tocopherol. *J. Agr. Food Chem.* **44**, 2030-2036 (1996).
Huang,S-W. and Frankel,E.N. Antioxidant activity of tea catechins in different lipid systems. *J. Agr. Food Chem.* **45**, 3033-3038 (1997).
Huang,S-W., Frankel,E.N., and German,J.B. Effects of individual tocopherols and tocopherol mixtures on the oxidative stability of corn oil triglycerides. *J. Agr. Food Chem.* **43**, 2345-2350 (1995).
Huang,S-W., Frankel,E.N., German,J.B., and Aeschbach,R. Antioxidant activity of carnosic acid and methyl carnosate in bulk oils and in oil-in-water emulsions. *J. Agr. Food Chem.* **44**, 2951-2956 (1996).
Huang,S-W., Hopia,A., Frankel,E.N., and German,J. B. Antioxidant activity of α-tocopherol and Trolox in different lipid substrates: bulk oils vs oil-in-water emulsions. *J. Agr. Food Chem.* **44**, 444-452 (1996).
Huang,M-T., Ho,C-T. and Lee,C.Y. (1992) *Phenolic Compounds in Food and Their Effects on Health. II. Antioxidants and Cancer Prevention*, ACS Series 507, American Chemical Society, Washignton, DC.
Hudson,B.J.F. Editor. *Food Antioxidants* (1990) Elsevier Applied Science, London.
Husain,S.R., Terao,J. and Matsushita,S. Effect of browning reaction products of phospholipids on autoxidation of methyl linoleate. *J. Am. Oil Chem. Soc.* **66**, 1457-1460 (1986).
Ingold,K.U. Inhibition of the autoxidation of organic substances in the liquid phase. *Chem. Rev.* **61**, 563-589 (1961).
Lambelet,P., Saucy,F. and Löliger,J. Radical exchange reactions between vitamin E, vitamin C and phospholipids in autoxidizing polyunsaturated lipids. *Free Rad. Res.* **20**, 1-10 (1994).
Lea,C.H. On the antioxidant activities of the tocopherols. II. Influence of substrate, temperature and level of oxidation. *J. Sci. Food Agric.* **11**, 212-218 (1960).
Lea,C.H. and Ward,R.J. Relative antioxidant activities of the seven tocopherols. *J. Sci. Food Agric.* **10**, 537-548 (1959).
Löliger,J. Natural antioxidants, in *Rancidity in Foods*, pp. 89-107 (1983) (edited by J.C. Allen, and R.J. Hamilton), Applied Science, London.
Löliger,J. Natural antioxidants for the stabilization of foods, in *Flavor Chemistry*, pp. 302-325 (1989) (edited by D. B. Min and T. Smouse), American Oil Chemists' Society, Champaign, IL.
Löliger,J., Lambelet,P., Aeschbach,R. and Prior,E.M. Natural antioxidants: From radical mechanisms to food stabilization, in *Food Lipids and Health*, pp. 315-344 (1996) (edited by R.E. McDonald and D.B. Min), Marcel Dekker, Inc., New York.
Löliger,J. and Saucy,F. A synergistic antioxidant mixture. *European patent* 0326829 (1989); Process for protecting a fat against oxidation. *U.S. Patent* 5,427,814 (1995).
Mehta,A.C. and Seshadri,T.R. Flavonoids as antioxidants. *J. Sci. Industr. Res.* **18B**, 24-28 (1959).
Meyer,A.S. and Isaksen,A. Application of enzymes as food antioxidants. *Trends Food Sci. Technol.* **6**, 300-304 (1995).
Namiki,M. Antioxidants/antimutagens in food. *Critical Rev. Food Sci. Nutr.* **29**, 273-300 (1990).
Niki,E. Antioxidants in relation to lipid peroxidation. *Chem. Phys. Lipids* **44**, 227-253 (1987).

Olcott,H.S. and Van der Veen,J. Comparison of antioxidant activities of tocol and its methyl derivatives. *Lipids* **3**, 331-334 (1968).

Porter,W.L., Black,E.D. and Drolet,A.M. Use of polyamide oxidative fluorescence test on lipid emulsions: contrast in relative effectiveness of antioxidants in bulk versus dispersed systems. *J. Agric. Food. Chem.* **37**, 615-624 (1989).

Rice-Evans,C., Miller,N.J. and Paganga,G. Structure-antioxidant activity relationships of flavonoids and phenolic acids. *Free Radic. Biol. Med.* **20**, 933-956 (1996).

Rice-Evans,C. and Miller,N.J. Total antioxidant status in plasma and body fluids. *Methods in Enzymology* **234**, 279-293 (1994).

Schwarz,K. Antioxidative Constituents of *Rosmarinus officinalis* and *Salvia officinalis*. III. Stability of phenolic diterpenes of rosemary extracts under thermal stress as required for technological processes. *Z. Lebensm. Unters. Forsch.* **195**, 104-107 (1992).

Scott,G. Antioxidants *in vitro* and *in vivo*. *Chem. Britain* **21**, 648-653 (1985).

Scott,G., editor. *Atmospheric Oxidation and Antioxidants*, Vol. I, II, III (1993), Elsevier, Amsterdam.

Shahidi,F. and Wanasundara,P.K.J. Phenolic antioxidants. *Crit. Rev. Food Sci. Nutr.* **32**, 67-103 (1992).

Sims,R.J. and Fioriti,J.A. Antioxidants as stabilizers for fats, oils, and lipid containing foods, in *CRC Handbook of Food Additives*, 2nd Ed. Vol. II, pp. 13-56, (1977)(edited by T.E. Furia, CRC Press, Inc., Boca Raton, Florida.

Sherwin, E.R. Antioxidants for food fats and oils. *J. Am. Oil Chem. Soc.*, **49**, 468-472 (1972).

Sherwin, E.R. Antioxidants for vegetable oils. *J. Am. Oil Chem. Soc.*, **53**, 430-436 (1976).

Witting, L.A. Vitamin E as food additive. *J. Am. Oil Chem. Soc.*, **52**, 64-68 (1975).

Yi., O-S., Meyer, A.S., and Frankel, E.N. Antioxidant activity of grape extracts in a lecithin liposome system. *J. Am. Oil Chem. Soc.*, **74**, 1301-1307 (1997).

CHAPTER 9

OXIDATION IN MULTIPHASE SYSTEMS

Lipid oxidation in multi-component foods is an interfacial phenomenon affecting prooxidant and antioxidant constituents that is greatly influenced by the interactions between the lipid and water components and the interface between them. Chapter 8 discussed how the relative activities of lipophilic and hydrophilic antioxidants vary between bulk oil systems and oil-in-water emulsions. The classical mechanism of inhibited lipid oxidation does not predict the changes in antioxidant effectiveness between solutions and emulsion systems. The same interfacial phenomena influence the effects of prooxidant metal catalysts, which can exist either in the free form or as complexes with proteins. Model systems have been used extensively to simulate foods or biological samples in research on lipid oxidation and its control. Although research on model systems is important to develop chemical principles that would be applicable to many foods, working with a variety of model systems other than the "real" foods can be misleading by oversimplifying the interfacial interactions of multiple components.

The term "interfacial oxidation" refers to the complex interaction between constituents in multiphase lipid systems in either promoting or inhibiting lipid oxidation. Interfacial oxidation is a "surface" reaction dependent on the rate of oxygen diffusion and its interactions with unsaturated lipids, metal initiators, radical generators and terminators, all of which are distributed in different compartments of colloidal systems.

Lipid oxidation is not well understood in systems in which the fat is dispersed as emulsion droplets. This interfacial oxidation affects a large number of foods, which exist partially or entirely in the form of emulsions. Examples include milk, cream, cheese, mayonnaise, margarine, butter, ice cream, soups, sauces and baby foods. For a better understanding of the lipid oxidation in multiphase systems, the following basic questions should be considered:
a) How is the mechanism of lipid oxidation affected by the physicochemical environment of the lipids?
b) What are the differences between lipid oxidation in bulk fats and in emulsified fats?
c) How are prooxidant and antioxidant constituents in multiphase systems related to their concentrations in different phases?

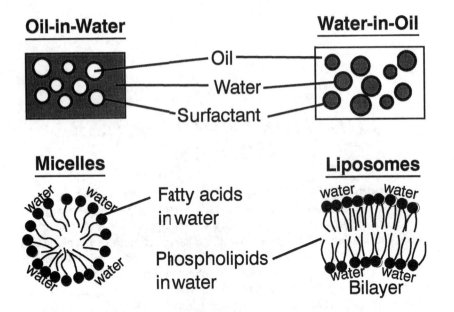

Figure 9-1. Four classical types of emulsion consisting of a distinct continuous phase and a discontinuous or dispersed phase.

d) What are the effects of surface-active compounds on the concentrations and interactions of prooxidants and antioxidants at oil-water interfaces and their activities?

Knowledge on the sites of prooxidant and antioxidant action in multi-component systems is essential to better predict the oxidative stability of complex foods and biological systems. Foods of improved quality may be developed if the association and driving forces of prooxidants and antioxidants can be controlled in multiphase systems.

A. NATURE OF COLLOIDAL SYSTEMS

An emulsion consists of two liquids such as oil and water in which one is dispersed in the other in the form of small droplets. Emulsions are thermodynamically unstable because energy is required to increase the surface between the oil and water phases. The dispersion of one liquid into another is achieved by the energy obtained from vigorous or ultrasonic stirring, or from homogenization. A mixture of oil and water is homogenized by forcing the mixture through a small valve opening to create small oil droplets. However, on standing with time the dispersion will separate into layers of different densities. To prepare emulsions that are stable on standing for a reasonable time, it is necessary to use *emulsifiers*, also referred to as *surfactants*, which are surface-active substances that have an affinity for the surface of droplets by forming a film or membrane covering the droplets to prevent them from

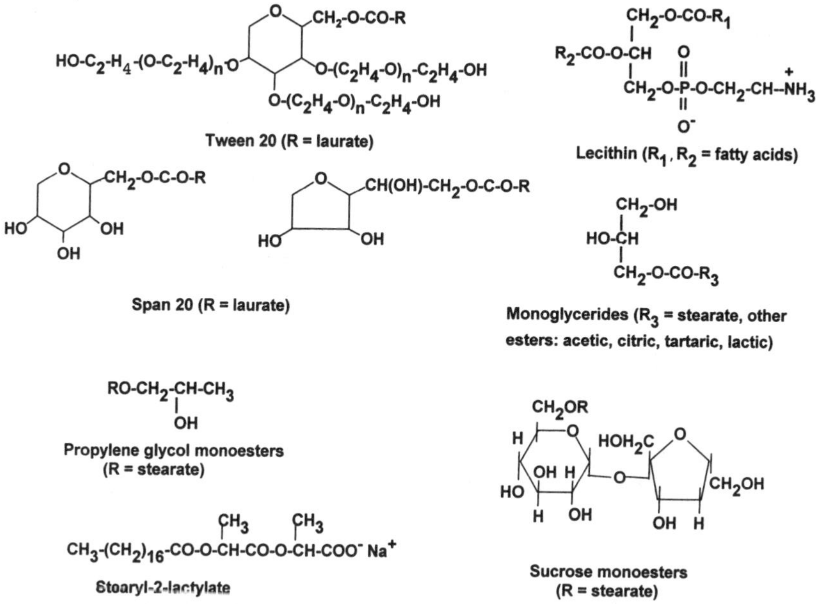

Figure 9-2. Common food emulsifiers: Tweens* (polyoxyethylene mono fatty acid esters of sorbitan or sorbitol anhydrides), Spans* (partial fatty acid esters and sorbitol anhydrides), lecithin (phosphatidylcholine), monoglycerides (monoacylglycerols), propylene glycol mono fatty acid esters (1,2-propanediol esters), sucrose mono fatty acid esters, stearyl-2-lactylate. *Tween/Span 20, monolaurate; Tween/Span 40, monopalmitate; Tween/Span 60, monostearate.

aggregating on standing. Emulsifiers are "amphiphilic" molecules consisting of a polar group oriented in one phase, and a non-polar group oriented in the other phase. In milk, the native milk fat globules can be decreased in size by homogenization under pressure to eliminate flocculation and cream formation.

We may consider four classical types of emulsion consisting of a distinct continuous phase and a discontinuous or dispersed phase (Figure 9-1):

i. *Oil-in-water emulsions* in which the oil droplets are dispersed in the continuous water phase. This dispersed phase is stabilized by an emulsifier composed of a lipophilic or hydrophobic part, consisting of a long chain alkyl residue with good solubility in the oil phase, and a hydrophilic part, consisting of a dissociable group or a number of hydroxyl groups with good solubility in the water phase. In an immiscible oil-in-water system, the emulsifier is located at the interface, where it decreases the interfacial tension, and facilitates a fine dispersion of one phase in the other. Common food emulsifiers include proteins (casein, whey, soy or egg), phospholipids (soy or egg lecithin), or surfactants (polyoxyethylene sorbitan monoesters named Tweens, or fatty acids) (Figure 9-2). Examples of oil-in-water food emulsions include milk, mayonnaise, salad dressing, cream and soups.

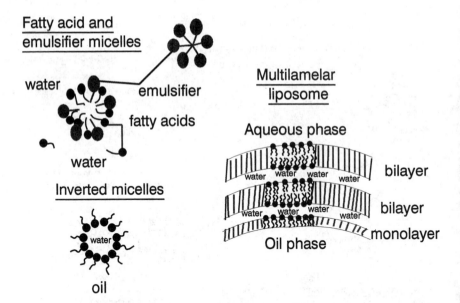

Figure 9-3. Structural pictures of emulsified fatty acids (with Tween 20), multilamellar liposome of aqueous lecithin and triglycerides, and inverted micelles.

ii. *Water-in-oil emulsions* in which water droplets are dispersed in the continuous oil phase. To stabilize water-in-oil emulsions, we need an emulsifier with a relatively stronger hydrophilic group and weaker lipophilic group, and predominantly soluble in the water phase (examples: monoglycerides and sorbitan fatty acid esters named 'Spans') (Figure 9-2). Examples of water-in-oil food emulsions include butter, margarine and spreads. Butter and margarines are unique water-in-oil emulsions because they are stabilized by the presence of crystals in the continuous phase.

iii. *Micelles*: When free fatty acids are dispersed in water they become arranged in the form of micelles, in which the polar head-groups lie mainly in the aqueous phase, and the non-polar tails lie mainly in the oil phase. This arrangement minimizes the free energy (or work content) of the system. Emulsified fatty acids can be either charged or ionic (example: sodium dodecyl sulfate, SDS) or uncharged (example: Tweens). Emulsified fatty acids can also be solubilized in surfactant micelles such as Tweens (Figure 9-3). Micelles can undergo rapid dynamic changes in size and shape involving the entering and leaving of emulsifier molecules within their structure.

Fatty acids ↔ Fatty acid micelles ↔ Emulsifier micelles ↔ Emulsifier

At high oil concentrations, the ternary mixture of oil, water and emulsifier can also exist as "inverted" micelles with the hydrocarbon chains of fatty acids or emulsifier (*e.g.* Tweens) located on the outside and the polar head-groups in the middle (Figure 9-3).

Ionic Emulsifier

Nonionic Emulsifier

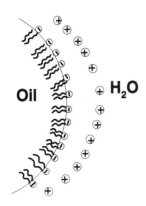
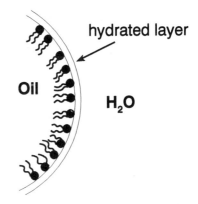

Figure 9-4. Stabilization of oil-in-water emulsion with an ionic emulsifier by electrostatic double layer, and with a nonionic emulsifier by formation of hydrated layer. Adapted from Belitz and Grosch (1987).

Micellar systems of free fatty acids are not common in foods, which are composed mainly of triacylglycerols or phospholipids. However, the presence of small amounts of micellar free fatty acids and moisture can greatly influence the oxidative stability of various foods by solubilizing and facilitating the transfer of non-polar molecules and their reactions in the aqueous phase. Emulsification of triacylglycerols in the presence of small amounts of free fatty acids leads to the formation of mixed micelles and emulsions. These mixed colloidal systems may form during fat frying, and lipid oxidation may be either promoted by the interactions of micellar free fatty acids with metal catalysts, or retarded by the interactions with hydrophilic antioxidants (Chapter 11, Section C).

iv. *Liposomes*: When dispersed in water, phospholipids spontaneously form multi-layers consisting of *bilayers* between the water phase and the innermost layer next to the oil phase being in the form of a monolayer (Figures 9-1 and 9-3). The surface-active phospholipid molecules arrange themselves with the non-polar tails located away from the water, and the polar head groups located in contact with the water. These systems have been commonly used to serve as models of meat and fish products, as well as cell membranes. The lipid bilayers of membranes containing phospholipids are well protected against oxidation by lipophilic antioxidants such as tocopherol. These lipophilic antioxidants can be regenerated by reaction with reducing agents such as ascorbic acid dissolved in the water phase. The interactions between mixtures of lipophilic and hydrophilic compounds provide efficient antioxidant systems in foods and biological materials (Section C.2).

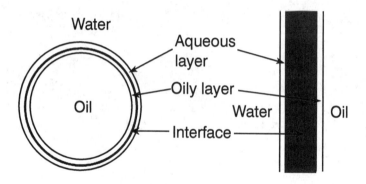

Interface = oil + water + emulsifier

Figure 9-5. Diagrammatic representation of the interface components in an oil-in-water emulsion system.

B. FOOD LIPID EMULSIONS

Oil-in-water emulsions are stabilized by *ionic emulsifiers*, which are oriented on the interface with alkyl residues solubilized in the oil droplets, and charged end groups oriented in the aqueous phase (Figure 9-4). Counter ion groups form an electrostatic double layer, which prevents aggregation of the oil droplets. *Nonionic emulsifiers* are oriented on the surface of the oil droplets with the polar end projected into the aqueous phase. An anchored hydrated layer is formed around the polar groups thus stabilizing the droplets by preventing their coalescence. Nonionic emulsifiers are most commonly used because they do not lose their surface activity by interaction with salt in foods.

The various molecules in an emulsion system become distributed according to their polarity and surface activity between the different phases, which include the oil phase, the water phase and the interfacial region. The interface is a narrow region between the phases that consists of a mixture of oil, water and emulsifier (Figure 9-5). Non-polar molecules are distributed mainly in the oil phase, polar molecule in the water phase, and amphiphilic molecules in the interface. The properties of emulsions are also affected by their dynamic nature because the droplets can undergo collisions and the emulsifier molecules can exchange at different rates between the interface and either the water or the oil phases.

Stable oil-in-water emulsions can also be obtained by dispersing polar lipids such as phospholipids into triglycerides and then emulsifying the oil in water. The presence of charged phosphatidylcholine components of phospholipids improves the stabilization of the emulsions. In most of these systems, the polar phospholipids form a separate phase at the interface where they form lamellar bilayers and a monolayer separated by triglyceride oil, between the outer water phase and the inner triglyceride oil phase (Figure 9-3).

1. Synthetic Emulsifiers and Surfactants

Numerous nonionic synthetic emulsifiers are used in the food industry. These materials are usually partial esters of medium- or long-chain acids (C-12 to C-22) with polyhydric alcohols such as glycerol, propylene glycol, sucrose, or sorbitan (Figure 9-2). Monoglycerides are usually made by glycerolysis (transesterification) by heating triglyceride mixtures with glycerol (30% by wt) in the presence of an alkaline catalyst (sodium hydroxide or sodium methoxide), or by direct esterification of fatty acids and glycerol. Animal fats or palm oil or hydrogenated vegetable oils are used to increase the level of saturated fatty acids. The product is a mixture of mainly 1- and 3-monoglycerides (about 45-60%), diglycerides (about 36-43%) and triglycerides (about 6-12%). Commercial monoglycerides are available either as mixed glycerides (42-46%) or as concentrates following distillation (90% or higher plus 5% diglycerides). These commercial products are used to increase emulsion stability and volume and texture in breads, cake mixes, sauces, ice cream, frozen desserts, cereals, margarine and peanut butter. In doughs, monoglycerides interact with starch to increase dough strength and improve the texture of baked products.

The properties of monoglycerides can also be modified by esterification of one of the hydroxyl groups with acetic, lactic, fumaric, tartaric or citric acids. Similar emulsifiers are synthesised by reacting propylene glycol with saturated fatty acids to give a mixture of mono- and diacyl esters which can be molecularly distilled to prepare concentrates containing about 90% propyleneglycol monostearate.

Other food emulsifiers are made by transesterification of fatty acid methyl esters with polyglycerols, sorbitan, sucrose and stearoyl lactate (Figure 9-2). Sugar, sorbitan fatty esters and polyoxyethylene sorbitan esters are used to stabilize oil-in-water food emulsions. Optimum emulsification is achieved with combinations of emulsifiers. Margarines are made with a combination of monoglycerides and lecithin, and cake mixes with a mixture of monoglycerides and propylene glycol monoesters.

2. Protein Emulsifiers

Proteins are one of the most important natural food emulsifiers because they are surface-active and are effective in preventing coalescence of oil-in-water emulsions. Protein-stabilized oil-in-water emulsions are complicated because protein molecules have various conformations at the oil/water interface. Whey proteins such as β-lactoglobulin adsorb at the oil-water interface without changing their native conformation, whereas milk plasma proteins such as β-casein become unfolded and provide further stabilization by the presence of negative charges (Figure 9-6). These types of adsorption can either be reversible to different extents or be irreversible. The unfolding of certain proteins and exposing their peptide chain at the oil-water interface would be expected to greatly influence lipid oxidation in food emulsions (Chapter 10).

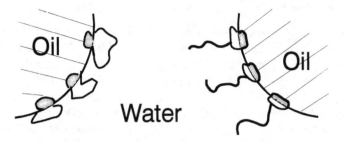

Figure 9-6. Schematic representation of adsorption of globular proteins such as β-globulin (left) and β-casein (right) at the interface of oil-in-water emulsions. Adapted from Larsson (1994) with permission.

Proteins can be displaced in food emulsions by other surface-active molecules or can exchange with other proteins. Homogenization of milk is an important method to control lipid oxidation by changing the interfacial composition of the fat globules. In the native state in raw milk, the fat globules are surrounded by a complex mixture of lipoproteins, phospholipids, diacylglycerols, sterols and enzymes. By homogenization, the milk fat membrane becomes resurfaced by adsorption of casein from the aqueous phase to form a protective layer that reduces oxidation. Lipid oxidation in emulsions can thus be inhibited to different degrees according to the partitioning of proteins between the emulsion interface and the aqueous phase, the composition of the interfacial membrane and the conformation of the protein molecules at the surface (Figures 9-5 and 9-6).

3. Phospholipid Emulsifiers

Phospholipids, also commonly referred to as "lecithins," (see Introduction) are important surface-active substances in the production of food emulsifiers and are generally derived commercially either from soybeans or egg yolk. These substances and several derivatives are used generally in combination with other emulsifiers and stabilizing polymers, such as proteins, starches and gums, and with monoglycerides. For oil-in-water food emulsions, phospholipids are modified by acetylation or hydroxylation to increase their hydrophilic properties. The ethanol-soluble fraction of lecithin rich in phosphatidylcholine (about 65%) is suitable to stabilize oil-in-water emulsions, and the ethanol-insoluble fraction rich in phosphatidylethanolamine (about 33%) and phosphatidylinositol (about 63%) is suitable to stabilize water-in-oil emulsions.

Phospholipids are generally considered to have an antioxidant effect in foods, but this activity is confounded by the formation of reducing browning material at elevated temperatures (Chapter 8). Phosphatidylethanolamine appears to have synergistic activity in mixtures with natural tocopherols and synthetic antioxidants. This synergistic activity is often related to the metal scavenging ability of phospholipids.

4. Mixtures of Emulsifiers

A mixture of emulsifiers can improve the stability of colloid food systems by strengthening the film membrane and by interaction between emulsifiers at the interface. Combinations of oil-soluble and water-soluble surfactants (Spans and Tweens) are often used in food formulations to produce more stable oil-in-water emulsions. Mixtures of phospholipids and proteins in foods undergo complex interactions during various thermal processing that can either promote or retard oxidation. At elevated temperatures phospholipids can produce antioxidant substances by the browning reaction, and proteins can denature to release reducing sulfhydryl groups that increase the oxidative stability of heated foods (*e.g.* evaporated milk).

C. LIPID OXIDATION IN EMULSIONS

Lipid oxidation in multiphase systems is greatly influenced by the nature of the interface, which in turn is affected by the composition of the oil and water phases. In foods, the membranes surrounding the emulsion droplets consist of surface-active substances or proteins or both. In addition to enhancing the physical stability of emulsion droplets, the film or membrane surrounding the droplets affect lipid oxidation processes in many ways.

i. *Protection of lipids.* The emulsifier provides a protective barrier to the penetration and diffusion of metals or radicals that initiate lipid oxidation. Food lipid radicals that can be generated by the action of metal catalysts in the aqueous phase markedly influence the oxidation of emulsions. Oxygenated decomposition products of lipid oxidation are more surface-active than the starting lipid and may become distributed in the interface and leach into the water phase. This "partitioning" of oxidation products promotes interactions with catalytic metals, and decreases the effectiveness of hydrophilic antioxidants (Chapter 8).

ii. *Emulsifier concentration.* The membrane increases the effectiveness of packing of emulsifier molecules. Higher concentrations of emulsifier can increase the oxidative stability of emulsified lipids against oxidation (Tables 9-1 and 9-2). Higher surfactant concentrations may cause tighter packing at the oil-water interface and the membrane becomes a more efficient barrier against diffusion of lipid oxidation initiators. At concentrations above the critical micellar concentration, a portion of polar emulsifiers, such as Tween-20, form micelles in the aqueous phase. These systems consist of an interface where the major part of the surfactant is located, and emulsifier micelles within the aqueous phase that are not located at the interface (Figure 9-3). Water-soluble and polar antioxidants can be solubilized effectively in bulk oil systems by formation of inverted micelles in the presence of low concentrations of moisture and emulsifiers such as lecithin. Ascorbic acid was thus solubilized effectively in oils in the presence of small amounts of water and emulsifiers such as lecithin. Synergistic effects were demonstrated with mixtures of

TABLE 9-1.

Effect of different factors on lipid oxidation of a safflower oil,emulsion at room temperature.[a]

Emulsifier %	Induction period[b]	pH	Induction period[b]	Sucrose %	Induction period[b]
0.5	22	6.2	14	0	3
1	28	7.2	25	17	3.5
2	38	8.2	49	33	5
4.0	47			50	10
				58	18
				67	40

From: Sims et al. (1979).
[a] Emulsion contained 25% safflower oil homogenized with sodium stearoyl-2-lactylate.
[b] Induction period in days to absorb 20 meq oxygen per kg oil.

ascorbic acid, lecithin and δ-tocopherol (Table 9-3). However, it is not clear whether the effects observed were due to solubilized ascorbic acid alone or due to the combination of ascorbic acid and lecithin necessary to prepare the inverted micelles. Phospholipids are generally useful as synergists in reinforcing the antioxidant activity of phenolic compounds (Chapter 8).

iii. *Sugars and amino acids.* Aqueous solutions of certain sugars and amino acids inhibit lipid oxidation in emulsions by scavenging free radicals in solution in the water phase. Some amino acids such as methionine, lysine and threonine retard lipid oxidation in emulsions, other amino acids such as histidine can promote lipid oxidation. However, these effects are very dependent on the type of emulsifier used, the presence of metals and the pH of the emulsion system. For example, the prooxidant activity of histidine was

TABLE 9-2.

Prooxidant effect of histidine on oxidation of methyl linoleate emulsions.[a, b]

Histidine ± metal[a]	Induction period, hr[a]	Emulsifier[b]	Induction period, hr[b]	Concentration SDS[b]	Induction period, hr[b]
Control	38	Span 20	24	0.002 M	0.6
Cu^{2+}	18	Ethenoxylated tetradecanol	13	0.004 M	1.1
Fe^{2+}/Fe^{3+}	8	Tween 20	10	0.008 M	2.4
Histidine	3	K palmitate	1		
Histidine + Cu^{2+}	3	SDS	0.5		
Histidine + Fe^{3+}	1				
Histidine + Fe^{2+}	0				

From Saunders et al. (1962)[a] and Coleman et al. (1964).[b] Induction period in hr to absorb 20 mmoles oxygen per mole ester.
[a] Methyl linoleate (0.0024 mole) emulsified with 0.1% sodium dodecyl sulfate (SDS) oxidized at 30°C.
[b] Methyl linoleate (0.1 M) homogenized with different emulsifiers (0.002 M) in the presence of histidine (0.01 M) oxidized at 30°C.

TABLE 9-3.
Effect of ascorbic acid solubilized in lecithin and water on antioxidant activity of δ-tocopherol in fish oil containing 0.1% water.[a]

Sardine oil [b]	Induction period, days [c]
Control	2
+ L + 0.2% δ-tocopherol (LT)	6
+ LT + 0.01% AA	24
+ LT + 0.02% AA	30
+ LT + 0.03% AA	32

[a] From Han *et al.* (1991). Sardine oil oxidized at 30°C.
[b] Aqueous solutions of ascorbic acid (AA) were injected into the oil containing 0.3% lecithin (L).
[c] Estimated as days to reach a peroxide value of 10.

greatly accelerated in the presence of ferrous or ferric ions (Table 9-2). Phosphate buffers retarded lipid oxidation in these emulsions.

iv. *Effect of pH*. Lipid oxidation is affected by pH in oil-in-water emulsions and liposomes. In general, lipid oxidation is slowest at high pH's and the rate is accelerated as the pH decreases (Table 9-1). This acceleration may be due to solubilization of metal catalysts. Lipid oxidation is also affected to varying degrees by pH according to the charges of different emulsifiers, as discussed below.

v. *Electric charge of the membrane*. Electrostatic attraction or repulsion by the membrane environment influences significantly the effectiveness of prooxidants and antioxidants. The rate of oxidation is much higher in emulsions prepared with ionic emulsifiers (SDS, potassium palmitate) than with nonionic emulsifiers (Span 20, ethenoxylated tetradecanol, Tween 20) (Table 9-2). Oxidation is accelerated in the emulsions stabilized by anionic emulsifiers, such as SDS. In these emulsions, electrostatic attraction occurs between the negatively-charged oil-water interface, and the positively-charged metal ions present either as trace impurities or added. Metals in the water phase become hydrated and more reactive with polar hydroperoxides and water soluble radicals (*e.g.* ˙OH, ˙OOH) at the oil-water interface.

vi. *Proteins*. As food emulsifiers, proteins have a complex interfacial effect on lipid oxidation. After saturating the oil-water interface, proteins at higher concentrations remain in the aqueous phase. Proteins can either retard or promote lipid oxidation in oil-in-water emulsions. By increasing the viscosity of the membrane surrounding the droplets, proteins may restrict the penetration and diffusion of radical initiators into the lipid phase. At certain concentrations when proteins are dispersed in the water phase they may either scavenge free radicals or be preferentially oxidized and retard lipid oxidation. Proteins

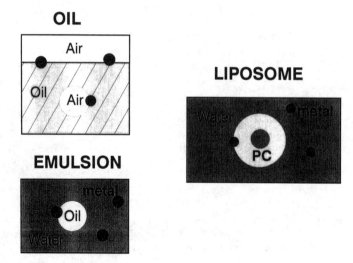

Figure 9-7. Metal catalysis in bulk oil and emulsion systems.

containing reducing sulfhydryl groups are particularly effective as antioxidants in milk, meat and fish. However, in the presence of metals some milk proteins may promote lipid oxidation (Chapter 10).

1. Metal Catalysis

The effects of metal catalysts as initiators of lipid oxidation in the presence of hydroperoxides were discussed in Chapter 1. Metal catalysis is not straight forward in various complex multiphase systems and the effects of a number of factors are difficult to unravel.

An important question to consider is where does metal catalysis takes place in multiphase systems? In bulk oils systems, the hydrophilic metals would be oriented in the air-oil interface to catalyse lipid oxidation (Figure 9-7). In emulsions and liposomes, the metals would be in solution in the aqueous phases and oriented in either the oil-water or phospholipid-water interfaces, where they may have an affinity for the hydrated layer around the droplets (Figure 9-4).

Metals become potent catalysts of lipid oxidation by forming complexes through the carboxylic acid groups of any free fatty acids present in oils. In liposome systems, metal catalysis may proceed by a different mechanism in the non-polar acyl chain moiety of the phospholipid molecule from that in the polar phosphate region. Different pathways for decomposition of lipid hydroperoxides occur also between bulk systems and emulsions. Water in emulsion systems may facilitate the breakdown of alkoxy radicals formed by decomposition of hydroperoxides. However, the effect of water in foods of varying moisture content shows a complex anomalous relationship that will be discussed in Chapter 10.

Metal ions decompose emulsified linoleic acid hydroperoxides at different rates according to the pH of the buffer (Table 9-4). Ferrous iron (Fe^{2+}) was 14 times more effective in decomposing linoleic hydroperoxides than ferric iron

TABLE 9-4.
Effect of metals and heme compounds on the decomposition of linoleic acid hydroperoxides at 23°C and two pH's.[a]

Metals ions	Relative rates		Heme compounds	Relative rates	
	pH 7	pH 5.5		pH 7	pH 5.5
Fe^{3+}	1	92	Hematin	4000	40000
Fe^{3+} + ascorbate	80	500	Hematin + ascorbate	64000	50000
Fe^{2+}	14	960	Methemoglobin	5000	76000
Cu^{2+}	0.2	1.6	Cytochrome c	2600	39000
Cu^{2+} + ascorbate	2200	6200	Oxyhemoglobin	1200	
Co^{2+}	600	0.2	Myoglobin	1120	

[a] From O'Brien (1969). Linoleic acid hydroperoxides (3.2 x 10^{-5} M) emulsified in separate buffers. Relative rates were calculated with different concentrations of each catalysts and expressed as moles of metal ions or heme per min x 10^{-6}; ascorbate concentration: 5.7 x 10^{-5} M.

(Fe^{3+}). Copper was much less effective than iron in decomposing linoleic acid hydroperoxides. The activity of both iron and copper for this decomposition reaction was increased significantly by the addition of ascorbic acid and by lowering the pH from 7.0 to pH 5.5.

Heme proteins containing Fe^{2+} and hemin proteins containing Fe^{3+} are widely distributed in animal tissue. Lipid oxidation is rapidly accelerated by hemoglobin, myoglobin and cytochrome C in fish, poultry and cooked meat (Chapter 10). Some heme compounds are much more active than iron ions in decomposing linoleic acid hydroperoxides (Table 9-4). These iron-protein complexes are thus much more efficient initiators of lipid oxidation. This higher activity can be attributed to a more favorable interaction of the iron-protein complexes at the micellar interface of the hydroperoxide acids.

Thermal denaturation of metmyoglobin (Mmb) containing Fe^{3+} causes structural changes to the heme protein that influence its prooxidant activity. Thus, heating hemin and metmyoglobin at 80°C produces considerable prooxidant activity, which decreases at high concentrations (Table 9-5). This thermal treatment is accompanied by the formation of free iron. By heating, the heme protein unfolds into several intermediate species, which eventually polymerize and precipitate.

Mmb → modified Mmb → Fe → polymerized Mmb → precipitate

Thermal modifications of heme proteins and the resulting oxidative processes play an important part in the oxidative stability of heat-processed meat (Chapter 10).

The effects of metal chelators may be very complex in multiphase systems. Chelators vary in their metal affinities and charge and in their solubility in lipids; they partition to different degrees between the lipid and water phase. Thus, in aqueous systems EDTA can reduce lipid oxidation by changing the

TABLE 9-5.
Prooxidant activity of free iron, hemin and metmyoglobin, before and after heating, in linoleic acid emulsions.[a]

Concentration	Fe(II)Cl$_2$	Hemin	Metmyoglobin
Before heating			
2 μM	0.028	0.172	0.275
4 μM	0.049	0.079	0.29
8 μM	0.045		0.246
After heating[b]			
2 μM	0.036	0.227	0.114
4 μM	0.037	0.167	0.155
8 μM	0.02		0.182

[a] From Kristensen and Andersen (1997). Oxygen consumption expressed in μM/sec.
[b] Heated to 80°C at 5°C/min.

location and by reducing the oxidizing capacity of iron. However, in the presence of reducing agents, complex formation with EDTA may also promote lipid oxidation in certain systems by producing alkoxyl radicals from hydroperoxides and initiate the radical chain. The effects of antioxidants and reducing agents in the presence of different forms of metal complexes have important consequences on lipid oxidation of multiphase systems, therefore.

2. Antioxidants

a. Initiators. Several dispersed systems have been used as models to study the interfacial behavior of various antioxidants on lipid oxidation induced with different artificial radical initiators. Quantitative kinetic studies of autoxidation are dependent on the use of an initiator (In) that decomposes into two radicals, which in solution are held together briefly in a cage of solvent molecules. The interactions of these radicals can retard or prevent a portion from diffusing into the bulk of the solution, and this phenomenon is known as the *cage effect* (see Chapter 1.B). Autoxidation chains are started only by the fraction of radicals that escape from the solvent cage. The radicals produced below are represented by a bar over the radical pairs for its caged nature.

In the decomposition of diazo initiators the number of free radicals in the bulk solutions is estimated to be less than two for each molecule of initiator consumed. This cage effect is dependent on the solvent viscosity. The more viscous the solvent, the more difficult it will be for the radicals to escape through the walls of the solvent cage.

$$\text{In} \longrightarrow \overline{\text{R}^\cdot \ \text{R}^\cdot} \xrightarrow{\text{diffusion}} 2\text{R}^\cdot \xrightarrow{\text{O}_2} 2\text{ROO}^\cdot$$

TABLE 9-6.
Effect of copper (II), azo initiators and BHT on oxidation of phosphatidylcholine (PC) liposomes at 37°C. [a]

Oxidation promoters [b]	Rate of oxidation [c]		Induction period min [d]
	PCOOH	Oxygen	
Control	7.8	7.4	
Cu (II)	24.3	21.5	
Cu (II) + BOOH	32.3	34.7	55
Cu (II) + BOOH + BHT			150
AAPH	23.5	30.5	65
AAPH + BHT			130
AMVN	31.2	35.9	40
AMVN + BHT			52

[a] From Yoshida and Niki (1992).
[b] Cu (II), 100 μM copper (II); BOOH, 3 mM t-butyl hydroperoxide; BHT, 1 μM. AAPH, 2mM 2,2'-azobis(2-amidinopropane) dihydrochloride; AMVN, 2mM 2,2'-azobis(2,4-dimethylvaleronitrile).
[c] Rates: d[PCOOH]/dt and -d[O_2]/dt, in nM per sec.
[d] Time necessary for the formation of 100 μM phosphatidylcholine hydroperoxides.

In contrast to solution, the higher viscosity of a lecithin liposome decreases the efficiency of free radical initiation, and retards their autoxidation. Artificial azo initiators have a very low efficiency (about 9%) when solubilized in the bilayer phase. For this reason, the oxidizability of lecithin dispersed in a liposome is much lower than in solution. Although the solvent cage effect may be unique to diazo initiators commonly used in kinetic studies, and is not necessarily relevant to food lipid systems, metal initiators which are relevant to foods and biological systems, may also be affected by solvent cage effects because of the hydrated layer in emulsions (Figure 9-4).

Antioxidants behave differently when oxidation is initiated with metals as compared to azo radical initiators. Thus, BHT was a more effective antioxidant in a phosphatidylcholine liposome oxidized with copper (II) in the presence t-butyl-hydroperoxide than in the same liposome oxidized with AAPH or with AMVN (Table 9-6). The inhibition by BHT was better with the water-soluble initiator AAPH than the oil-soluble initiator AMVN. This result suggests that radicals produced in the water phase by copper (II) were trapped by BHT in the same way as the radicals produced by AAPH. The rates of oxygen absorption in this liposome system were increased three fold with 100 μM copper (II) compared to four fold with 2 mM AAPH and about five fold with 2 mM AMVN. Such results must be interpreted with caution, because at sufficiently high concentrations metals react stoichiometrically and promote chain termination, a condition that is not observed at low and trace concentrations in multiphase food systems. These model studies may,

OIL   **EMULSION**

● Hydrophilic ＞ ● Lipophilic   Lipophilic ＞ Hydrophilic
   Antioxidants   Antioxidants

Figure 9-8. Interfacial distribution of hydrophilic and lipophilic antioxidants in bulk oil vs oil-in-water emulsion systems. From Frankel *et al.* (1994).

therefore, not be particularly applicable to food lipid systems because lipid oxidation is initiated predominantly by trace amounts of metals in the presence of small amounts of hydroperoxides.

b. Emulsions. The contrasting behavior of different antioxidants in bulk oil versus oil-in-water emulsions was discussed in Chapter 8. Antioxidant activity varies in different multiphase systems which differ in the distribution of the lipid phase. In oil-in-water emulsions, hydrophilic antioxidants are generally less effective than lipophilic antioxidants, whereas in bulk oil systems hydrophilic antioxidants are more effective. Thus, the order of activity of α-tocopherol versus Trolox and ascorbic acid versus ascorbyl palmitate, is reversed when comparing bulk oil and emulsified systems (Table 8-9, Figures 8-7 and 8-8). This phenomenon was explained in relation to the affinities of the antioxidants toward the air-oil interfaces in bulk oil and the water-oil interfaces in emulsions. In the corn oil bulk system, the hydrophilic antioxidants Trolox and ascorbic acid are more protective against oxidation by being oriented in the air-oil interface; the lipophilic antioxidants α-tocopherol and ascorbyl palmitate are less protective by remaining in solution in the oil phase where they are present at low concentrations (Figure 9-8). In contrast, in the oil-in-water emulsions, the lipophilic antioxidants are oriented at the oil-water interface by virtue of their surface activity, and are more protective against oxidation than the hydrophilic antioxidants, which are dissolved and become diluted in the water phase.

c. Liposomes. The behavior of different types of antioxidants in emulsions, referred to as the *polar paradox*, is based on the phenomenological observation that non-polar antioxidants are more effective on polar lipids in emulsions, while polar antioxidants are more effective on non-polar lipids. In a lecithin liposome oxidized with hematin, the activities of different antioxidants showed a linear relation with polarity based on the R_f values in

TABLE 9-7.
Relative effectiveness of antioxidants in lecithin liposomes versus polarity by thin-layer chromatography (TLC). [a]

Antioxidants	Relative efficiency [b]	R_f silica TLC
BHT	14.7	0.88
BHA	11.1	0.68
Propyl gallate	8.0	0.23
TBHQ	2.6	0.43
Quercetin	2.6	0.16
Methyl gallate	1.9	0.18
Caffeic acid	0.3	0.21
Gallic acid	0.3	0.06

[a] From Porter et al. (1989).
[b] Relative efficiency = $(IP_a / IP_c) - 1$, where IP_a is induction period with added antioxidant, and IP_c is induction period without added antioxidant.

silica gel thin-layer chromatography (Table 9-7). The lipophilic BHA, BHT, and propyl gallate were much more active antioxidants in protecting polar phospholipids in this liposome system than the hydrophilic TBHQ, caffeic and gallic acids.

A number of kinetic studies were carried out to elucidate the interaction between α-tocopherol and ascorbic acid in liposomes. By using either water-soluble or oil-soluble diazo radical initiators, the effect of radicals can be compared when they are produced either in the phospholipid bilayer or in the aqueous phase. Oxidation of a soybean phosphatidylcholine liposome in the presence of radicals produced in the water phase with a water-soluble radical initiator [2,2'-azobis(2-amidinopropane) dihydrochloride, (AAPH)], showed an induction period with both α-tocopherol (vitamin E) and ascorbic acid (vitamin C) (Figure 9-9). With a mixture of α-tocopherol and ascorbic acid, the length of the induction was close to the sum of the individual induction periods. This result indicates an additive effect in suppressing oxidation by both vitamin E and vitamin C. Ascorbic acid apparently traps radicals in the water phase. When oxidation was induced by radicals produced in the lipid phase with an oil-soluble radical initiator [2,2'-azobis(2,4-dimethylvaleronitrile), (AMVN)] incorporated into the membrane, ascorbic acid alone had no effect while α-tocopherol had a greater effect because it is lipophilic. However, the mixture of α-tocopherol and ascorbic acid had a synergistic effect by increasing the induction period. Similar results were reported with liposomes of dilinoleoyl phosphatidylcholine. This effect can be explained by assuming that the radicals initiated in the lipid phase are first trapped by α-tocopherol within the liposome, and the resulting α-tocopherol radicals become oriented at the bilayer surface and are reduced by ascorbic acid in the aqueous phase to regenerate α-tocopherol.

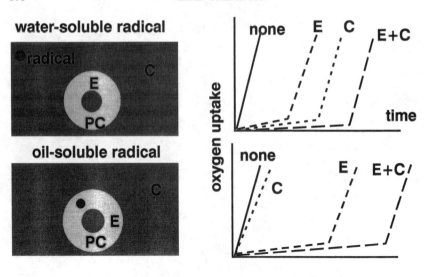

Figure 9-9. Effect of vitamin E (E) and vitamin C (C) and their mixtures on the oxidation of a soybean phosphatidylcholine (PC) liposome in the presence of a water-soluble radical initiator (AAPH) and oil-soluble radical initiator (AMVN). From Niki et al. (1985).

The relative antioxidant activity of polyphenolic flavonoids in green tea changed significantly according to the test system. In egg yolk phosphatidylcholine liposomes, using a water-soluble AAPH initiator, epicatechin gallate had the highest antioxidant activity, followed by epicatechin and quercetin, and α-tocopherol was the least active. In contrast, the reverse trend was obtained with methyl linoleate oxidized in hexane-isopropanol solution using a lipid-soluble AMVN initiator. In this solution, α-tocopherol

TABLE 9-8.

Effect of catechins, quercetin and α-tocopherol on the oxidation of phosphatidylcholine (PC) liposomes or solutions initiated by AAPH [a]

Antioxidants 10 μM	Liposomes[b]		Solutions [c]	
	Induction period, min	Decomposition time, min	Antioxidant efficiency (k_{inh} / k_p)	Inhibition time (t_{inh})
Epicatechin gallate	350	220	130	23
Epicatechin	290	110	135	27
Quercetin	250	100	280	13
α-Tocopherol	160	50	1700	5

[a] From Terao et al. (1994).
[b] PC oxidized with water-soluble AAPH; Induction period for the formation of 0.25 mM phosphatidylcholine hydroperoxides; Decomposition time for the complete loss of antioxidant.
[c] Hexane-isopropanol solution oxidized with oil-soluble AMVN.

Egg yolk PC liposome: 37°C, Initiator: AAPH (water-soluble)
ECG > EC > Q >> α-T

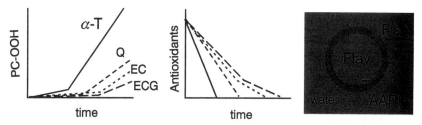

Me linoleate solution (hexane:isoprOH:EtOH)
Initiator: AMVN (lipid soluble)
α-T >> Q > EC = ECG

Figure 9-10. Relative antioxidant activities of epicatechin gallate (ECG), epicatechin (EC), quercetin (Q), and α-tocopherol (α-T) on the oxidation of egg yolk phosphatidylcholine liposome in the presence of a water-soluble initiator (AAPH) and in hexane-isopropanol solution in the presence of a lipid-soluble initiator (AMVN). From Terao et al. (1994). FLAV, flavonoids.

had the highest activity, followed by quercetin, and epicatechin, which had the same activity as epicatechin gallate (Table 9-8; Figure 9-10). In liposomes, flavonoids associate with the surface to protect the polar phospholipids, while α-tocopherol remains within the membrane and is less protective. Also, α-tocopherol was the least stable in the liposome system and disappeared first during oxidation, followed by quercetin, epicatechin and epicatechin gallate.

The activity of antioxidants is greatly enhanced by increasing the electrostatic attraction between the lipid surface and antioxidants. The charge of phospholipid liposomes can thus greatly affect the activity of charged antioxidants. The charge of liposomes was changed by adjusting the pH of two types of phospholipids. The negatively-charged Trolox had no antioxidant activity toward the zwitterionic (±) liposomes prepared with dilinoleoyl-phosphatidylcholine (DLPC) at pH 11 and with the negatively-charged dilinoleoyl-phosphatidylglycerol (DLPG) at pH 11 and 7, due to electrostatic repulsion between the surface charge and the antioxidant charge (Table 9-9). Conversely, Trolox became a much more active antioxidant toward the positively charged DLPC liposome at pH 4 due to electrostatic attraction. Similarly, Trolox was active by electrostatic attraction in the positively-charged DPLC in the presence of stearyl amine at pH 7. Trolox was also active in dimyristoyl-phosphatidic acid in the presence of linoleic acid when it was positively charged at pH 2, but not when it was negatively charged at pH 7. Trolox was more active than the positively charged quaternary ammonium analog of α-tocopherol, hydroxychroman substituted with -CH-$(CH_2)_2$-N^+-$(CH_3)_3$ [6-hydroxy-2,5,7,8- tetramethyl-2-N,N,N-trimethyl-

TABLE 9-9.
Effect of charge of liposome on the antioxidant activity of Trolox.[a]

Liposome (charge) [b]	pH	Inhibition period, min	Inhibition rate constant, k_{inh} [c]
	Trolox (-)		
DLPC (±)	11	No	—
DLPC (±)	7	123	2.98
DLPC (+)	4	97	5.8
DLPC/SA (+)	7	83	4.16
DLPG (-)	7/11	No	—
DLPG (±)	4	109	1.22
DMPA/LA (+)	2	74	5.55
DMPA/LA (-)	7	No	—
	MDL (+) [d]		
DLPC (-)	7	170	0.95
DLPG (-)	7	96	1.62

[a] From Barclay and Vinqvist (1994). Liposome oxidized at 37°C in the presence of lipid-soluble initiator azo-bis (2-amidinopropane hydrochloride) (ADVN).
[b] DLPC, dilinoleoyl-phosphatidylcholine; SA, stearyl amine; DLPG, dilinoleoyl-phosphatidylglycerol; DMPA, dimyristoyl-phosphatidic acid; LA, linoleic acid.
[c] Inhibition rate constants calculated from plots of oxygen uptake, in $M^{-1} s^{-1} \times 10^{-3}$.
[d] 6-Hydroxy-2,5,7,8-tetramethyl-2-N,N,N-trimethylethanaminium methyl-benzenesulfonate.

ethanaminium methyl-benzene sulfonate, (MDL™)], in the DLPC liposome at pH 7 (Table 9-9). Conversely, MDL was more active than Trolox in the negatively charged DLPG liposome at pH 7.0.

d. Micelles. Although triacylglycerols are the major components in foods, polyunsaturated fatty acids and their esters are most often used as model lipid substrates in studies of antioxidants. However, free fatty acids are amphiphilic, and dissociate in aqueous systems according to the pH to partially water-soluble anions that form micelles. This micelle formation is in marked contrast to the water-insoluble fatty acid esters and triacylglycerols that are dispersed in water as emulsion particles. Although at neutral pH the oxidizability of linoleic acid anions remains constant within a wide range of concentration in SDS micelles, the oxidizability of methyl linoleate increases with increasing concentrations because it becomes solubilized within the micelles at higher concentrations. Above the critical micellar concentration, polar emulsifiers form mixed micelles with the free fatty acids in the aqueous phase (Figure 9-3). Polar antioxidants such as Trolox are more active in such mixed micelle systems consisting of emulsifier and free fatty acids, than in oil-in-water emulsions of methyl linoleate or corn oil where they partition mainly in the water phase (Section 2.e.). For these reasons, linoleic acid cannot be used as a

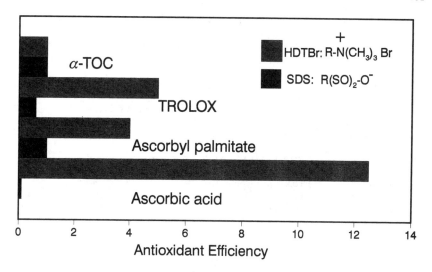

Figure 9-11. Relative antioxidant efficiencies of α-tocopherol (α-TOC), Trolox, ascorbic acid and ascorbyl palmitate on the oxidation of linoleic micelles stabilized with HDTBr and SDS. From Pryor et al. (1993).

representative model system for foods since antioxidant behavior in linoleic acid micelles will be significantly different from that in foods composed mainly of triacylglycerols.

The charge of fatty acid micelles affects markedly the efficiency of charged and uncharged antioxidants. The negatively charged ascorbic acid and Trolox were significantly more effective antioxidants in a micelle system of linoleic acid stabilized with a positively charged emulsifier such as hexadecyltrimethylammonium bromide (HDTBr), and much less effective when the system was stabilized with a negatively charged emulsifier, such as SDS (Figure 9-11). These charged antioxidants have an ionized carboxylate group which causes it to be repelled by the negatively charged SDS micelles. The stronger effect observed with ascorbic acid indicates that it is more readily attracted by the positively charged HDTBr micellar sphere where oxidation occurs. In this system the activity of the uncharged and lipophilic antioxidant α-tocopherol was not affected by the charge of the emulsifiers. The same trend was observed, but to a smaller extent, with the hydrophilic ascorbyl palmitate as with ascorbic acid. With ascorbyl palmitate there may be a balancing effect between electrostatic and polar affinities for the linoleic acid-water interface.

The lipophilic α-tocopherol and its hydrophilic analog Trolox behave quite differently in linoleic acid micelles compared to phosphatidylcholine liposomes. With micelles of linoleic acid in SDS initiated with a water-soluble initiator [2,2'-azobis-(2-amidino propane) dihydrochloride, ABAP], α-tocopherol was much less active than Trolox (Table 9-10). With phosphatidylcholine liposomes initiated with a lipid-soluble initiator AMVN, α-tocopherol and Trolox had

TABLE 9-10.
Effect of lipid systems on relative activities of hydrophilic versus lipophilic antioxidants.

Lipid systems [a]	Antioxidant trends	Ref.[b]
Bulk corn oil	Trolox > α-tocopherol, ascorbic acid > ascorbyl palmitate	(1)
Oil/water emulsion	α-tocopherol > Trolox, ascorbyl palmitate > ascorbic acid	(1)
Lecithin liposome	BHT > BHA > propyl gallate > TBHQ > gallic acid	(2)
DLPC liposome	Trolox = α-tocopherol	(3)
PC liposome	epicatechin gallate > epicatechin > quercetin > α-tocopherol	(4)
LA / SDS micelles	Trolox >> α-tocopherol	(5)
SDS- micelles	α-tocopherol = ascorbyl palmitate > Trolox > ascorbic acid	(5)
HDTBr+ micelles	ascorbic acid > Trolox > ascorbyl palmitate > α-tocopherol	(5)
Solution [c]	α-tocopherol >> quercetin > epicatechin = epicatechin gallate.	(4)

[a] Abbreviations: DLPC, dilinoleoyl phosphatidylcholine; PC, phosphatidylcholine (egg); LA, linoleic acid; SDS, sodium dodecyl sulfate; HDTBr, hexadecyltrimethyl ammonium bromide.
[b] References: (1) Frankel et al. (1994), (2) Porter et al. (1989), (3) Barclay et al. (1990), (4) Terao et al. (1994), (5) Pryor et al. (1988).
[c] hexane-isopropanol (1:1, v:v).

about the same antioxidant activity. Linoleic acid and SDS form mixed micelles in the aqueous phase in which the polar antioxidant Trolox equilibrates more rapidly and becomes more effective than in liposomes. Also, Trolox can efficiently trap radicals from the water-soluble initiator ABAP.

In summary, the relative activities of various antioxidants vary widely according to the site of action in different systems, the charge and solubility of components in micelle compared to liposome systems and solutions (Table 9-10). Antioxidant activity is thus significantly affected by the physical state of different multiphase systems.

e. Partition Studies. In multiphase systems antioxidants partition between several environments, including the aqueous phase, lipid phase and surfactant-enriched environment, such as the interface and micelles. Antioxidants partition into the different phases according to their relative affinities toward these phases and their relative amounts. The partition behavior of different antioxidants in relation to their concentration and location in different phases are expected to change their effectiveness in food and biological systems. However, the literature on this subject is relatively sparse.

The concentration of Trolox in the water phase was lower in mixtures of linoleic acid and water and of methyl linoleate and water than in the mixture of corn oil triglycerides and water (Figure 9-12). The partition coefficient of Trolox between linoleic acid or methyl linoleate and water is twice as high (6.4) as between corn oil and water (3.8). Trolox has, therefore, a much lower affinity toward the triglyceride than toward either linoleic acid or methyl linoleate. In the corresponding oil-in-water emulsions, the concentration of Trolox in the water phase decreased because of its affinity toward the

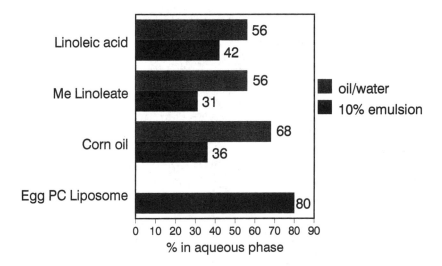

Figure 9-12. Partition of Trolox in different oil/water mixtures and in Tween 20 emulsions. From Huang et al. (1996), and Barclay et al. (1994).

emulsifier Tween-20. The concentration of Trolox in the water phase was lower in the systems prepared with corn oil triacylglycerols and methyl linoleate than those with linoleic acid. In egg lecithin liposome, the bulk of Trolox (80%) resided in the aqueous phase.

In corn oil-water mixtures, more than 90% of the hydrophilic antioxidants catechin, gallic acid, propyl gallate and rosmarinic acid partitioned into the aqueous phase, compared to 68% for Trolox (Figure 9-13). The concentration of these antioxidants decreased markedly in the corresponding oil-in-water

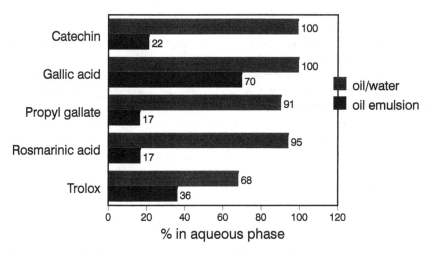

Figure 9-13. Partition of antioxidants in different oil/water mixtures and in Tween 20 emulsions. From Huang et al. (1997).

TABLE 9-11.
Partition of antioxidants between different surfactant environments in emulsion systems.[a]

Antioxidants	Surfactant	% in water phase	% in oil phase	% in surfactant phase
Ferulic acid	Tween 20	48.6	8.9	42.5
Ferulic acid	SDS micelles	54.1	9.9	36
Ferulic acid	DTABr	51.4	9.4	39.2
Caffeic acid	Tween 20	54.6	1.5	43.9
Caffeic acid	SDS micelles	75.9	2.1	22
Caffeic acid	DTABr	42.2	1.2	56.6
Propyl gallate	Tween 20	42.9	9.6	47.5
Propyl gallate	SDS micelles	57.6	12.9	29.5
Propyl gallate	DTABr	19.4	21.1	59.5
Gallic acid	Tween 20	68.2	1.9	29.9
Gallic acid	SDS micelles	82.1	2.4	15.5
Gallic acid	DTABr	62.9	1.8	35.3
Trolox	Tween 20	26.7	23.6	49.7
Trolox	SDS micelles	38.6	34.4	27
Trolox	DTABr	26.4	23.6	50
Catechin	Tween 20	48.1	0.9	51
Catechin	SDS micelles	84	1.6	14.4
Catechin	DTABr	54.8	1.1	44.1

[a] From Schwarz et al. (1996). Proportions of antioxidants in the oil and surfactant phases were calculated on the assumption that the oil phase is identical to the bulk oil in biphasic systems.

Abbreviations: Tween 20, polyoxyethylene sorbitan monolaurate; SDS, sodium dodecylsulfate; DTABr, dodecyltrimethylammonium bromide.

emulsions, in the order propyl gallate and rosmarinic acid, catechin, Trolox and gallic acid. The order of this decrease in partition in the water phase between oil-water mixtures and the corresponding emulsions can be attributed to the association or affinity of the antioxidants with the Tween 20 emulsifier.

In emulsified corn oil systems, the hydrophilic antioxidants favored the interface or surfactant enriched environment compared to the oil phase (Table 9-11). All these hydrophilic antioxidants showed lower affinity toward the SDS micelle systems because of the lower bonding capacity of the negatively charged SDS compared to the other surfactants. The bulk of caffeic acid, gallic acid and catechin resided in the aqueous phase of SDS micelles. Caffeic acid, propyl gallate and gallic acid showed the highest proportion in the aqueous phase with the positively charged DTABr surfactant. These hydrophilic antioxidants thus favor the interface or surface-enriched environments compared to the oil phase. In the corn oil system emulsified with Tween 20, a significant proportion of this emulsifier was estimated to be not only in the oil-water interface (26-68%) but also formed micelles (32-74%). The low antioxidant activities of Trolox and other hydrophilic antioxidants (rosmarinic acid, catechin, propyl gallate, gallic acid and tea catechins) in oil-in-water emulsions were attributed to their partition into the Tween 20 micelles and into the aqueous phase.

The oxidative stability of food lipids varies according to their colloidal location because of their exposure to prooxidants and antioxidants. The partition behavior of antioxidants may provide, therefore, an important tool for the delivery of specific inhibitors in highly oxidizable lipid surfaces. In foods, antioxidants partition between non-aqueous and aqueous phases depending on the solvent properties of the phases. Partition of antioxidants into the non-aqueous phase can exert an important effect on activity. Surfactant can significantly increase the solubility of lipophilic antioxidants in the interface and exert a large effect on activity. Other factors that should be considered include the relative stability of antioxidants and their anomalous behavior in different colloidal phases. The role of discrete phases in lipid oxidation must now be recognized for a better understanding of the oxidative stability of food lipids, but the exact details still remain unclear.

BIBLIOGRAPHY

Barclay,L.R.C. Model biomembranes: quantitative studies of peroxidation, antioxidant action, partitioning, and oxidative stress. *Can. J. Chem.* **71**, 1-16 (1993).
Barclay,L.R.C., Baskin,K.A., Dakin,K.A., Locke,S.J. and Vinqvist,R. The antioxidant activities of phenolic antioxidants in free radical peroxidation of phospholipid membranes. *Can. J. Chem.* **68**, 2258-2269 (1990).
Barclay,L.R.C. and Vinqvist,M.R. Membrane peroxidation: Inhibiting effects of water-soluble antioxidants on phospholipids of different charge types. *Free Radical Biol. Med.* **16**, 779-788 (1994).
Belitz,H.-D. and Grosch,W. *Food Chemistry*, 1987. Springer Verlag, Berlin.
Boyle,E. and German,J.B. Monoglycerides in membrane systems. *Crit. Rev. Food Sci. & Nutrition.* **36**, 785-805 (1996).
Charalambous,G. and Doxastakis,G., editors. *Food Emulsifiers: Chemistry, Technology, Functional Properties and Applications*. Developments in Food Science, Vol. 19, 1989. Elsevier, Amsterdam.
Coleman,J.E., Hampson,J.W. and Saunders,D.H. Autoxidation of fatty materials in emulsions. II. Factors affecting the histidine-catalyzed autoxidation of emulsified methyl linoleate. *J. Am. Oil Chem. Soc.* **41**, 347-351 (1964).
Coupland,J.N. and McClements,D.J. Lipid oxidation in food emulsions. *Trends Food Science & Technol.* **7**, 83-91 (1996).
Coupland,J.N., Zhu,Z., Wan,H., McClements,D.J., Nawar,W.W. and Chinachoti,P. Droplet composition affects the rate of oxidation of emulsified ethyl linoleate. *J. Am. Oil Chem. Soc.* **73**, 795-801 (1996).
Dickinson,E. and Stainsby,G. *Colloids in Foods*. Applied Science Publishers, London. 1982.
Doba,T., Burton,G.W. and Ingold,K.U. Antioxidant and co-antioxidant activity of vitamin C. The effect of vitamin C, either alone or in the presence of vitamin E or a water-soluble vitamin E analogue, upon the peroxidation of aqueous multilamellar phospholipid liposomes. *Biochim. Biophys. Acts* **835**, 298-303 (1985).
Frankel,E.N., Huang,S-H., Kanner,J. and German,J.B. Interfacial phenomena in the evaluation of antioxidants: Bulk oils vs Emulsions. *J. Agric. Food Chem.* **42**, 1054-1059 (1994).
Fritsch,C.W. Lipid oxidation - the other dimensions. Inform 5, 423-436 (1994).
Goñi,F.M. and Alonso,A. Studies of phospholipid peroxidation in liposomes, in *CRC Handbook of Free Radicals and Antioxidants in Biomedicine*, Volume III, pp. 103-122 (1989) (edited by J. Miquel, A.T. Quintanilha and H. Weber), CRC Press, Boca Raton, Florida.
Gunstone,F.D. *Fatty Acid and Lipid Chemistry*, pp. 234-236. (1996) Blackie Academic & Professional, London.
Han,D., Yi,O-S. and Shin,H-K. Solubilization of vitamin C in fish oil and synergistic effect with vitamin E in retarding oxidation. *J. Am. Oil Chem. Soc.* **68**, 740-743 (1991).
Huang,S-W., Frankel,E.N., German,J.B. and Aeschbach,R. Partition of selected antioxidants in corn oil-water model systems. *J. Agr. Food Chem.* **45**, 1991-1994 (1997).

Huang,S-W., Hopia,A., Frankel,E.N. and German,J.B. Antioxidant activity of α-tocopherol and Trolox in different lipid substrates: Bulk oils vs oil-in-water emulsions. *J. Agr. Food Chem.* **44**, 444-452 (1996).

Ingold,K.U. Metal catalysis, in *Lipids and their Oxidation*, pp. 93-121 (1962) (edited by H.W. Schultz, E.A. Day and R.O. Sinnhuber), The Avi Publishing Co., Westport, CN.

Kristensen,L. and Andersen,H.J. Effect of heat denaturation on the pro-oxidative activity of metmyoglobin in linoleic acid emulsions. *J. Agric. Food Chem.* **45**, 7-13 (1997).

Larsson,K. *Lipids - Molecular Organization, Physical Functions and Technical Applications.* (1994). The Oily Press Ltd., Dundee, Scotland.

Maiorino,M., Zamburlini,A., Roveri,A. and Ursini,F. Copper-induced lipid peroxidation in liposomes, micelles, and LDL: which is the role of vitamin E. *Free Rad. Biol. Med.* **18**, 67-74 (1995).

Niki,E., Kawakami,A., Yamamoto,Y. and Kamiya,Y. Oxidation of Lipids. VIII. Synergistic inhibition of phosphatidylcholine liposome in aqueous dispersion by vitamin E and vitamin C. *Bull. Chem. Soc. Jpn.* **58**, 1971-1975 (1985).

O'Brien,P.J. Intracellular mechanisms for the decomposition of a lipid peroxide. I. Decomposition of a lipid peroxide by metal ions, heme compounds, and nucleophiles. *Can. J. Biochem.* **47**, 485-492 (1969).

Porter,W.L., Black,E.D. and Drolet,A.M. Use of polyamide oxidative fluorescence test on lipid emulsions: Contrast in relative effectiveness of antioxidants in bulk versus dispersed systems. *J. Agric. Food Chem.* **37**, 615-624 (1989).

Pryor,W.A., Strickland,T. and Church,D.F. Comparison of the efficiencies of several natural and synthetic antioxidants in aqueous sodium dodecyl sulfate micelle solutions. *J. Am. Chem. Soc.* **110**, 2224-2229 (1988).

Pryor,W.A., Cornicelli,J.A., Devall,L.J., Tait,B., Trivedi,B.K., Witiak,D.T. and Wu,M. A rapid screening test to determine the antioxidant potencies of natural and synthetic antioxidants. *J. Org. Chem.* **58**, 3521-3532 (1993).

Riisom,T., Sims,R.J. and Fioriti,J.A. Effect of amino acids on the autoxidation of safflower oil in emulsions. *J. Am. Oil Chem. Soc.* **57**, 354-359 (1980).

Sanders,D.H., Coleman,J.E., Hampson,J.W., Wells,P.A. and Riemenschneider,R.W. Autoxidation of fatty materials in emulsions. I. Pro-oxidant effect of histidine and trace metals on the oxidation of linoleate esters. *J. Am. Oil Chem. Soc.* **39**, 434-439 (1962).

Schaich,K.M. Metals and lipid oxidation. Contemporary issues. *Lipids* **27**, 209-218 (1992).

Schwarz,K., Frankel,E.N. and German,J.B. Partition behaviour of antioxidative phenolic compounds in heterophasic systems. *Fett/Lipid* **98**, 115-121 (1996).

Terao,J., Piskula,M. and Yao,Q. Protective effect of epicatechin, epicatechin gallate, and quercetin on lipid peroxidation in phopholipid bilayers. *Arch. Biocem. Biophys.* **308**, 278-284 (1994).

Yoshida,Y. and Niki,E. Oxidation of phosphatidylcholine liposomes in aqueous dispersions induced by copper and iron. *Bull. Chem. Soc. Japan* **65**, 1849-1854 (1992).

Yoshida,Y. and Niki,E. Oxidation of methyl linoleate in aqueous dispersions induced by copper and iron. *Arch. Biochem. Biophys.* **295**, 107-114 (1992).

CHAPTER 10
FOODS

Lipid oxidation is the major form of deterioration in foods even when the lipid content is very small. The rancidity developed in foods causes changes in quality that affect their flavor, odors, taste, color, texture and appearance. Thus, lipid oxidation is a decisive factor in the useful storage of food products. The changes due to oxidation occur at different stages of food preparation, ranging from the raw materials, through processing, packaging, storage, cooking, and various retail or large scale applications. An appreciation of various oxidation factors in food processing is necessary to control and minimize the impact of lipid oxidation on food quality.

A. RAW MATERIALS

1. Metals and Metallo-Proteins

Traces of catalytic metals are present in enzymes and bound with proteins. *Heme* (Fe^{2+}) and *hemin* (Fe^{3+}) proteins occur in many plant and animal foods. In animal tissues, lipid oxidation is catalysed by hemoglobin, myoglobin and cytochrome C. Heme catalysis is of widespread significance in meats, and poultry, frozen fish, milk and plant foods by causing rancidity and the destruction of natural tocopherols and carotenoids. Important heme and hemin proteins in plant foods include peroxidase and catalase. These heme compounds are much more active than the free metal ions in catalysing decomposition of lipid hydroperoxides (Table 9-4). The resulting alkoxyl radical intermediates are active in initiating lipid oxidation. However, when horseradish peroxidase is heated, the prooxidative activity of the denatured proteins is increased significantly (Table 10-1). Therefore, the mixture of free iron and heat-denatured proteins becomes an effective non-enzymatic catalyst of lipid oxidation.

2. Enzymatic Oxidation in Plant Foods

Various hydrolytic and oxidative enzymes affect lipid oxidation in raw materials when tissues are disrupted or injured. Enzyme action occurs sequentially with the initial hydrolysis of triacylglycerols or lipolysis catalysed by *lipases* producing free fatty acids. The polyunsaturated free fatty acids are

TABLE 10-1.
Effect of temperature on activity of horseradish peroxidase and on non-enzymatic lipid oxidation.[a]

Temperature (°C)[b]	Enzyme activity (%)	non-enzyme activity (nmol O_2/min)
25	100	20
50	100	25
75	80	40
100	75	85
125	30	130

[a] From Eriksson (1987).
[b] Heating time 2 min.

then oxidized by *lipoxygenases*, which represent important oxidizing enzymes found in plant raw materials, and in animal tissues. In plant foods, lipases and lipoxygenases are responsible for the formation of characteristic flavors. Lipolysis occurs readily after crushing or grinding oilseeds and vegetables. The resulting off-flavor production can be prevented by heat inactivation of lipases and lipoxygenases. For example, raw peanuts can be stored without lipid oxidation for over one year if they remain whole and intact. However, if the peanuts are ground, they undergo flavor deterioration readily due to enzyme activation. On the other hand, if the peanuts are dry-roasted to inactivate enzymes, they are oxidatively stable.

Lipoxygenases (Lox) are selective towards polyunsaturated fatty acids containing the *cis,cis*-1,4-pentadiene moiety to produce either the 9(S)-hydroperoxide, 13(S)-hydroperoxide, or a mixture of both from linoleic and linolenic acids. These enzymes contain an iron atom in their active center. They are activated by hydroperoxides, and the Fe^{2+} is oxidized to Fe^{3+}, according to a scheme in which the pentadiene radical of linoleic acid becomes bound to the enzyme, reacts with oxygen, and the peroxyl radical formed is reduced by the enzyme to produce a hydroperoxide after reaction with a proton.

$$Lox\text{-}Fe^{2+} + LOOH \longrightarrow Lox\text{-}Fe^{3+} + LH \longrightarrow Lox\text{-}Fe^{2+}\text{-}L^{\cdot}$$

$$Lox\text{-}Fe^{2+}\text{-}L^{\cdot} + O_2 \longrightarrow Lox\text{-}Fe^{2+}\text{-}LOO^{\cdot} + H^+ \longrightarrow LOOH$$

Several isoenzymes of lipoxygenase are known, including L-1 and L-2 from soybeans, with pH optima of 9.0 and 6.5, respectively. Isoenzymes L-1 with high pH optima prefer anionic free fatty acids, and L-2 with pH optima near neutral prefer neutral substrates such as triacylglycerols. L-1 produces the 13-hydroperoxide stereoselectively from linoleic acid, while L-2 produces a mixture of 9- and 13-hydroperoxides from linoleic acid and linoleic glycerides. Thus, isoenzyme L-2 does not require prior lipolysis of triacylglycerols to be

TABLE 10-2.
Formation of volatile compounds produced by different lipoxygenases.

Lipoxygenases	Substrates	Main volatiles [a]	Ref. [b]
Soy L-1	18:2 n-6	hexanal, t2,t4-decadienal	(1)
Soy L-2	18:2 n-6	hexanal, t2,t4-decadienal, t2,c4-decadienal, t-2-heptenal, t-2-octenal	(1)
Soy L-1	18:3 n-3	t-2-hexenal, propanal, pentenal	(2)
Soy L-2	18:3 n-3	propanal, t2,c4-heptadienal, t2-pentenal, t-2-hexenal, 2,4,6-nonatrienal, 3,5-octadien-2-one	(3)
Pea L	18:2 n-6	hexanal, 2,4-decadienal, 2-heptenal, 2-octenal, pentanal, 2,4-nonadienal	(2)
Pea L	18:3 n-3	t2,c4-heptadienal, propanal, 2-pentenal, acetaldehyde, crotonaldehyde, 2-hexenal	(2)
Potato L	18:3 n-3	2,6-nonadienal, 2-hexenal, propanal, 3,5-octadien-2-one, 2,4-heptadienal	(4)
Beans L	18:3 n-3	2-hexenal, propanal, 2,4-heptadienal	(4)
Peanut L	18:2 n-6	pentane, hexanal	(5)

[a] t, *trans*; c, *cis*.
[b] (1) Fischer and Grosch (1977), (2) Grosch (1968), (3) Grosch and Laskawy (1975), (4) Grosch et al. (1976), (5) Singleton et al. (1976).

active in foods. L-2 can also cooxidize carotenoids and chlorophyll into colorless products. This property of L-2 is used to bleach flour (Section F.1.).

3. Enzymatic and Non-Enzymatic Volatile Decomposition Products

The hydroperoxides produced by lipoxygenases undergo cleavage, either by hydroperoxide *lyases* or non-enzymatically, and other reactions including isomerization, and rearrangement to produce aldehydes, C-18 and short-chain alcohols, vinyl ethers, oxo acids (or aldehyde acids), epoxy-hydroxy esters, and dimers that contribute to rancid flavors. Hydroperoxides from both isoenzymes L-1 and L-2 produce a wide spectrum of volatile aldehydes from lipids that are similar to those formed by non-enzymatic autoxidation (Chapter 3). From linoleic acid, isoenzyme L-1 yields mainly hexanal and *trans*-2,*trans*-4-decadienal, while L-2 forms hexanal, *trans*-2-heptenal, *trans*-2-octenal, *trans*-2,*trans*-4-nonadienal and *trans*-2-*cis*-4-decadienal (Table 10-2). From linolenic acid, L-1 produces mainly *trans*-2-hexenal, propanal and pentenal, and L-2 produces the same aldehydes plus *trans*-2,*cis*-4-heptadienal, 2,4,6-nonatrienal, and 3,5-octadien-2-one.

Hydroperoxides from lipoxygenase isoenzymes in peas, potatoes and beans produce similar aldehyde mixtures from linoleic and linolenic acids. Peanut lipoxygenase forms pentane and hexanal. Under anaerobic conditions, lipoxygenase reactions produce oxo acids, dimeric fatty acids and oxygenated dimers that may be derived by free radical intermediates. In many of these decomposition reactions, it is difficult to distinguish between enzymatic and non-enzymatic processes, which may occur either sequentially or

TABLE 10-3.
Flavor scores and gas chromatographic analyses of soybean oils. [a]

Oils	Storage 60°C	Flavor score [b]		2,4-decadienal	
		Heated [c]	Unheated	Heated [c]	Unheated
Crude	0-time	6.4	6.6	20	20
	24 days	6.5	5.9	190	270
Degummed	0-time	6.5	6.7	20	20
	8 days	6.4	6.5	210	410
Refined	0-time	7.0	6.2	80	60
	4 days	6.0	6.4	620	430
Bleached	0-time	6.6	5.1	30	30
	2 days	6.3	5.6	140	150
Deodorized	0-time	8.1	8.5	6	8
	8 days	6.2	6.2	200	250
Defatted flour [d]	0-time	6.3	5.2	—	—

[a] From Frankel et al. (1988).
[b] Flavor score: 1-10; 10, bland; 1, strong. Crude and partially processed oil tasted after dilution 95:5 with good quality soybean oil; deodorized oils tasted undiluted.
[c] Dehulled, cracked soybeans heated with steam for 2 min.
[d] Tasted as 2% dispersion in water.

simultaneously. After heat inactivation, lipoxygenases and other enzymes can act as non-enzymatic metallo-protein catalysts that may be activated under reducing conditions. During crushing and solvent extraction of oilseeds, metallo-proteins may dissociate, and the free metals may bind to any free fatty acids and catalyse lipid oxidation during processing.

To prevent flavor deterioration due to the action of lipoxygenases and other enzymes, unprocessed plant foods such as beans and peas are blanched before canning or freezing to inactivate these enzymes. These heat treatments must be carried out under mild conditions to avoid undesirable changes in texture and loss of protein bioavailability.

Before solvent extraction, soybeans are cracked, heated (65-75°C) and the moisture content adjusted (10-11%) before flaking. This heat treatment is not sufficient to inactivate the lipases and lipoxygenase, which requires temperatures of 100-110°C. Treating soybean flakes with steam for 2 min inactivated L-1 and L-2 and produced higher initial flavor scores of refined and bleached oils and higher flavor scores after storage of the crude and bleached oils (Table 10-3). The crude, degummed, bleached and deodorized oils from the heated beans produced less 2,4-decadienal after storage at 60°C than the oils from unheated beans. The treatment of beans with steam was also important to produce defatted flour of higher flavor score and quality than the unheated beans.

Toasting of soybeans by heating at 100-110°C to inactivate lipoxygenase and antinutritional factors (*e.g.* trypsin inhibitor) is important to improve the

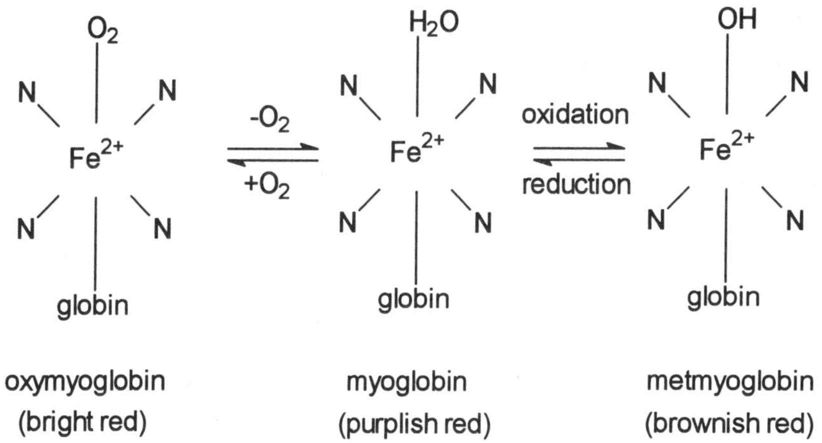

FOODS   191

$$\underset{\text{globin}}{\underset{\text{oxymyoglobin (bright red)}}{\overset{O_2}{\underset{N}{\overset{N}{\diagdown}}\underset{N}{\overset{N}{\diagup}}Fe^{2+}\underset{N}{\overset{N}{\diagup}}\underset{N}{\overset{N}{\diagdown}}}}} \underset{+O_2}{\overset{-O_2}{\rightleftharpoons}} \underset{\text{globin}}{\underset{\text{myoglobin (purplish red)}}{\overset{H_2O}{\underset{N}{\overset{N}{\diagdown}}\underset{N}{\overset{N}{\diagup}}Fe^{2+}\underset{N}{\overset{N}{\diagup}}\underset{N}{\overset{N}{\diagdown}}}}} \underset{\text{reduction}}{\overset{\text{oxidation}}{\rightleftharpoons}} \underset{\text{globin}}{\underset{\text{metmyoglobin (brownish red)}}{\overset{OH}{\underset{N}{\overset{N}{\diagdown}}\underset{N}{\overset{N}{\diagup}}Fe^{2+}\underset{N}{\overset{N}{\diagup}}\underset{N}{\overset{N}{\diagdown}}}}}$$

Figure 10-1. Schematic representation of the heme complex of myoglobin: iron bound to the nitrogen atoms of four pyrole rings in the porphyrin molecule and to the globin protein. From Cross *et al.* (1986).

quality of soybean protein products used for either animal feed or human consumption. However, this thermal enzyme inactivation is not carried out in the conventional processing of soybean oil. Therefore, the soybean flakes must be solvent extracted without delay to minimize free fatty acid and peroxide formation in the extracted crude oil and to produce a finished oil of improved oxidative and flavor stability.

4. Enzymatic and Non-Enzymatic Oxidation in Animal Foods

Three major forms of metal-containing systems catalyse lipid oxidation in raw meat and seafood, namely heme iron, non-heme iron and microsomal enzymes. Myoglobin is the predominant hemoprotein responsible for the color of red meat. In myoglobin, the iron atom is bound covalently to the nitrogen atoms of the four pyrrole rings of the porphyrin structure, and one nitrogen of a histidine molecule of the globin protein (Figure 10-1). The sixth bond orbital reacts with oxygen or water, involved in oxidation and reduction. The role of myoglobin in lipid oxidation in meat is still unclear. Microsomal enzymes in raw meat and fish catalyse lipid oxidation by maintaining iron in the reduced form. The relative activity of these metal-containing systems is strongly affected by heat treatments. The non-enzymatic catalytic system becomes more significant at higher temperatures and in cooked meat.

Many enzymes in milk are known to initiate lipid oxidation. *Trypsin* shows a protective effect that may be attributed to partial proteolysis increasing the copper-chelating capacity of milk. *Xanthine oxidase* is a flavin enzyme found as a significant component of the milkfat globule membranes that generates superoxide radicals (O_2^-). This reactive oxygen species is formed when oxygen takes up one electron, and occurs as an anion radical acting as a reducing agent. Under acidic conditions, $O_2^{-\cdot}$ has free radical activity producing the hydroperoxy radical ($HO_2^{\cdot}$), which undergoes dismutation to produce hydrogen peroxide:

$$O_2^{-\cdot} + H^+ \longrightarrow HO_2^{\cdot} + H^+ \longrightarrow H_2O_2 + O_2$$

Superoxide dismutase catalyses the dismutation of $O_2^{-\cdot}$ to produce H_2O_2, a reaction that occurs in animal and plant tissues. The generation of H_2O_2 takes place in turkey muscle tissues, and increases during aging at low temperatures. By removing superoxide from tissues, superoxide dismutase can have antioxidant activity in the presence of catalase to remove H_2O_2 (Chapter 8.E).

Hydrogen peroxide produced by reduction of $O_2^{-\cdot}$ is relatively innocuous and has limited reactivity in the absence of metals. However, in the presence of Fe^{2+}, H_2O_2 produces the very reactive hydroxyl radical ($HO^{\cdot}$) by the so-called Fenton reaction. Iron compounds complexed with phosphate esters (adenosine di- and triphosphate, ADP and ATP) can decompose H_2O_2 to form free radicals. The presence of the ADP-Fe complex in plant and animal foods may trigger the reduction of H_2O_2 to form the very reactive and potentially damaging hydroxyl radicals.

$$ADP\text{-}Fe^{3+} + O_2^{-\cdot} \longrightarrow ADP\text{-}Fe^{2+} + O_2$$

$$ADP\text{-}Fe^{2+} + H_2O_2 \longrightarrow HO^{\cdot} + OH^-$$

Hydroxyl radicals react non-selectively with all organic molecules. There is evidence that hydroxyl radicals are generated in muscle homogenates by heating, and in the presence of iron-ascorbate, but their importance with respect to lipid oxidation is unclear.

The autoxidation of iron-heme proteins such as myoglobin, that are abundant in muscle tissues, produces *metheproteins*, the ferric states of these proteins. The interaction of hydrogen peroxide with metmyoglobin (MetMb) or methemoglobin (MetHb) generates activated heme proteins, in which the iron is converted to a higher oxidation state by the formation of a porphyrin (P) cation radical, $P\text{-}Fe^{+4}=O^{\cdot}$, a *ferryl* species considered to be more selective than the hydroxyl radical, that could initiate lipid oxidation in muscle foods. The mechanism of lipid oxidation by hemoglobin and myoglobins involves activation of the iron catalyst by the formation of ferryl ion, as follows:
The ferryl ion can initiate lipid oxidation by hydrogen abstraction to produce a lipid radical plus a proton. Ascorbic acid and other reducing compounds in turkey muscle cytosol effectively inhibit lipid oxidation promoted by ferryl ions in membranes.

$$P\text{-}Fe^{+3} + H_2O_2 \longrightarrow P\text{-}Fe^{+3}\overset{H}{\underset{|}{\text{-}O\text{-}OH}}$$

$$P\text{-}Fe^{+3}\overset{H}{\underset{|}{\text{-}O\text{-}OH}} \longrightarrow P\text{-}Fe^{+4}\text{-}O^{\cdot} + H_2O$$

$$P\text{-}Fe^{+4}\text{-}O^{\cdot} \longrightarrow P^{+}\text{-}Fe^{+4}{=}O$$

Free iron and copper ions catalyse redox reactions in muscle lipids. *Ferritin* is a protein that can hold up to 4,500 ions of iron per mole, and the main iron storage protein in the cells. Iron can be released from ferritin by reducing agents. The formation of superoxide as a reductant may thus be a main pathway for the synthesis of myoglobin in muscle cells by mitochondria. A significant amount of iron is released from ferritin during storage of turkey muscle (Table 10-4). Most of the copper present in animal tissues is inactivated by chelation to proteins in the form of ceruloplasmin, albumin, carnosine and other histidine-dipeptides. These chelating compounds are good inhibitors of lipid oxidation in meat and fish. Carnosine present in muscle tissues can also reduce ferryl to oxymyoglobin.

The 12-lipoxygenase and 15-lipoxygenase have been identified in fish gill tissue and in chicken muscle by their production of 12-hydroperoxide and 15-hydroperoxide from arachidonic acid and other polyunsaturated fatty acids in fish. The substrate dependency and product specificity toward other long-chain polyunsaturated fatty acids in fish (20:5 n-3, 22:4 n-6, 22:5 n-3) indicate that the 12-lipoxygenase is specific towards the methyl rather than the carboxyl end and can be described as an n-9 lipoxygenase. The hydroperoxides produced by these enzymes may be decomposed in the presence of metal catalysts to form volatile carbonyls that cause off-flavors in fish and frozen chicken meat.

TABLE 10-4.
Loss of iron from ferritin isolated from turkey muscle after storage at 4°C.[a]

Storage time (days)	Dark muscle	Light muscle
0	1.50	0.55
2	1.14	0.42
9	0.60	0.08

[a] From Kanner (1992). Fe-ferritin in $\mu g/g$ wet tissue.

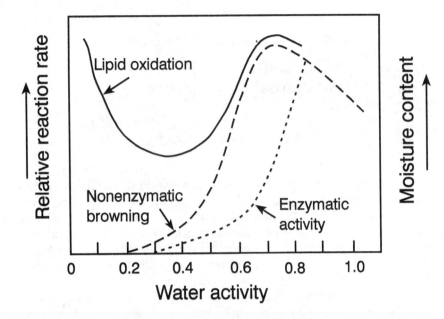

Figure 10-2. Food stability as a function of water activity. Adapted from Labuza (1971).

B. PROCESSED FOODS

A number of processing steps affect the oxidative stability of foods, including heat treatment, drying, fermentation, storage and packaging. The two most important factors due to processing include changes in water activity and the formation of non-enzymatic browning materials at elevated temperatures by the interactions of sugars or lipid oxidation products with proteins and amino acids. These compounds act as oxidation inhibitors, and their formation is influenced by the water activity, which affects the quality of foods and their nutritional value.

1. Water Activity

Water is a key factor affecting lipid oxidation in foods during storage by a complex series of events. The water activity of a food is a more appropriate physicochemical property than water content because in heterogeneous systems it determines the distribution of water between different components. Water activity (Aw) or the thermodynamic availability of water is defined as:

$$Aw = p / p_o$$

where p = partial pressure of water in food, and p_o = vapor pressure of water

Water activity is often used as a predictor of food quality, but problems arise from the inherent non-equilibrium nature of foods, and the kinetics of changes in moisture content must be followed as a function of temperature. The concept of water activity is rigorously applicable only in dilute aqueous solutions at equilibrium.

Lipid oxidation is high at low water activity (0.1-0.2) but decreases as water activity is increased to a minimum, below the monolayer value up to 0.2-0.4, and then increases again above 0.4, reaching a maximum in the range 0.6-0.8 (Figure 10-2). This "food stability map" is problematic because the relative rate of lipid oxidation, which is a kinetic property, cannot be plotted against water activity, which is an equilibrium thermodynamic function. The equilibrium nature of water activity should be clearly distinguished from the kinetics of lipid oxidation.

Several mechanisms have been suggested to explain the effect of water activity on lipid oxidation in foods. When dry, the metal catalysts are most active. At increasing moisture levels, lipid oxidation may decrease as a result of hydrogen bonding between water and hydroperoxides, thus retarding their decomposition into initiating radicals, and by decreasing the prooxidant activity of metal catalysts by hydration and changes in their coordination sphere. Metal salt hydrates are formed that are less lipid soluble and less active. At high water activity levels, lipid oxidation may be accelerated by increasing the diffusion and mobilization of the metals in the aqueous phase and allowing more intimate association with the lipid-water interface where polar hydroperoxides concentrate. Above a water activity of 0.7, lipid oxidation may decrease again, apparently as a result of dilution of the metal catalysts. Accordingly, foods of intermediate moisture would be very susceptible to oxidation.

Water has a profound effect on interactions between lipid oxidation products and other food components, including heme compounds, amino acids and proteins. The oxidative stability of several foods is greatly affected by their water activity. For example, walnuts are oxidatively stable at a water activity of about 0.3, and become less stable at higher (0.45-0.54) and lower (0.26) water activities. In another example, potato chips oxidize much slower at a water activity of 0.4 compared to lower water activities, reaching a maximum rate at a water activity of 0.01. Foods that contain their natural moisture such as refrigerated cooked meats are readily oxidized. Meat proteins may also oxidize, and this oxidation is often accelerated by increasing the water content. The type of rancidity developed in foods in the dry state may be different compared to the normal moisture state. Carbonyls produced by lipid oxidation can also interact with proteins in the presence of moisture and lead to browning. Similarly, the effectiveness of antioxidants and metal chelators is affected by the moisture content of foods (Section C.3.). Therefore, precautions are necessary in the preparation and storage of foods susceptible to lipid oxidation by using either packaging materials of low permeability to water vapor, or vacuum packaging, to reduce changes in the food water activity.

Although the concept of water activity and its effect on food stability is theoretically limited in its predictive capability, it continues to prove

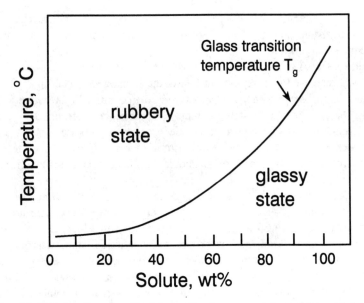

Figure 10-3. State diagram of a binary food system. Adapted from Nelson and Labuza (1992).

empirically useful. To get a better understanding of the relationship between moisture content and the rate of lipid oxidation, a new approach has recently been advanced by considering foods as polymers or a mixture of macromolecules. The glass transition theory, used successfully to understand the physicochemical properties of synthetic polymers, is now being exploited by food scientists to explain the complex relationship between water content and stability. Like polymers, foods can exist in either "glassy" or "rubbery" states according to their temperature and moisture content. The *glass transition temperature* (T_g) is the temperature at which amorphous regions of polymers change from a glassy state to a rubbery state. In the glassy state there is less free volume, or space in the food matrix that is not taken by the biopolymer chains of the macromolecules, than in the rubbery state. The diffusion of a small molecule like oxygen through the food matrix is hindered by the limited free volume present in the glassy state (Figure 10-3). This temperature-composition state diagram shows zones of different stabilities, where the glass transition temperature is dependent on solute composition and water content.

According to the glass transition theory, an important property of water is to act as an excellent plasticizer, which swells and increases the free volume of the food matrix. An increase in moisture results in increased diffusion and molecular mobility in a multi-component food system, accelerating lipid oxidation by increasing the number of catalytic sites. The molecular mobility decreases in solutes of increasing molecular weight. In polymer science, the rate of chemical reactions increases at an exceedingly high value by raising the temperature above the glass transition temperature range. In the glassy state, diffusion in the food matrix is restricted and reaction rates are diminished.

Accordingly, to increase food stability, formulations should be aimed at either bringing the matrix into the glassy state or increasing its glass transition temperature. Freeze-dried or spray-dried sugar-containing foods remain in the glassy state at low moisture conditions. At high moisture conditions, the "glasses" can collapse, and mobility allows the glasses to be converted to an amorphous or rubbery state that facilitates the diffusion of oxygen and accelerates oxidation. Water thus acts as a plasticizer that facilitates diffusion of oxygen to lipids entrapped in the food matrix. The prediction of food stability is thus complicated by the interactive influences of moisture, water activity, the state of the water and other factors influencing the transition from glassy to rubbery states in moist foods.

According to the glass transition theory, lipids are dispersed in the free volume of the food matrix composed of carbohydrates and protein polymers. In the rubbery state, the lipids react readily with oxygen and become oxidized. In the glassy state, however, the lipids are stable to oxidation because they are encapsulated and there is no free volume. The glass transition temperature, which determines when the food matrix changes from one state to the other, increases with a decrease in moisture and water activity. In many foods, the difference in oxidizability of different portions of lipids may be due to their different transitions between rubbery and glassy states. Therefore, to control lipid oxidation in foods it would be necessary to vary the formulation or processing conditions to minimize or eliminate the oxidizable lipid portions. However, this goal is difficult to achieve because the factors that determine how much of the lipids can be stabilized are not clear. To better control lipid oxidation, a fuller understanding is needed of the influence of moisture on the mechanism of transition between different states in various foods. The following section discusses how water activity also strongly affects food proteins and their interactions with oxidized lipids and sugars.

2. *Interactions of Lipids, Proteins and Sugars*

Sugars and carbonyl compounds interact with amino acids or proteins in a sequence of complex reactions known as the *Maillard reaction* or as *non-*

$$R-CHO + H_2N-R_1 \longrightarrow R-\underset{OH}{\underset{|}{C}H}-NH-R_1 \underset{+H_2O}{\overset{-H_2O}{\rightleftharpoons}} R-CH=N-R_1$$

D-glucose $\qquad\qquad\qquad\qquad\qquad\qquad\qquad\qquad$ imine

$$+H_2O \updownarrow -H_2O$$

$$R-CH_2NH-R_1$$

D-glucosylamine

enzymatic browning. The browning products from this reaction have a marked influence on lipid oxidation. They generally retard lipid oxidation in foods, and contribute to meat flavors. Lipid oxidation products can also react with proteins and amino acids, leading to the loss of essential amino acids with impact on the oxidative stability and the nutritional quality of foods.

The interaction of amino compounds with reducing sugars proceeds by the addition of a carbonyl group to a primary amino group of an amino acid, peptide or protein, by water elimination through an intermediate imine, cyclizising to a glycosylamine (N-glycoside).

N-glycosides are readily formed in foods containing a reducing sugar in the presence of proteins, amino acids and peptides, at elevated temperatures, or at low water activity. The glucosylamine undergoes rearrangement to a 1-amino-1-deoxyketose by a series of reactions known as the *Amadori rearrangement*, through an endiol-enaminol intermediate:

$$-\underset{\underset{OH}{|}}{CH}-CH=NH-R \rightleftharpoons -\underset{\underset{OH}{|}}{C}=CH-NH-R \rightleftharpoons -\underset{\underset{O}{\|}}{C}-CH_2-NH-R$$

$$\text{1-amino-1-deoxy-ketose}$$

The Amadori products are found in dried fruits, dehydrated vegetables and milk products. In milk powder, the reaction between the ε-amino group of proteins and lactose leads to the formation of N-alkyl-1-amino-1-deoxy-lactulose. Color formation by the Maillard reaction is accompanied by aroma generation during cooking, frying, baking and roasting. These reactions lead also to the loss of essential amino acids, undesirable discoloration and development of objectionable flavor compounds in dehydrated foods during storage, or heat processing (milk sterilization, pasteurization and condensation, meat roasting). The reaction between water-soluble reducing sugars and amino acids generally occurs only in foods of water activity in the range above the monolayer (0.3-0.7). If foods of intermediate moisture contents are heated at sufficiently high temperatures, the rate of the browning reaction can be sufficiently high that lipid oxidation is inhibited effectively by the Maillard reaction products. This process is an effective way of stabilizing certain foods like evaporated milk, if the browning color does not detract from the acceptability of the product. When considered undesirable in certain foods, the Maillard reaction can be inhibited by lowering the pH, avoiding high temperatures and the critical range of water activity (0.7-0.8) during processing and storage (Figure 10-2), or by using formulations containing non-reducing sugars, and sulfite as an additive.

The antioxidant properties of Maillard reaction products are attributed to the formation of *reductone* (enaminone) structures that have both reducing and metal complexing properties. These compounds can also act as peroxide destroyers by reducing hydroperoxides into stable allylic alcohols. Reductones are intermediate compounds produced by the thermal reactions of reducing sugars that retain a carbonyl group in the vicinity of an endiol group. In the presence of amino compounds, reductones are produced under mild conditions. Reductones consist of vicinal diol compounds undergoing enolization and oxidation to the dehydro compounds.

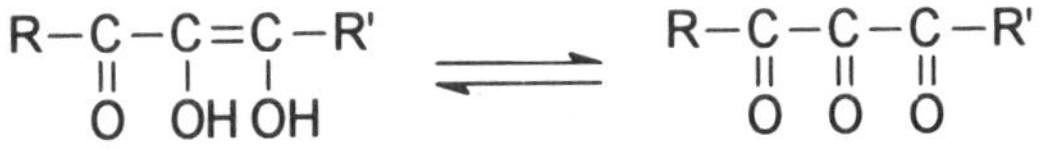

$$R-\underset{\underset{O}{\|}}{C}-\underset{\underset{OH}{|}}{C}=\underset{\underset{OH}{|}}{C}-R' \rightleftharpoons R-\underset{\underset{O}{\|}}{C}-\underset{\underset{O}{\|}}{C}-\underset{\underset{O}{\|}}{C}-R'$$

At pH below 6, reductones are resonance stabilized as anions.
Ascorbic acid is a reductone characterized by strong reducing properties under acidic conditions and at low temperatures. Basic amino acids (histidine, lysine and arginine) produce the most effective antioxidant products with sugars. Fractions from histidine-glucose reaction mixtures have good antioxidant properties. A range of Maillard interaction products of amino acids and sugars are known to have antioxidant activity, and may be useful in food processing to retard oxidation in foods. Various heat treatments can improve the stability of bakery and cereal products and inhibit oxidation in milk products.

$$R-\underset{\underset{O}{\|}}{C}-\underset{\underset{OH}{|}}{C}=\underset{\underset{O^-}{|}}{C}-R' \rightleftharpoons R-\underset{\underset{O^-}{|}}{C}=\underset{\underset{OH}{|}}{C}-\underset{\underset{O}{\|}}{C}-R'$$

Lipid oxidation products can also interact with proteins and amino acids and can affect the flavor deterioration and nutritive value of food proteins. Peroxyl radicals are very reactive with labile amino acids (tryptophane, histidine, cysteine, cystine, methionine, lysine and tyrosine) undergoing decarboxylation, decarbonylation and deamination. Methionine is oxidized to a sulfoxide; combined cysteine is converted to cystine to form combined thiosulfinate (Figure 10-4). Aldehydes, dialdehydes and epoxides derived from the decomposition of hydroperoxides react with amines to produce imino *Schiff* bases (R-CH=N-R'). Schiff bases polymerize by aldol condensation producing dimers and complex high-molecular weight brown macromolecules

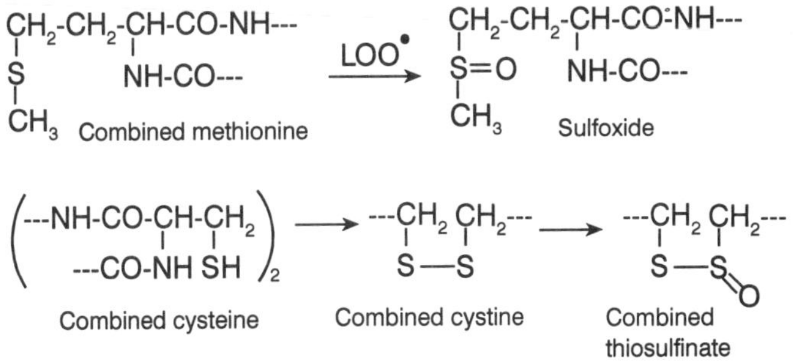

$$\underset{\text{Combined methionine}}{\underset{|}{\overset{|}{\text{CH}_2\text{-CH}_2\text{-CH-CO-NH---}}}\atop\underset{\text{CH}_3}{\overset{|}{\text{S}}\quad\text{NH-CO---}}} \xrightarrow{\text{LOO}^\bullet} \underset{\text{Sulfoxide}}{\underset{|}{\overset{|}{\text{CH}_2\text{-CH}_2\text{-CH-CO-NH---}}}\atop\underset{\text{CH}_3}{\overset{|}{\text{S=O}}\quad\text{NH-CO---}}}$$

$$\left(\begin{array}{c}\text{---NH-CO-CH-CH}_2\\ \text{---CO-NH SH}\end{array}\right)_2 \longrightarrow \underset{\text{Combined cystine}}{\begin{array}{c}\text{---CH}_2\text{ CH}_2\text{---}\\ |\quad\;\;|\\ \text{S---S}\end{array}} \longrightarrow \underset{\substack{\text{Combined}\\\text{thiosulfinate}}}{\begin{array}{c}\text{---CH}_2\text{ CH}_2\text{---}\\ |\quad\;\;|\\ \text{S---S}_{\searrow\text{O}}\end{array}}$$

$$\underset{\text{Combined cysteine}}{}$$

Figure 10-4. Interaction products between combined methionine and combined cysteine with lipid hydroperoxides.

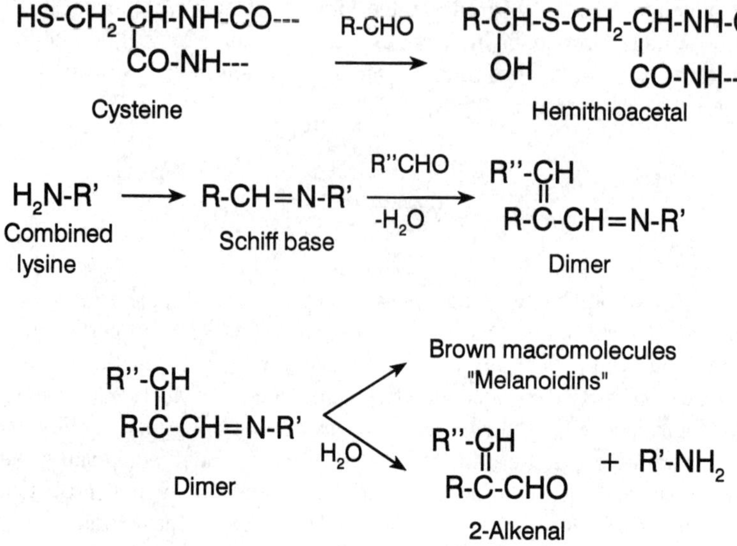

HS-CH$_2$-CH-NH-CO---   R-CHO   R-CH-S-CH$_2$-CH-NH-CO---
         | | |
         CO-NH--- $\longrightarrow$ OH   CO-NH---
Cysteine   Hemithioacetal

                            R"CHO   R"-CH
H$_2$N-R' $\longrightarrow$ R-CH=N-R' $\xrightarrow{-H_2O}$ ‖
Combined   Schiff base   R-C-CH=N-R'
lysine   Dimer

                                    Brown macromolecules
R"-CH   ↗   "Melanoidins"
  ‖
R-C-CH=N-R' $\xrightarrow{H_2O}$
                               ↘   R"-CH
   Dimer   ‖   + R'-NH$_2$
                                    R-C-CHO
                                    2-Alkenal

Figure 10-5. Interaction products between cysteine and aldehydes, and between a Schiff base and combined lysine.

known as *melanoidins* that are not well characterized (Figure 10-5). These polymeric brown materials are unstable and generate new volatiles by scission of the macromolecule or dehydration, that affect the flavor characteristic of foods during cooking and processing. Hydroperoxides react with proteins to form protein free radicals which cause the polymerization of the peptide chain to form protein-protein cross-links (Figure 10-6). The polymerization of lipids and peptide chains leads to browning of food proteins and changes in food texture and rheological properties.

Numerous studies have been published on the reactions of mixtures of lipid oxidation products and pure amino acids or proteins as model systems. The

$$LOOH + PH \longrightarrow LO^\cdot + P^\cdot + H_2O$$

$$PH + LO^\cdot \longrightarrow P^\cdot + L-OH$$

$$P^\cdot \xrightarrow{P} P-P^\cdot \xrightarrow{P} P-P-P^\cdot$$

$$P^\cdot + P^\cdot \longrightarrow P-P$$

Figure 10-6. Protein-protein crosslink formation by interaction of protein radicals (P) with lipid hydroperoxides.

TABLE 10-5.
Lipid oxidation in a whey protein-linolenate/linoleate model system stored under different condition.[a]

Temp. °C	Aw	Moisture g/kg	Oxygen mol/mol	Hydroperoxides[b]		Ethane[b]	
				7 days	14 days	7 days	14 days
20	0.90	170	4	5500	2900	3.6	6.6
37	0.90	170	1	2100	1000	5.9	6.6
37	0.90	170	4	1100	500	6.0	6.7
37	0.67	70	4	1600	750	8.2	9.0
37	0.30	20	4	1500	900	0.5	10.8
55	0.90	170	4	0	0	7.3	7.8

[a] Nielsen et al. (1985). Whey protein (from skimmed milk) was mixed with a mixture of methyl linolenate and linoleate in a hand mixer, the mixture was milled after spraying with water, and stored in enameled aluminum tins sealed under air.

[b] μmol/g lipid

interactions involved in such complex mixtures are often difficult to understand and to relate to real food systems. Several factors influenced lipid oxidation of a model system consisting of mixture of methyl linolenate and linoleate with whey proteins during storage in sealed tins under air (Table 10-5). Storage temperature had the most significant effect on hydroperoxide formation, with the highest amount formed after one week at 20°C, and no hydroperoxides left after the same period at 55°C. More hydroperoxides were produced as the water activity was decreased, and surprisingly, when the oxygen level was reduced. Ethane formation (from decomposition of linolenate hydroperoxides) in the headspace decreased at lower storage temperature and water activity. Losses in amino acids (lysine and tryptophan) increased at higher water activity, and decreased at lower oxygen levels. Tryptophan was more stable in this model system during storage.

The literature in this field is replete with a wide variety of model systems using for example cellulose, or silica, or amylopectin as solid supports, extruded mixtures of polyunsaturated fatty acids and esters with pure proteins, in the presence of humectants (to control moisture content) such as glycerol, or dextran as water-binding agent. Studies with such model systems are inherently problematic because they do not simulate real foods where oxidation is naturally inhibited by the separation of lipids from catalysts in different cellular compartments.

Changes in flavor result from the interactions of lipid-derived carbonyl compounds by aldolization with the amino groups of proteins. Undesirable flavors are produced when beef or chicken are fried in oxidized fats by the interaction of secondary lipid oxidation products with meat proteins. Cooked meat undergoes flavor deterioration on storage through the oxidized fat adhering to the food surface. The complex interaction products formed during actual cooking and frying of meat products are not well characterized, and it is not clear how they degrade into volatile compounds that affect flavor. The chemistry of interactions between lipid oxidation products and proteins is very complex and the impact on flavor deterioration is still not well understood.

C. MILK PRODUCTS

Oxidation of milk lipids is an important cause of flavor deterioration of dairy products that is often referred to as *Oxidative Rancidity*, to distinguish it from *Hydrolytic Rancidity* resulting from lipolysis. Milk products have complex compositions, physico-chemical properties, and contain natural prooxidants and antioxidants. Milk fat consists mainly of triacylglycerols (95-96%) existing naturally in the form of fat globules surrounded by a complex fat globule membrane composed of proteins (41%), phospholipids (30%), which are included in lipoproteins and glycoproteins, cholesterol (2%) and enzymes. The fatty acid composition of milk fat includes 63-69% saturated (24% palmitic, 13% stearic), 27-34% monounsaturated (25% oleic), and 2.5-3% polyunsaturated (2% linoleic) fatty acids. The milkfat globule membrane is a principal site of lipid oxidation because the main phospholipid components contain about 6% linoleic acid and are associated with prooxidants in a lipoprotein matrix.

Flavor defects in milk and cream are collectively referred to as the "oxidized flavor," in butter as "metallic" or "tallowy." Oxidized flavors are noticeable in stored milk at very low levels of oxidation, usually at a peroxide value less than 1. The development of oxidized flavor is most evident in fluid milk, cream and butter because they have mild and more delicate flavors.

Oxidation occurs initially in the polyunsaturated phospholipid fraction associated with the fat globule membrane, followed by the main triacylglycerol fraction. Different milks vary widely in susceptibility to oxidation, and oxidative deteriorations in milk are classified empirically as: i. *spontaneous*, developing oxidized flavors within 48 hr after milking without added metal catalyst; ii. *susceptible*, developing oxidized flavors within 48 hr after contamination with copper, and iii. *resistant (or non-susceptible)*, not developing oxidized flavors after 48 hr even after addition of copper or iron. Many hypotheses have been advanced for these differences in oxidative susceptibility, including the oxidation potential of milks, and the action of xanthine oxidase and lactoperoxidase, which is controversial. However, there are no substrates for these enzymes in milk. The action of various metallo proteins in milk may be confounded as enzymes. These metallo proteins act as powerful lipid oxidation catalysts in the presence of oxygen and redox systems involving ascorbic acid.

Although in fluid milk the phospholipid fraction is more susceptible to oxidation than the triacylglycerol fraction, in dry milk products, the triacylglycerol fraction is more susceptible to oxidation and the phospholipids act as antioxidants. Thus, solvent-extracted milkfat containing phospholipids is much more stable to oxidation than milkfat free of phospholipids, obtained by melting churned butter (also called butter oil). The susceptibility of milk

TABLE 10-6.
Effect of mixtures of ascorbic acid and copper on the formation of conjugated dienes (CD) during oxidation of linoleate.[a]

Ascorbic acid (M)	Copper (1.3 x M)	CD formation (k_i x 10^6)[b]
1.8×10^{-6}	10^{-7}	1.0
1.8×10^{-6}	10^{-3}	6.6
1.8×10^{-5}	10^{-7}	5.3
1.8×10^{-5}	10^{-3}	9.8
1.8×10^{-4}	10^{-7}	11.4
1.8×10^{-4}	10^{-5}	0
1.8×10^{-3}	10^{-7}	13.8
1.8×10^{-3}	10^{-6}	0

[a] From Haase and Dunkley (1969).
[b] Initial rates in mol/L/min

phospholipids to oxidation appears to be dependent on whether they are suspended in water or fat. This difference of oxidative stability influences the development of different flavor defects in various dairy products. With butter, which is a water-in-oil emulsion system containing an aqueous phase of phospholipids dispersed in fat, the phospholipids oxidize more readily than the triacylglycerol components.

1. Metal Catalysis

Metal catalysts are recognized as the most important factors in accelerating lipid oxidation in dairy products. Although copper occurs naturally in milk at a lower concentration (20-40 μg/L) than iron (100-250 μg/L), it is the most important catalyst for the development of oxidized flavors. These metals occur as complexes with proteins, and are associated with the fat globule membrane. Milks that develop oxidized flavors spontaneously contain more copper in the fat globule membrane. The most accepted mechanism of initiation of lipid oxidation in milk involves the catalytic action of lipoprotein-copper complexes at the fat globule membrane-aqueous interface. However, the integrity of this membrane as affected by processing must play an important part in determining the variation in oxidative susceptibility of milks. Metal chelators, such as ethylenediamine tetraacetic acid (EDTA), reduce the development of oxidized flavors in milk, but the use of these additive is not allowed in the U.S.A. and other countries.

Ascorbic acid, found in milk at a concentration of 10-20 mg/L, is also considered to be linked with the development of oxidized flavors in milk. Milk treated to remove ascorbic acid is more resistant to copper-catalysed lipid oxidation. The catalytic activity of copper is enhanced in the presence of ascorbic acid. However, when ascorbic acid is present in milk at sufficiently high concentrations (50-200 mg/L), it inhibits effectively the development of

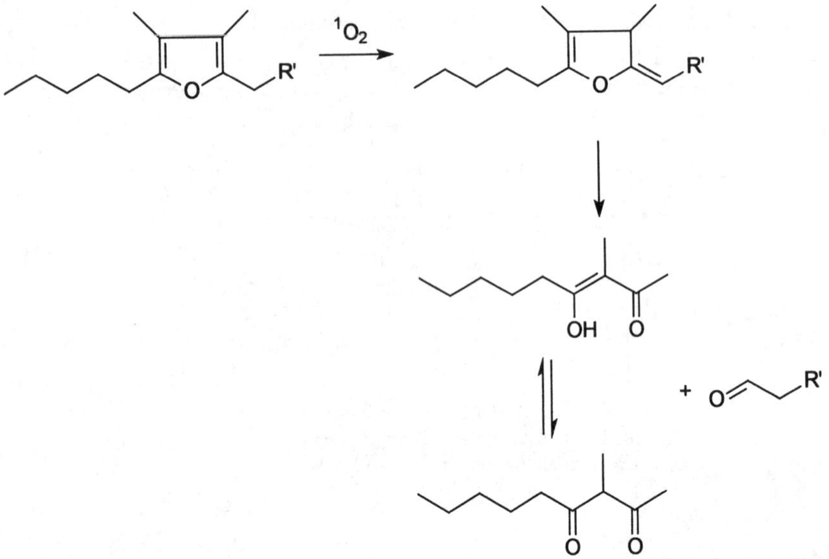

Figure 10-7. Proposed mechanism for the formation of 3-methylnonane-2,4-dione by photosensitized oxidation of furanoid fatty acids. From Grosch et al. (1992).

oxidized flavors. Combinations of ascorbate and copper have either prooxidant effects, at relatively low levels of ascorbate, or antioxidant effects at relatively high concentrations of ascorbate (Table 10-6). Several mechanisms may explain these effects, including conversion of cupric ions to the more active cuprous state, formation of an active copper-ascorbic acid-oxygen complex catalyst, sparing lipid oxidation by preferential oxidation of ascorbic acid and depleting the available oxygen, and reduction of hydroperoxides by ascorbate into stable and innocuous allylic alcohols. Clearly, more work is needed to clarify the paradoxical behavior of ascorbic acid. Other reducing agents such as thiols may also have dual pro- and antioxidant effects in the presence of copper, similar to those of ascorbic acid.

The ligands associated with copper and iron can have a profound effect on their catalytic activities. Significant portions of endogenous copper and iron are associated with proteins and the milk fat globule membrane. The formation of a copper-protein complex with the phospholipids of this membrane material has been suggested. The distribution of metals in various milk products and the effects of processing on their catalytic activities are influenced by a wide number of factors that are not clearly understood.

2. Light Oxidation

Exposure of milk and butterfat to light, especially sunlight, produces off-flavors, initially described as burnt, light-activated or sunlight flavor. On prolonged exposure to light, this flavor defect changes to the typical oxidized flavor of milk. The sunlight flavor originates in the serum proteins of milk and riboflavin is implicated as a potent water-soluble photosensitizer. On exposure to light in the presence of riboflavin, several amino acids (tryptophan, tyrosine, cysteine, and lysine) are destroyed in a low-density lipoprotein fraction associated with the milk fat globule membrane. Carbonyl compounds and methional, formed by degradation of methionine, are attributed to the development of sunlight flavor of milk, which is also related to thiols, sulfides, and disulfides in milk.

Recent studies of photooxidized butter and butter oil identified by aroma extract dilution analysis, 3-methylnonane-2,4-dione, a potent volatile compound derived from furanoid fatty acids (see Section C.4.) (Figure 10-7). Six different furanoid fatty acids were established as dione precursors, and were found in various samples of butter made from either sweet cream (116-476 mg/kg) or from sour cream (153-173 mg/kg), or from butter oil (395 mg/kg). Similar precursors of the dione were identified in stored boiled beef and vegetable oils. This flavor defect arising by photooxidation of butter or butter oil is apparently different from the light-activated flavor in milk, that involves the interaction of sulfur-containing proteins and riboflavin. However, more sensory comparisons are needed to distinguish between these two flavor defects due to light oxidation.

3. Antioxidants

α-Tocopherol is the only important natural antioxidant in milk, which contains an average of 25 $\mu g/g$ milkfat and 44 $\mu g/g$ of fat globule membrane. The content of this antioxidant varies according to the feed of the animal and the season of the year. The oxidative stability of milk correlates well with the α-tocopherol level in milk, and especially in the lipids of the fat globule membrane. Supplementing the ration of animals with various forms of tocopherols provides an effective control measure against lipid oxidation in milk. However, direct addition of emulsified α-tocopherol to milk is a more efficient measure of controlling oxidized flavor in milk than ration supplementation. Several synthetic antioxidants (BHA, BHT, propyl gallate) and metal chelators (citric acid, phosphoric acid and salts of ethylenediamine tetraacetic acid) have proven effective in inhibiting lipid oxidation, but the use of these compounds is not legally permitted in dairy products in the U.S.A. and other countries.

Ascorbic acid acts as an antioxidant in milk at normally low copper concentrations. However, during storage the concentration of ascorbic acid decreases continuously and is depleted by consuming dissolved oxygen. Milk is also protected from lipid oxidation by casein and other proteins capable of binding and inactivating copper and iron. One mechanism suggested for the improved oxidative stability of homogenized milk is the resurfacing of the fat globules with casein providing a protective membrane (see Section 5.d). Casein has an effective stabilizing action in preventing lipid oxidation in milk products. Its protective effect can be attributed to its binding capacity for copper and its association with the oil-water interface around the fat globule

TABLE 10-7.
Aldehydes found in autoxidized dairy products.[a,b]

Products	n-Alkanals	2-Alkenals	2,4-Alkadienals
Butter phospholipids	C_2 to C_4	C_5 to C_{11}	C_8, C_9
Butter oil	C_1 to C_{10}	C_3 to C_{11}	C_7, C_{11}
Skimmilk	C_2, C_6	C_4 to C_{11}	C_6 to C_{11}
Nonfat dry milk	C_1, C_2, C_5-C_{10}, C_{12}, C_{14}		
Dry whole milk	C_1 to C_3, C_5-C_{10}, C_{12}	C_5 to C_{11}	
Fluid whole milk	C_5-C_{10}	C_6-C_{11}	C_8-C_{12}
Butter [b]	C_5-C_{12}	C_5 to C_{11}	C_7, C_9, C_{10}, C_{11}

[a] From Day (1966) and Parks (1974).
[b] Other carbonyls: *cis*-4-heptenal, *trans,cis*-2,6-nonadienal, 2,5-octadienal, 2,4,6-nonatrienal, 2,4,7-decatrienal, 1-penten-3-one, 1-octen-3-one, *trans,cis*-3,5-octadien-3-one, 3,5-undecadien-3-one, 1-octen-3-one, 1-octen-3-ol

membrane. Heating milk products to activate sulfhydryl groups in proteins is another method of increasing the oxidative stability of dairy products.

4. Flavor Compounds

The volatile carbonyl compounds developed from milkfat oxidation are detectable at extremely low levels (parts per billion). These compounds are much more easily perceived in milk where they have much lower threshold values (0.004-0.1) than in oil systems (0.1-2.5). Because flavor compounds are very easily perceived in milk products, they can be detected at extremely low levels of oxidation. Fluid milk becomes rancid at peroxide values less than one.

TABLE 10-8.
Effect of storage of butter oil on increase in the dilution factor of carbonyl compounds. [a]

Compounds	Fresh	Stored
hexanal	16	64
cis-3-hexenal	64	128
1-octen-3-one	128	1024
cis-1,5-octadiene-3-one	32	256
nonanal	0	64
cis-2-nonenal	64	128
trans-2-nonenal	32	256
trans,trans-2,4-nonadienal	8	128
trans,trans-2,4-decadienal	32	64

[a] From Widder *et al.* (1991). Storage for 42 days at room temperature.
Dilution factor = dilution necessary to still detect compounds by GC olfactometry (using a sniffing port).

TABLE 10-9.
Changes in dilution factors of flavor compounds formed in photooxidized butter oil. [a]

Compounds	fresh butter oil	photooxidized butter oil
1-hexen-3-one	128	64
hexanal	<32	32
cis-3-hexenal	64	64
cis-4-heptenal	64	64
1-octen-3-one	128	256
cis-2-nonenal	64	64
trans-2-nonenal	32	128
3-methylnonane-2,4-dione	0	256
trans,trans-2,4-decadienal	32	128
trans-4,5-epoxy-trans-2-decenal	0	128

[a] From Grosch et al. (1992). Butter oil exposed to fluorescent light for 48 hours.

Aldehydes found in various dairy products include those derived from the hydroperoxides of polyunsaturated fatty acid components, and further oxidation products of polyunsaturated aldehydes (Table 10-7). Flavors imparted by these carbonyl compounds are described as "oxidized, fishy, metallic, painty and tallowy". A powerful technique known as *aroma extract dilution analysis* was used to determine potent odor and flavor compounds in various milk and food products. This method determines the odor activity of volatile compounds in an extract eluted from a high-resolution capillary gas chromatograph (GC) by detection with a sniffing port (SP). The odor activity of a compound is expressed as the *flavor dilution factor*, which is the ratio of its concentration in the initial extract to its concentration in the most dilute extract in which the odor can be detected by GC-SP. By this technique, the sweet buttery flavor of fresh butter was attributed mainly to a mixture of diacetyl, acetic acid, butyric acid, guaiacol, vanillin, cis-6-dodecen-γ-lactone, δ-octalactone, δ-decalactone and various carbonyl compounds formed by lipid oxidation. The effect of storage of butter oil on the dilution factor of carbonyl compounds was determined by the technique of aroma extract dilution analysis (Table 10-8). On storage at room temperature the dilution factor of carbonyl compounds increased, with 1-octen-3-one showing the most significant increase. Antioxidants (BHA, BHT, α- and γ-tocopherols) effectively inhibited the formation of cis- and trans-2-nonenal, cis-4-heptenal and 1-octen-3-one in butter oil stored at 35°C in the dark. Butter oil exposed to fluorescent light developed an off-flavor described as "green, strawy and fatty," which was attributed mainly to the production of 3-methylnonane-2,4-dione, trans-4,5-epoxy-trans-2-decenal and an increase in concentration of trans-2-nonenal and trans,trans-2,4-decadienal (Table 10-9). The formation of 3-methylnonane-2,4-dione was shown to be derived from a mixture of furanoid fatty acids occurring naturally in minor amounts (Figure 10-7).

TABLE 10-10.
Oxygen consumption by pasteurized milk during storage at 7°C in the dark for 6 days. [a]

Consumed by:	Oxygen (mg/L)
Bacteria	1.48
Ascorbate	1.41
Peroxidation	0.003
Sulfhydryl groups	0.14
Total	4.10

[a] From Allen and Joseph (1983). 1.07 mg/L accounted for by loss through septum and other sources.

Not all of the potent volatile compounds are derived from lipid oxidation, including a number of lactones that come from naturally occurring hydroxy fatty acids, diacetyl and vanillin in butter oil (from melted butter). The concentrations of the mixtures of carbonyl compounds exceed the flavor threshold values for individual aldehydes, and the "oxidized flavor" results from a combination of volatile compounds.

5. Factors Affecting Lipid Oxidation

Lipid oxidation during processing and storage of milk products may influence prooxidants, antioxidants and the oxygen content of the final products.

a. Storage temperature. The effect of storage temperature on oxidative deterioration of dairy products is anomalous, because milk develops oxidized flavors more readily and shows higher TBA values at 0 to 4°C than at 4 to 20°C. Differences in activation energy between peroxide formation (10.3 kcal/mole) and increase in TBA values (13.9 kcal/mole) suggest that these methods measure two different events in the sequence of lipid oxidation. Initial hydroperoxide formation requires less energy than hydroperoxide decomposition into TBA-reactive products. Another explanation offered for the effect of storage temperature relates to the lowering of the oxygen content by bacterial activity at higher temperatures. However, other types of bacterial deteriorations would be expected at the higher storage temperatures.

b. Oxygen levels. The oxygen content of pasteurized milk varies from 9 mg/L at 20°C to 8 mg/L at 25°C, and storage in the dark consumes 4.1 mg/L after six days (Table 10-10). Much more of the oxygen is consumed by bacteria and oxidation of ascorbic acid than oxidation of sulfhydryl groups or autoxidation.

The development of oxidized flavors in milk, butter oil and dry milk powder is effectively inhibited by removal of dissolved oxygen and by deaeration. The shelf-life of dry whole milk is significantly extended by vacuum treatment, replacing air oxygen and packing with inert gas. Other effective techniques to remove oxygen from milk powder include the use of a mixture of glucose and catalase as an oxygen scavenger (see Chapter 8.E) in the presence of calcium as a desiccant, or the oxygen-absorbing mixture, Na_2SO_3 and $CuSO_4$, enclosed in porous paper pouches, or by using a mixture of nitrogen and hydrogen in the presence of a palladium or a platinum catalyst. Dried milk stored in nitrogen shows slower development of off-flavors than unprotected samples.

c. **Heat treatments.** Pasteurization of milk increases its susceptibility to oxidized flavors by increasing the copper content of the cream phase. However, heating milk to higher temperatures decreases this oxidative susceptibility because of the formation of sulfhydryl reducing compounds, derived from β-lactoglobulin, that impart a "cooked" flavor to milk.

d. **Homogenization.** The homogenization of milk, which disrupts the natural fat globule integrity, retards lipid oxidation and the development of oxidized flavors effectively. Several mechanisms have been advanced to explain this stabilization effect: (i) redistribution of catalytic metals, which may involve complex exchange between the fat globule membrane as a prooxidant catalyst, and casein in milk plasma as an inactive metal-protein complex, (ii) migration of the phospholipids into either the serum phase or the interior of the fat globules, (iii) redistribution of the phospholipids in milk, (iv) denaturation of proteins with increase in available sulfhydryl groups, (v) structural changes or decrease of the copper-protein complex that inactivates ascorbic acid in promoting lipid oxidation. Although considerable work has been published in this area, details of the structural changes resulting from homogenization of milk are still not clear.

D. MEAT AND POULTRY PRODUCTS

The lipids in lean beef consist of about 2 to 4% triacylglycerols and 0.8 to 1% phospholipids containing 44% polyunsaturated fatty acids, which are mainly subject to oxidation. Oxidation occurs initially in the phospholipids of cellular and subcellular membranes which are in close proximity to the heme catalysts of the mitochondria and microsomes. A typical fatty acid composition of beef membrane phospholipids includes 22% 18:2, 2% 18:3, 15% 20:4, less than 1% 20:5 and 2% 22:6. Chicken and turkey muscle are more susceptible to oxidation than beef because of their higher polyunsaturated phospholipid fraction and relatively low levels of natural tocopherols. Red poultry muscles oxidize faster than white because of higher phospholipid and iron contents.

Raw meat is resistant to oxidation when refrigerated. Lipid oxidation proceeds rapidly when the meat structure is disrupted by grinding, chopping or by heating, exposing the phospholipids to oxygen, enzymes, heme pigments and metal ions. When the cellular and biochemical integrity of muscle tissues is lost during food processing, there is a shift in the balance between prooxidants and antioxidants in favor of prooxidant activity. Pigment changes occurring by oxidation of the red oxymyoglobin and myoglobin to the brown metmyoglobin are important reactions affecting quality of meat (Figure 10-1).

Rancidity development in meat is distinguished between "normal oxidation," due to oxidation of the triacylglycerol fraction, and rapid oxidation referred to as the "warmed-over flavor," attributed mainly to the phospholipids. Normal oxidation occurs during long-term refrigeration storage, while the development of warmed-over flavor takes place in meat after cooking and varies at different stages of lipid oxidation. Rancidity due to lipid oxidation occurs also in uncooked meat when the muscle membranes are broken by mechanical separation and restructuring. The interactions of carbonyl compounds from lipids in the browning Maillard reaction impart desirable flavor attributes to cooked meat. Lipid oxidation in meat is also associated with deterioration of the desirable beef flavor by formation of other flavor defects described as "cardboard" or "painty."

1. Iron Catalysts

Iron can be found either in the free form, non-heme iron, or bound to proteins. The free catalytically active iron represents 2 to 4% of the total muscle iron in beef, lamb, pork and chicken. Protein-bound iron exists in the form of soluble proteins, myoglobin, ferritin and other iron-complexed proteins (transferrin, ovotransferrin, lacto-transferrin). Cooking of meat extracts causes a release of iron from heme proteins. Ferritin releases ferrous iron in the presence of reducing agents such as ascorbate and thiols. Lipid oxidation is also catalysed in mixtures of ferritin isolated from meat extracts in the presence of ascorbate and cysteine. As in milk, depending on the level of catalytically active iron in meat, ascorbate can act either as a prooxidant at low concentrations by reducing iron to a low valence state, or as an antioxidant at high concentrations by scavenging lipid free radicals.

Although the free iron is catalytically active, the relative importance of protein-bound iron is being debated. Lipid oxidation in cooked meat is not only attributed to the release of catalytically active non-heme iron, but also to the disruption of cell membranes in meat that brings the polyunsaturated lipids in close contact to the catalysts. Unfortunately, in many studies using the TBA test to determine the effects of different forms of iron on lipid oxidation in meats, the results must be interpreted with caution because the TBA reaction is significantly affected by iron and other metals, and is subject to serious interference by other factors. In addition to products of lipid oxidation, TBARS are also formed from proteins and nucleic acids and other non-lipid components in meat tissues that confound the results of the TBA test.

There is increasing evidence that degradation of proteins is associated with the development of the warmed-over flavor. Selective methods to measure protein oxidation may be useful as a complementary approach in evaluating oxidative deterioration of meat products. The determination of *protein carbonyls* (as dinitrophenyl hydrazine derivatives) is now used for this purpose. More reliable methods are needed to determine specific products of lipid oxidation acting as precursors of flavor compounds to establish the relative contribution of heme and non-heme iron to the development of rancidity in various meat products.

2. Enzymatic Lipid Oxidation

Lipid oxidation in subcellular fractions can be mediated by enzyme systems in muscle microsomes that maintain iron in the ferrous form by reduced nicotinamide adenine dinucleotide (NADH). However, this redox system may not be enzymatic because unlike lipoxygenase, no specific lipid oxidation products have been identified. Ascorbate and other reducing agents may have the same effects in the presence of heme-protein complexes. On the other hand, the presence of 15-lipoxygenase in chicken muscle may be responsible for oxidative deteriorations in uncooked chicken meat during frozen storage. Phospholipases inhibit lipid oxidation, apparently by forming iron complexes with the free fatty acids liberated by hydrolysis of phospholipids in the membrane.

Lipid oxidation in meats can also be accelerated by depletion of glutathione peroxidase and glutathione transferase, which decompose hydroperoxides by reduction into hydroxy compounds in the presence of glutathione. Lipoxygenases can participate in co-oxidation reactions at low oxygen levels that produce volatile products affecting the flavor quality of meat. The role of lipoxygenase in initiating lipid oxidation in meat systems is not clear yet and more detailed studies are needed to assess the significance of this enzyme.

3. Volatile Flavor Compounds

The interactions of lipid hydroperoxides and secondary products with proteins and amino acids have a considerable impact on the flavor of meat products during processing, cooking and storage. Hydroperoxide radicals react with sulfur and amine and amino acids, and secondary oxidation products with thiols from cysteine (Figures 10-4 and 10-5). Non-enzymatic browning reaction products and Schiff bases formed from the interactions of aldehydes with proteins are unstable and produce complex mixtures of volatile compounds that affect the flavor characteristics of meat products during cooking and processing.

A large number of volatile compounds have been identified in beef fat (Table 10-11). The flavor significance of individual volatile compounds can be calculated on the basis of their threshold values. Thus, two of the carbonyl compounds formed in the smallest concentrations, *cis*-4-heptenal and *trans,trans*-2,6-nonadienal, were the most flavor significant. On the other hand, the most abundant carbonyl, 2-decenal, was the least flavor significant.

A direct and more advanced analytical approach to screening important

TABLE 10-11.
Carbonyl compounds from beef fat.[a]

Carbonyls (in order of concentration)	Flavor threshold	Flavor significance
2-decenal	5.5	12
2-undecenal	4.2	11
2-nonenal	0.1	5
2-octenal	1.0	10
hexanal	0.15	6
2-heptenal	0.63	9
heptanal	0.042	3
nonanal	0.32	8
octanal	0.068	4
2,4-decadienal	0.28	7
cis-4-heptenal	0.001	1
trans-4-heptenal	0.32	8
trans,trans-2,6-nonadienal	0.001	1
trans,cis-2,6-nonadienal	0.002	2

[a] From Frankel (1984). Flavor threshold in ppm.

volatile compounds in cooked beef is based on the aroma extract dilution analysis (Section C.4.). By this technique 35 compounds were identified as having dilution factors greater than 4 (Table 10-12). This technique evaluates compounds on the basis of their relative aroma intensities. The most significant aroma compounds with dilution factors greater than 512 included, 2-methylfuran-3-thiol, methional, trans-2-nonenal, trans,trans-2,4-decadienal, β-ionone and bis-(2-methyl-3-furyl)-disulfide. 2-Methylfuran-3-thiol had an odor threshold of 0.0025-0.01 ng/L, and bis-(2-methyl-3-furyl)-disulfide, an odor threshold of 0.0007-0.0028 ng/L.

The effects of lipid oxidation in cooked meat is to produce directly a number of volatile compounds known to be derived from lipid hydroperoxides (e.g. alkanals, alkenals, alkadienals, and ketones), and indirectly decrease the levels of volatile compounds associated with desirable meaty odors (furans with sulfur substituents). The most important source of these furan compounds is the Amadori rearrangement and Maillard reactions between carbohydrates and amino acids, and further reactions with ammonia and hydrogen sulfide (see Section B.2).

The volatiles identified in cooked chicken were the same as those of beef, but with higher levels of volatiles derived from unsaturated lipids (2,4-decadienal, and γ-dodecalactone), and lower beef aroma compound bis-(2-methyl-3-furyl)-disulfide. Chicken meat has a higher level of linoleic acid than beef and would be expected to contribute more volatile lipid decomposition products.

4. Antioxidants

Muscle tissues contain a multi-component antioxidant system consisting of lipid-soluble compounds (α-tocopherol, ubiquinone), water-soluble compounds

TABLE 10-12.
Volatile flavor compounds with high aroma values from cooked beef.[a]

Flavor dilution factor	Volatile compound	Description
512	2-methylfuran-3-thiol	meaty, sweet
	methional	cooked potato
	trans-2-nonenal	tallowy
	trans,trans-2,4-decadienal	fried potato
	bis(2-methyl-3-furyl)disulfide	meaty
256	2-acetyl-1-pyrroline	roasted, sweet
	1-octene-3-one	mushroom
128	2-acetylthiazole	roasted
	trans,trans-2,4-nonadienal	fatty
64	2-octanone	fruity, musty
	trans-2-octenal	fruity, fatty, tallowy
	2-decanone	musty, fruity
	2-dodecanone	musty, fruity
32	trans-2-heptenal	fatty, tallowy
	1-cis-5-octadien-3-one	geranium, metallic
16	hexanal	green
	trans-2-hexenal	green
	2-heptanone	fruity, musty
	heptanal	green, fatty, oil
	trans-2-cis-6-nonadienal	cucumber
	2-undecanone	tallowy, fruity
	2-tridecanone	rancid, fruity, tallowy
8	1-octen-3-ol	mushroom
	2-nonanone	fruity, musty
	nonanal	tallowy, green
	2,4-decadienal	fatty
4	2-methyl-3-(methylthio)furan	sulfurous
	2-acetythiophene	sulfurous, sweet

[a] from Gasser and Grosch (1988).

(ascorbate, histidine-dipeptides) and enzymes (superoxide dismutase, catalase, glutathione peroxidase). Lipid oxidation in meats can be effectively controlled by the use of various phenolic compounds derived from spice extracts, by vitamin E supplementation of animal diets, and by processing of cured meat with sodium nitrite. Various natural antioxidant formulations containing mixtures of tocopherols, ascorbyl palmitate and citric acid show synergistic effects in stabilizing cooked and frozen meat. Synthetic antioxidants, BHA, TBHQ, propyl gallate (see Chapter 8) and combinations with citric acid, ascorbic acid or phosphates are also effective formulations used to retard lipid oxidation in meat. A number of metal chelators (EDTA, sodium pyrophosphate, tripolyphosphate, citric acid), and reducing agents (ascorbic acid, isoascorbic acid and their salts, sulfur dioxide) are effective protective agents in improving the sensory quality and color stability of stored meat products.

As in milk, ascorbic acid has either prooxidant activity at low concentrations (up to 250 ppm) by reducing iron, or antioxidant activity at higher levels (500 ppm) by acting as an oxygen scavenger. Higher concentrations of ascorbic acid

(500-1000 ppm) may be detrimental by the formation of hydrogen peroxide in air that may not be removed by catalase in meat, leading to the destruction of myoglobin with bleaching of the meat color (Figure 10-1) and other quality deteriorations.

Supplementation of vitamin E in the diet has a significant effect in increasing the oxidative and color stability of poultry, beef and pork products. Dietary incorporation of α-tocopherol is more effective in stabilizing meat than exogenous addition of this antioxidant. Incorporation of α-tocopherol into the membrane lipids by the diet is considered to be an effective means of extending the shelf-life of restructured and precooked meat products. The greater storage stability of chicken than turkey meat is attributed to the higher level of natural tocopherols in chicken meat. Vitamin E supplementation is also an effective means of reducing the formation of cholesterol oxidation products (cholesterol oxides) in raw and cooked veal muscle. Cholesterol oxides in foods are regarded as deleterious because of their association with health disorders (see Chapter 12).

Sodium nitrite used as a curing agent in processing cured meat is an effective inhibitor of lipid oxidation. The addition of nitrites to other meats (cooked ground beef, chicken and pork) retards oxidation, and inhibits the development of warmed-over flavor of stored cooked meat, and of rancidity in raw beef and chicken during storage. Nitrites produce a red color in cured meat due to the formation of nitrosylmyoglobin, which has antioxidant activity. The stability of cured meat decreases as the nitrosylmyoglobin becomes oxidized to produce metmyoglobin. A number of mechanisms have been suggested for the antioxidant activity of nitrite; it inhibits heme-catalysed lipid oxidation by forming a complex with the iron porphyrins, stabilizes the polyunsaturated lipids in membranes, chelates catalytic metals, forms inactive complexes with non-heme iron and copper, or low-molecular weight iron fractions, produces nitrosated amines that possess antioxidant activity, stabilizes cell membranes, and scavenges free radicals by nitric oxide. The disadvantages of using nitrite in meat include the formation of undesirable N-nitrosamines, particularly in cooked bacon, altering the desirable color of myoglobin, and consumer objections.

A number of natural antioxidants show potential applications for stabilizing meat products. Rosemary extracts containing potent antioxidants, including carnosic acid and carnosol (see Chapter 8), are effective in stabilizing cooked pork, beef, chicken and turkey meat. In cooked ground beef, rosemary extracts are effective in controlling the development of warmed-over flavor during storage. Many plant extracts and spices containing flavonoids and polyphenols reduce flavor deterioration in various ground meat products. However, the flavor associated with these plant extracts may limit their usefulness in muscle foods. Phytic acid and carnosine (histidine-containing dipeptide), obtained from cereal and meat by-products, are effective inhibitors of lipid oxidation by several mechanisms, including metal inactivation and free radical quenching. Uric acid obtained from the decomposition of adenosine triphosphate in muscle also inhibits lipid oxidation by the same mechanisms. However, the importance of uric acid as an endogenous antioxidant in muscle foods is not clear. Various protein concentrates from soybeans, cottonseed and peanuts inhibit lipid oxidation in muscle foods. In addition to their iron binding activity, these crude extracts contain complex polyphenolic flavonoids that have potent antioxidant activity.

5. Other Control Measures

Several factors can be manipulated to control and reduce flavor deterioration in meat due to lipid oxidation. Factors related to raw material include vitamin E content and age, while factors related to processing include addition of antioxidants, heat treatment and packaging. Heating and grinding raw meat accelerate lipid oxidation. In general, it is important to maintain the integrity of heated meat products to retard flavor deterioration in meat from lipid oxidation.

Oxidative deteriorations of meat increase with increasing heating temperatures and heating time. However, above 100°C browning Maillard reaction products are formed at the surface of meat that inhibit lipid oxidation and the development of warmed-over flavor. Unfortunately, this heat treatment changes the water-holding capacity and the appearance of the meat which may not be desirable.

Packaging at reduced oxygen pressure and vacuum packaging are effective means of prolonging the oxidative stability of meat products such as precooked chill-stored, or freezer stored, sliced products. Modified-atmosphere (nitrogen or carbon dioxide) packaging is also an effective but not commonly used measure to increase the stability of meat products.

Many controversial questions remain unanswered in evaluating the best measures to control flavor deterioration in processed meat. The effect of heating meat is explained by different mechanisms that accelerate the oxidation of phospholipids, by liberating iron from myoglobin and other metallo-proteins, by transforming myoglobin from an antioxidant to a prooxidant species, by disrupting in cell membranes the compartmental separation between the phospholipids and prooxidant metal catalysts, by thermal decomposition of hydroperoxides, and by denaturing antioxidant enzymes. Progress in this field has been hampered by the use of a number of over simplified model systems to study muscle tissue oxidation under unrealistic conditions of oxidation (*e.g.* heme catalysts activated with hydrogen peroxide and halides, combinations of iron and ascorbic acid). Transferring results obtained with such simplistic systems to the more complex meat systems may be questionable. The general and indiscriminate use of the TBA test to measure lipid oxidation is particularly problematic in complex foods such as meats because they contain a multitude of prooxidant and antioxidant substances that seriously interfere with this test (see Chapter 5.E).

E. FISH PRODUCTS

Fish lipids are highly susceptible to oxidation because they contain high levels of polyunsaturated fatty acids (30-40%), including eicosapentaenoic acid (EPA) (5-12%) and docosahexaenoic acid (DHA) (1-2%). These unique fatty acids distinguish fish lipids from other plant and animal lipids. Fish also contain prooxidants that generate lipid free radical initiators, and natural antioxidant systems. These pro-and antioxidant systems are affected by many factors, including storage temperature, oxygen concentration, pH, metal contamination, the use of salt, light, and processing causing cellular damage. Many of the same mechanisms of oxidative deterioration discussed for meat products (Section D) apply in varying degrees to fish products. Flavor deterioration in fish by lipid oxidation is the major problem in quality deterioration during processing and storage.

The polyunsaturated fatty acids in phospholipids associated with the fish membranes oxidize faster than the triacylglycerols found in fat deposits. Lipid oxidation in fish is influenced by several factors discussed below.

1. Metal Catalysts

Iron is implicated in several stages of the lipid oxidation chain. Both heme (6-15 ppm) and non-heme iron (10 ppm) present in fish tissues catalyse lipid oxidation. Iron complexed to metabolites, referred to as low-molecular weight iron, and iron complexed with myoglobin and hemoglobin, referred to as high-molecular weight iron, participate in the initiation and propagation of lipid oxidation in fish. Increased rates of lipid oxidation in cooked fish muscles indicate the release of non-heme iron during cooking. Other prooxidant components present in fish include: mitochondria, nicotinamide adenine dinucleotide phosphate (NADP), lipoxygenase, ascorbate, ferritin and transferrin. Heme pigments are abundant in dark muscle of fish, which is also richer in unsaturated lipids and thus more susceptible to oxidation than white muscle.

As in meat, in cooked fish the myoglobin is activated into the catalytic metmyoglobin form, which accelerates the development of rancidity. On prolonged heating, the active heme iron is decomposed and the liberated free iron causes non-heme catalysis of lipid oxidation. Metmyoglobin can be activated in the presence of hydrogen peroxide to produce the hypervalent ferryl ion (Section A.4.). However, there is no direct evidence showing the hydrogen peroxide-activated heme compounds in fish. Hemoprotein oxidation is increased by subjecting fish to freeze-thaw treatment which activates hemoproteins. The relative importance of hydrogen peroxide activation of metmyoglobin and the formation of free iron is not clear. The inhibition of hemoprotein-catalysed oxidation by EDTA in cooked fish indicates exposure of the iron to the lipid substrate resulting from heating.

The inhibition of lipid oxidation from both heme and non-heme iron is difficult. In the early post mortem stages, fish tissues have sufficient reducing capacity to activate free iron. This reducing capacity of fish tissues is lost on storage and the ferric form of iron remains inactive as a catalyst. However, during conditions of prolonged storage, myoglobin is oxidized to metmyoglobin (Figure 10-1), and hemoglobin to methemoglobin with iron in the ferric state. These oxidized hemes may be activated by reacting with hydrogen peroxide to the ferryl state.

Metal catalysis in fish is further complicated by interfacial phenomena between multiphase and compartmentalized reactions in fish systems. New understanding of interfacial oxidation in complex fish tissues may be developed by partition studies with heme and non-heme iron by the same methodology used in partition studies with antioxidants (Chapter 9, C. 2.e.).

2. Enzyme Oxidation

Although lipoxygenases are recognized catalysts of lipid oxidation in plant tissues, in fish like meat tissues both enzymatic and non-enzymatic oxidation can take place. In general, lower temperatures favor enzymatic oxidation and higher temperatures promote non-enzymatic oxidation. This issue is confounded, however, because under conditions where enzymes are thermally inactivated, the denatured metallo-proteins provide active catalysts for the breakdown of hydroperoxides causing flavor deterioration. In stored fish muscle, enzyme-catalysed lipid oxidations are active for longer periods than non-enzymatic metal-catalysed oxidations. During frozen storage microsomal lipid oxidation is not significant, but after thawing, enzyme activity becomes an important factor for quality deterioration.

Lipoxygenases occurring in gill and skin tissues of fish produce conjugated diene hydroperoxides that may degrade either by glutathione peroxidase to produce stable hydroxy fatty acids, or non-enzymatically by metal-heme catalysis to produce alkoxyl radicals that undergo cleavage into aldehydes and carbonyl compounds causing rancidity. The 12-lipoxygenase and 15-lipoxygenase in fish gill tissue produce the 12- and 15-hydroperoxides from arachidonic acid and other polyunsaturated fatty acids in fish (Section A.4). The major volatile compounds produced from the oxidation of arachidonic acid and EPA by 12-lipoxygenase were 1-octen-3-ol, 2-octenal, 2-nonenal, 2-nonadienal, 1,5-octadien-3-ol and 1,5-octadien-1-ol.

Gill lipoxygenase can be thermally inactivated above 60°C with a resulting improvement in shelf-life stability of fish. However, heating increases non-enzymatic oxidation also, and this may exceed the oxidation due to lipoxygenase.

3. Volatile Compounds in Stored Fish

Oxidized fish oils, rich in n-3 polyunsaturated fatty acids, produced volatile compounds more readily than oxidized vegetable oils, rich in linoleic acid. Activation energy for the formation of propanal from fish oils was lower than for the formation of hexanal from vegetable oils. A mixture of aldehydes

TABLE 10-13.
Various carbonyl volatile compounds derived by lipid oxidation in fish tissues.[a]

Compounds	Origin (PUFA)[b]	Conc. (ppb)	Threshold (ppb)	Descriptions
4-heptenal	n-3	1-10	1	creamy
2,4-heptadienal	n-3	1-10	10	rancid hazel nuts
2-hexenal	n-3	1-10	17	green grass
2,4,7-decatrienal	n-3	1-10	150	oxidized fish oil
1-octen-3-ol	n-6	10-100	10	mushroom, melon
1,5-octadien-3-ol	n-3	10-100	10	mushroom, seaweed
2,5-octadien-1-ol	n-3	1-10	—	mushroom, seaweed
1,5-octadien-3-one	n-3	0.1-5	0.001	mushroom
2-nonenal	n-6	0-25	0.08	cucumber
2,6-nonadienal	n-3	0-35	0.01	cucumber

[a] From Hsieh and Kinsella (1989); ppb, parts per billion.
[b] n-3 PUFA = 18:3 n-3, 20:5 n-3 and 22:6 n-3
n-6 PUFA = 18:2 n-6, 20:4 n-6.

contributed to the characteristic odors and flavors of oxidized fish, described as "rancid, painty, fishy and cod-liver oil like" (Table 10-13). Oxidation of unsaturated fatty acids in fish was related to the formation of 2-pentenal, 2-hexenal, 4-heptenal, 2,4-heptadienal and 2,4,7-decatrienal. The fishy or trainy characteristic of fish oil was attributed to 2,4,7-decatrienal. Studies of volatiles from boiled trout after storage showed significant increases in potent volatiles by aroma extraction dilution analysis (Table 10-14). Volatiles with the highest odor impact included 1,5-octadien-3-one, 2,6-nonadienal, 3-hexenal, and 3,6-

TABLE 10-14.
Effect of storage on potent odor compounds from boiled trouts analysed by aroma extraction dilution analysis.[a]

compounds	concentration[b]		odor activity value[c]	
	fresh	stored[d]	fresh	stored[d]
butane-2,3-dione	268	363	54	73
penatane-2,3-dione	140	167	28	33
cis-3-hexenal	1.6	13	53	430
hexanal	37	98	4	9
cis-4-heptenal	2.8	6.0	46	100
methional	5.8	7.5	145	188
1-octen-3-one	0.6	0.9	60	90
1-cis-5-octadien-3-one	0.36	0.64	900	1600
cis,cis-3,6-nonadienal	<1.2	14.0	<24	280
trans-2-nonenal	2.7	5.0	34	62
trans,cis-2,6-nonadienal	8.0	10.5	400	525
trans,cis-2,4-nonadienal	4.7	6.4	78	107

[a] From Milo and Grosch (1993).
[b] μg / kg boiled fish.
[c] odor activity value = concentration / odor threshold.
[d] stored 17 weeks at $-13°C$ before boiling.

TABLE 10-15.
Effect of storing raw and baked minced carp at -18°C on free fatty acids (FFA), thiobarbituric acid (TBA) and carbonyl values.[a]

Storage (weeks)	Raw			Baked		
	FFA	TBA	Carbonyl	FFA	TBA	Carbonyl
0	229	0.44	18.9	224	2.11	17.5
2	259	0.41	47.5	233	0.96	24.8
4	258	0.42	38.1	213	0.93	16.7
6	281	0.49	61.6	252	0.96	50.7
8	336	0.44	40.7	241	1.11	32.7

[a] From Flick et al. (1992). TBA and carbonyl values in μmoles/100 g wet sample.

nonadienal. 3,6-Nonadienal and 3-hexenal were considered to contribute most to the "fatty, fishy flavor" in stored boiled fish.

4. Fish Processing

The fish processing industry is subject to many biological and environmental factors causing undue fluctuations in the quality of raw materials, which are extremely perishable. Many complex post-mortem changes occur in fish muscles that disintegrate the membrane and affect lipid oxidation by mechanisms that are not well understood. A decrease in adenosine triphosphate (ATP) may initiate lipid oxidation by reducing nucleosides, conversion to hypoxanthine and change of xanthine dehydrogenase to xanthine oxidase. A decrease in reducing compounds such as ascorbate, NADH, and glutathione may affect lipid oxidation by changing the relative catalytic activity of heme iron and free iron, and the release of iron from ferritin.

a. Effect of temperature. The storage life of fish is limited by the formation free fatty acids below 0°C due to hydrolysis of triacylglycerols by lipases and of phospholipids by phospholipases. Tissues containing free fatty acids are more readily oxidized and rancidity increases by long term cold and frozen storage. Both free fatty acids and peroxide values increase in fish during frozen storage. By cooling just below the freezing point (-4°C), rancidity is accelerated in fish by a complicated process involving the removal of free water by crystallization resulting in the concentration of catalytic metals. Other unfavorable changes that affect the quality of fish include protein denaturation, causing toughening of the fish muscle and yellow-brown discoloration, attributed to non-enzymatic browning due to the formation of lipid-protein complexes. The rate of oxidation in fish decreases rapidly at temperatures below -18°C. The development of rancidity in stored fish can be retarded by glazing with a thin layer of ice.

Cooking fish decreases the rate of rancid development on storage. This effect is attributed to the destruction of lipoxygenase, formation of water-soluble antioxidants, and destruction of heme compounds. Baked fish developed lower free fatty acids and carbonyl values but higher TBA values than raw fish after storage at -18°C for 8 weeks (Table 10-15). The TBA and carbonyl values reached a maximum and then decreased after 6 weeks storage, apparently because of further decomposition of carbonyl compounds. However, these determinations did not necessarily reflect the degrees of flavor deterioration which needed to be determined by sensory analyses.

b. Antioxidants. Lipid oxidation is retarded in fish by synthetic (BHA, BHT, TBHQ) and natural antioxidants (tocopherols, flavonoids) and by metal chelators (EDTA, ascorbate, phosphate, citrate, carnosine). Ascorbic acid retards the oxidation of herring during frozen storage, but may promote oxidation in cooked fish. Flavonoids are effective antioxidants in prolonging the shelf-life of ground fish. EDTA and antioxidants inhibit enzyme-catalysed lipid oxidation (superoxide dismutase, catalase, and peroxidases) by removing iron and reducing hydrogen peroxide. Antioxidants are more effective in minced fish where they become more readily incorporated with the oxidizable lipids than in whole fish.

c. Oxygen level and packaging. Other measures to retard development of rancidity in fish include minimizing exposure to air and using vacuum and opaque packaging. The flavor score of frozen stored fish can be improved and the shelf life can be extended by vacuum and controlled atmosphere packaging (nitrogen and carbon dioxide). These measures are often more effective in fish products than the use of antioxidants.

To improve our understanding of the complex oxidation reactions in meat and fish products, evaluations should be carried out with each product stored under realistic conditions. Lipid oxidation should be followed by several and more reliable methods than the TBA test. These methods should measure specific lipid oxidation products (*e.g.* hydroperoxides or conjugated dienes) acting as precursors of rancid flavors, and specific volatile products (*e.g.* hexanal or carbonyls) that serve as reliable indicators of flavor deterioration. Instrumental analyses of volatile oxidation products should be supported and confirmed by sensory evaluations.

F. CEREAL PRODUCTS

Cereals are more stable than other foods because they are low in total fat (2-5%) and contain relatively high levels of natural tocopherols (20-50 ppm α-, β- and γ-tocopherols). Cereals are also stabilized by antioxidant products formed during baking by the browning or Maillard reaction. Natural compartmentalization within plant cells and low water activity also contribute to the low susceptibility of cereal lipids to develop rancidity. As in other foods, since flavor deterioration in cereals is caused by minor amounts of lipid decomposition products, the amount of lipid in a product is much less important than its susceptibility to oxidation.

Cereal products from whole grain or those containing bran or germ components are more easily oxidized because they contain more lipoxygenases and more unsaturated lipids (40-60% linoleic acid and 1-10% linolenic acid). Recent popularity of whole grain products has increased the importance of rancidity and its control in cereal foods. Whole grains are stable during storage. However, after milling into flour the grain integrity is disrupted, lipids become subject to oxidation by exposure to hydrolytic and oxidative enzymes activated by air and moisture, and catalytic metal ions. Rancidity in cereal products results from hydrolytic and oxidative degradations, usually sequentially because hydrolytic products are more susceptible to oxidation.

1. Hydrolytic Rancidity

The hydrolysis of polyunsaturated lipids in cereals produces free fatty acids that undergo further enzymatic or non-enzymatic oxidation to form volatile and non-volatile undesirable flavor compounds. Lipoxygenases act mainly on free fatty acids, which are also more easily oxidized than those esterified as triacylglycerols. In addition, free fatty acids are detrimental to functional properties of cereal products.

Lipase activity is found mainly in the bran components, which readily accumulate free fatty acids during ambient storage. Bran lipase is most active at 17% moisture, in finely divided milled products; it is rapidly inactivated by heating at 100°C for 10 min. Lipase activity is high in other cereal grains including oats and rice.

2. Lipoxygenase Action

This enzyme oxidizes linoleic and linolenic acids rapidly in whole flour or milling products containing wheat germ or bran mixed with water. The initial hydroperoxides formed by lipoxygenases in stored wheat bran are converted to secondary products, mono- and tri-hydroxy fatty acids. These oxidation products causing bitter and rancid flavors are formed in higher concentrations in hydrated products than in dry raw materials. Rancid flavors develop rapidly on hydration.

The addition of small amounts of unheated soy flour to wheat dough improves its rheological properties in the presence of air, but the mechanism for this improvement in bread quality is not clear. For the production of white bread, soybean flour containing lipoxygenase can be used, in small amounts (1%), to bleach the yellow pigment by co-oxidation of the carotenoids. However, heat inactivation of lipoxygenase is necessary in the preparation of pasta products to preserve the desirable yellow color of flour from durum wheat rich in carotenoids. Lipoxygenases, present in the germ component of most cereals, produce a mixture of secondary oxidation products in ground grain suspended in water. The formation of trihydroxy fatty acids cause bitter and rancid flavors in hydrated unheated milled oats.

3. Non-Enzymatic Oxidation

In contrast to enzymatic oxidation, free radical non-enzymatic oxidation of cereal lipids is readily inhibited by antioxidants. Undamaged grains and cereal germ oils are oxidatively stable because of their tocopherol contents. However, tocopherols are readily decomposed during storage of cereal products. Oats contain other flavonoid compounds. Certain processing steps can increase the susceptibility of grains to oxidation by decomposing tocopherols, disrupting the grain, redistributing the lipids and exposing them to catalytic metals.

Iron is added as a supplement to baby and to adult cereals, and to wheat flour as a nutritional supplement. This source of iron, and any contamination from equipment during processing, provide the main catalyst for lipid oxidation in cereal products. Natural metallo-proteins in cereals can also be activated during thermal processing in the presence of moisture to catalyse lipid oxidation.

4. Factors Affecting Lipid Oxidation

a. Raw materials. Cereal products from grains obtained under wet pre-harvest conditions are more susceptible to rancidity due to physical damage resulting in cell disruption that promotes the activities of microbial and fungal lipases. Thermal inactivation of enzymes should be carried out at the lowest temperature to reduce susceptibility of the grains to non-enzymatic lipid oxidation and to avoid detrimental changes in functionality. Poor baking performance results from loss of gluten functionality by thermal denaturation of proteins.

b. Processing. Enzymes are more easily inactivated by heat at higher water activity. Thus, at higher moisture, oat lipoxygenase can be inactivated at lower temperature (64°C at 12% moisture) than at low moisture (103°C at 8% moisture). Enzymatic lipolysis can also be prevented at storage temperatures sufficiently low to solidify the oil. Moisture contents below 5% are recommended to minimize non-enzymatic rancidity in cereal products.

c. Antioxidants. Endogenous tocopherols are important to prevent rancidity during grain processing. Thermal processing designed to inactivate enzymes should be carried out at the lowest temperature necessary to minimize the oxidation of natural tocopherols. However, the tocopherol content of wheat flour decreases significantly during storage, and the lipids become more susceptible to oxidation. Added synthetic antioxidants such as BHA, TBHQ and natural antioxidants from rosemary extracts are effective in prolonging the shelf-life of dry cereal products. The presence of natural flavonoid antioxidants in oats and other cereals should be further exploited to minimize problems of rancidity.

d. Iron fortification. Infant cereals are usually fortified with iron because it is considered to be the major source of iron during the weaning period. Adult

TABLE 10-16.
Effect of iron fortification on pentane formation and odor evaluation of infant cereals stored for 3 months at 37°C.[a]

Iron source (mg Fe/100g)	Headspace pentane (ppm)	Odor score[b]
Wheat cereal		
No iron	2.3	6.7
Dry ferrous sulfate (10)	2.3	6.7
Hydrated ferrous sulfate (10)	15.1	3.0
Encapsulated ferrous sulfate (10)[c]	2.0	6.7
Fe pyrophosphate (10)	21.5	3.3
Electrolytic Fe (10)	1.9	6.3
Rice cereal		
No iron	0.2	7.3
Ferrous fumarate (18.5)	0.2	7.0
Ferrous succinate (18.5)	0.2	7.7
Mixed cereal		
No iron	2.7	7.7
Ferrous fumarate (18.5)	4.7	5.8
Ferrous succinate (18.5)	5.4	5.3

[a] From Hurrell et al. (1989).
[b] Odor score: 9, excellent; 5, acceptable; 1, very strongly oxidized.
[c] Encapsulated in hydrogenated soybean oil.

cereals are also often supplemented with iron for presumed "energy" and "health" effects. Unfortunately, the most bioavailable forms of iron are soluble iron compounds, such as ferrous sulfate or ferrous gluconate, that also promote lipid oxidation and cause flavor deterioration when added to stored cereals. Many less soluble and less bioavailable forms of iron (elemental iron, ferric pyrophosphate and ferric orthophosphate) are used commercially for economic reasons, and because they are relatively less active as oxidation catalysts. The prooxidant activity of soluble ferrous sulfate can be inhibited by encapsulation with either hydrogenated vegetable oils or mono- and diglycerides. Various iron complexes have proven to be suitable sources of iron fortification, including ferrous fumarate and ferrous succinate. (Table 10-16). These iron chelates are bioavailable and do not promote lipid oxidation during storage at 37°C for 3 to 6 months. For either economic or legal reasons, these forms of iron supplement are not commonly used in baby cereals. However, the frequent practice of iron supplementation of adult cereal products and flours may be questionable in view of the problems of rancidity development during storage and processing, and the possible nutritional and biological consequences of excessive iron in the body (Chapter 12).

BIBLIOGRAPHY

Allen,J.C. and Joseph,G. Deterioration of pasteurized milk on storage. *J. Dairy Res.* **52**, 469-487 (1985).
Belitz,H.-D. and Grosch,W. *Food Chemistry*, Chapter 4, *Carbohydrates*, pp. 201-256 (1987) Springer Verlag, Berlin.
Cross,H.R., Durland,P.R. and Seidemen,S.C. Sensory qualities of meat, in *Muscle as Food*, pp. 279-320 (1986) (edited by P.J. Bechtel), Academic Press, New York.
Day,E.A. Role of milk lipids in flavors of dairy products, in *Flavor Chemistry*, pp. 94-120 (1966) (edited by I. Hornstein), Advances in Chemistry Series, American Chemical Society, Washington, D.C.
Decker,E.A. and Hultin,H.O. Lipid oxidation in muscle foods via redox iron, in *Lipid Oxidation in Foods*, pp. 33-54 (1992) (edited by A.J. St. Angelo), American Chemical Society Symposium Series 500, Washington, D.C.
Eriksson,C.E. Oxidation of lipids in food systems, in *Autoxidation of Unsaturated Lipids*, pp.207-231 (1987) (edited by H.W.-S. Chan), Academic Press, London.
Fischer,K-H. and Grosch, W. Co-oxidation of linoleic acid to volatile compounds by lipoxygenase isoenzymes from soya beans. *Z. Lebensm. Unters.-Forsch.* **165**, 137-139 (1977).
Flick,G.J., Hong,G-P. and Knobl,G.M. Lipid oxidation of seafood during storage, in *Lipid Oxidation in Foods*, pp. 183-207 (1992) (edited by A. J. St. Angelo), American Chemical Society Symposium Series 500, Washington, D.C.
Forss,D.A., Angelini,P., Bazinet,M.L. and Merritt,C. Volatile compounds produced by copper-catalysed oxidation of butterfat. *J. Am. Oil Chem. Soc.* **44**, 141-143 (1967).
Frankel,E.N. Recent advances in the chemistry of rancidity of fats, in *Recent Advances in the Chemistry of Meat*, pp. 87-118 (1984) (edited by A.J. Bailey), The Royal Society of Chemistry, London.
Frankel,E.N., Warner,K. and Klein,B.P. Flavor and oxidative stability of oil processed from null lipoxygenase-1 soybeans. *J. Am. Oil Chem. Soc.* **65**, 147-150 (1988).
Galliard,T. Rancidity in cereal products, in *Rancidity in Foods*, pp. 109-130 (1983) (edited by J.C. Allen and R.J. Hamilton), Applied Science Publ., London.
Gardner,H.W. Decomposition of linoleic acid hydroperoxides. Enzymic reactions compared with non-enzymic. *J. Agric. Food Chem.* **23**, 129-136 (1975).
Gardner,H.W. Lipid enzymes: Lipases, lipoxygenases and "hydroperoxidases," in *Autoxidation in Food and Biological Systems*, pp. 447-504 (1980) (edited by M.G. Simic and M. Karel), Plenum Publ. Corp., New York.
Gasser,U. and Grosch,W. Identification of volatile flavour compounds with high aroma values from cooked beef. *Z. Lebensm. Unters.-Forsch.* **90**, 3-8 (1988).
German,J.B. and Kinsella,J.E. Lipid oxidation in fish tissue. Enzymatic initiation via lipoxygenase. *J. Agric. Food Chem.* **33**, 680-683 (1985).
German,J.B., Zhang,H. and Berger,R. Role of lipoxygenases in lipid oxidation in foods, in *Lipid Oxidation in Foods*, pp. 74-92 (1992) (edited by A. J. St. Angelo), American Chemical Society Symposium Series 500, Washington, D.C.
Gillatt,P.N. and Rossell,J.B. The interaction of oxidized lipids with proteins, in *Advances in Applied Lipid Research*, Vol. 1, pp. 65-118 (1992) (edited by F. B. Padley), Jai Press Ltd., London.
Gray,J.I. and Monahan,F.J. Measurement of lipid oxidation in meat and meat products. *Trends Food Sci. Technol.* **3**, 315-319 (1992).
Gray,J.I. and Pearson,A.M. Lipid-derived off-flavours in meat - formation and inhibition, in *Flavor of Meat and Meat Products*, pp. 116-143 (1994) (edited by F. Shahidi), Blackie Academic & Professional, London.
Grosch,W. Linoleic and linolenic acid as substrate for enzymatic formation of volatile carbonyl compounds in pea. *Z. Lebensm. Unters.-Forsch.* **137**, 216-223 (1968).
Grosch,W., Konopka,U.C. and Guth,H. Characterization of off-flavors by aroma extract dilution analysis, in *Lipid Oxidation in Foods*, pp. 266-278 (1992) (edited by A. J. St. Angelo), American Chemical Society Symposium Series 500, Washington, D.C.
Grosch,W. and Laskawy,G. Differences in the amount and range of volatile carbonyl compounds formed by lipoxygenase isoenzymes from soybeans. *J. Agric. Food Chem.* **23**, 791-794 (1975).
Grosch,W., Laskawy,G. and Fischer,K-H. Oxidation of linolenic acid in the presence of haemoglobin, lipoxygenase or singlet oxygen. Identification of the volatile carbonyl compounds. *Lebensmitt.-Wiss. U.-Technol.* **7**, 335-338 (1974).

Grosch,W., Laskawy,G. and Weber,F. Formation of volatile carbonyl compounds and cooxidation of β-carotene by lipoxygenase from wheat, potato, flax, and beans. *J. Agric. Food Chem.* **24**, 456-459 (1976).
Haase,G. and Dunkley,W.L. Ascorbic acid and copper in linoleate oxidation. 3. Catalysts in combination. *J. Lipid Res.* **10**, 568-576 (1969).
Hurrell,R.F., Furniss,D.E., Burri,J., Whittaker,P., Lynch,S.R. and Cook,J.D. Iron fortification of infant cereals: a proposal for the use of ferrous fumarate or ferrous succinate. *Am. J. Clin. Nutr.* **49**, 1274-1282 (1989).
Hsieh,R.J. and Kinsella,J.E. Oxidation of polyunsaturated fatty acids: Mechanisms products, and inhibition with emphasis on fish. *Adv. Food and Nutr. Res.* **33**, 233-341 (1989).
Kanner,J. Mechanism of non-enzymic lipid peroxidation in muscle foods, in *Lipid Oxidation in Foods*, pp. 55-73 (1992) (edited by A. J. St. Angelo), American Chemical Society Symposium Series 500, Washington, D.C.
Kanner,J., German,B.J. and Kinsella,J.E. Initiation of lipid peroxidation in biological systems. *CRC Rev. Food Sci. Nutr.* **25**, 317-364 (1987).
Karel,M. Chemical effects in food stored at room temperature. *J. Chem. Educ.* **61**, 335-339 (1984).
Karel,M. Lipid oxidation, secondary reactions, and water activity of foods, in *Autoxidation in Food and Biological Systems*, pp. 191-206 (1980) (edited by M.G. Simic and M. Karel), Plenum Publ. Corp., New York.
Labuza,T.P. Kinetics of lipid oxidation in foods. *CRC Critical Rev. Food Tech.* **2**, 355-405 (1971).
Labuza,T.P., McNally,L., Gallagher,D., Hawkes,J. and Hurtado,F. Stability of intermediate moisture foods. 1. Lipid oxidation. *J. Food Sci.* **37**, 154-159 (1972).
Labuza,T.P., Tannenbaum,S.R. and Karel,M. Water content and stability of low-moisture and intermediate-moisture foods. *Food Technology* **24**, 35-42 (1970).
Ladikos,D. and Lougovois,V. Lipid oxidation in muscle foods: A review. *Food Chem.* **35**, 295-313 (1990).
Mielche,M.M. and Bertelsen,G. Approaches to the prevention of warmed-over flavour. *Trends Food Sci. Technol.* **5**, 322-327 (1994).
Milo,C. and Grosch,W. Changes in the odorants of boiled trout (*Salmo fario*) as affected by the storage of the raw material. *J. Agric. Food Chem.* **41**, 2076-2081 (1993).
Mustakas,G.C., Albrecht,W.J., McGhee,J.E., Black,L.T., Bookwalter,G.N. and Griffin,E.L. Lipoxidase deactivation to improve stability, odor and flavor of full-fat soy flours. *J. Am. Oil Chem. Soc.* **46**, 623-626 (1969).
Nelson,K.A. and Labuza,T.P. Relationship between water and lipid oxidation rates, in *Lipid Oxidation in Foods*, pp. 93-103 (1992) (edited by A.J. St. Angelo), American Chemical Society Symposium Series 500, Washington, D.C.
Nielsen,H.K., Löliger,J. and Hurrell,R.F. Reactions of proteins with oxidizing lipids. 1. Analytical measurements of lipid oxidation and of amino acid losses in a whey protein-methyl linolenate model system. *British J. Nutr.* **53**, 61-73 (1985).
O'Connor,T.P. and O'Brien,N.M. Lipid oxidation, in *Advanced Dairy Chemistry*. Vol. 2 pp. 309-347 (1995) (edited by P. F. Fox), Chapman & Hall, London.
Parks,O.W. Autoxidation, in *Fundamentals of Dairy Chemistry, Second Edition*, pp. 240-272 (1974) (edited by B.H. Webb, A.H. Johnson and J.A. Alford), The Avi Publ. Co., Inc., Westport, CT.
Pearson,A.M. Physical and biochemical changes occurring in muscle during storage and preservation, in *Muscle as Food*, pp. 103-134 (1986) (edited by P.J. Bechtel), Academic Press, New York.
Richardson,T. and Korycka-Dahl,M. Lipid oxidation, in *Developments in Dairy Chemistry - 2. Lipids*, pp. 241-363 (1983) (edited by P.F. Fox), Applied Science Pub., London.
Singleton,J.A., Pattee,H.E. and Sanders,T.H. Production of flavor volatiles in enzyme and substrate enriched peanut homogenates. *J. Food Sci.* **41**, 148- (1976).
Slade,L. and Levine,H. Beyond water activity. Recent advances based on an alternate approach to the assessment of food quality and safety. *Crit. Rev. Food Sci. Nutr.* **30**, 115-360 (1991).
Tappel,A.L. Hematin compounds and lipoxidase as biocatalysts, in *Symposium on Foods: Lipids and their Oxidation*, pp. 122-138 (1962) (edited by H. W. Schultz, E. A. Day and R. O. Sinnhuber), Avi Publ. Co., Inc., Westport, CT.
Widder,S., Sen,A. and Grosch,W. Changes in the flavour of butter oil during storage. *Z. Lebensm.-Unters. U-Forsch.* **193**, 32-35 (1991).

CHAPTER 11

FRYING FATS

Although the frying of foods is an important application of edible fats, it is one of the most complex and difficult processes to understand because of the multitude of reactions taking place and the complexity of the products formed. During frying, the water released from the food into the heated oil produces steam that causes hydrolysis of the fat. On the one hand, the resulting free fatty acids are more rapidly oxidized and promote thermal oxidation by solubilizing metal catalysts. On the other hand, the steam blankets the surface of the frying oil and moderates the rate of lipid oxidation by reducing the availability of oxygen from the air. The steam stripping of volatile decomposition products also delays deterioration of the frying fat.

During frying, foods absorb oil in amounts varying from 5 to 40% by weight, and release some of their lipids into the frying fat. The composition and stability of the fat can thus be changed by the foods during frying. After frying chicken or fish, for example, the frying fat will contain chicken or fish lipids. An important step in commercial frying is the continuous addition of fresh oil to attain a steady state by replacing the oil absorbed by the food. The presence of food particles (*e.g.* breading and meat particles) will accelerate the thermal darkening process. The reactions of frying fats with food protein and carbohydrate components contribute to both desirable and undesirable flavors. The formation of desirable flavors, odors and texture is necessary for the fried foods to acquire their unique sensory characteristics. However, extensive thermal oxidative decomposition of fats during prolonged frying will not only reduce the sensory quality of the fried foods but also may diminish their nutritional value. Good frying practice is aimed, therefore, at carefully controlling the temperature, time and replacement oil during frying to optimize desirable sensory qualities of the fried foods, and to minimize the accumulation of thermal deterioration products causing off-flavors and potential safety problems.

A. CHEMISTRY OF FRYING

The oxidation of unsaturated fats is not only greatly accelerated at high temperatures, but the free radical mechanism is changed by the decrease in oxygen concentration in heated fats. At elevated temperatures, the oxygen availability is lower and becomes limiting (see Chapter 6.E). The alkyl radicals, formed by initiation (1), become more important because the rate of the oxygenation reaction (2) is diminished. The termination reactions (5-10) also become more important because the alkyl radicals condense into stable products of higher molecular weights.

Initiation: $\quad\quad LH \longrightarrow L^{\cdot}$ (1)

Propagation: $\quad L^{\cdot} + O_2 \longrightarrow LOO^{\cdot}$ (2)

$\quad\quad\quad\quad LH + LOO^{\cdot} \longrightarrow L^{\cdot} + LOOH$ (3)

$\quad\quad\quad\quad LOOH \longrightarrow LO^{\cdot}$ (4)

Termination: $\quad 2\, L^{\cdot} \longrightarrow L-L$ (5)

$\quad\quad\quad\quad LO^{\cdot} + L^{\cdot} \longrightarrow LOL$ (6)

$\quad\quad\quad\quad LOO^{\cdot} + L^{\cdot} \longrightarrow LOOL \longrightarrow LO^{\cdot} \longrightarrow L-L + LOL$ (7-10)

At temperatures above 100°C, the initial hydroperoxides formed by reaction (3) exist only transiently and decompose rapidly into a multitude of volatile and nonvolatile products. Two mechanisms have been postulated for the thermal oxidation of unsaturated fats: (a) *thermal decomposition* by direct interactions of radicals, when unsaturated fats are continuously heated at elevated temperatures, and (b) *induced decomposition* through the intermediacy of hydroperoxides, when unsaturated fats are subjected to intermittent heating. Under these conditions, hydroperoxides accumulating at the lower temperatures contribute more radicals by decomposition when the fats are reheated. It is generally assumed that intermittent heating of unsaturated fats is more destructive than continuous heating. However, the evidence for these mechanisms is far from clear due to the complexity of reactions products formed.

The main changes occurring during frying include three major reactions: oxidation, polymerization and hydrolysis. Lipid oxidation at the elevated temperatures of frying (180-200°C), in the presence of moisture generated from the food, produces a multiplicity of compounds that contribute to desirable and undesirable changes affecting the flavor and quality of foods. During the frying of foods additional products may also be derived from interactions between lipid oxidation compounds and food components (proteins and carbohydrates). Thermal lipid oxidation and hydrolysis produce complex mixtures of volatile, and non-volatile monomeric (cyclic and non-cyclic) and polymeric (polar and non-polar) substances.

1. Volatile Compounds

During frying a large mixture of volatile substances is produced by rapid

TABLE 11-1.
Volatile compounds identified in frying vegetable oils and in french fries.[a,b]

Compounds found in frying oils	Relative amounts [c]	Compounds found in french fries [b]
1-penten-3-ol	++	hexanal
pentanal	++	methyl pyrazine
heptane	++	2-hexenal
1-pentanol	+++	2-heptanone
hexanal	+++	nonane
octane	++	heptanal
furfural	++	2,5-dimethylpyrazine
trans-2-hexenal	++	2-heptenal
furfuryl alcohol	+++	2-pentylfuran
heptanal	++	octanal
2-acetylfuran	++	trans,trans-2,4-heptadienal
trans-2-heptenal	+++	2-octenal
5-methylfurfural	+++	nonanal
1-octen-3-ol	+++	2-nonenal
octanal	+++	decanal
2-pentylfuran	+++	2-decenal
trans-2-octenal	+++	trans,cis-2,4-decadienal
nonanal	+++	undecanal
trans-2-nonenal	+++	trans,trans-2,4-decadienal
4-oxo-nonanal	++	
trans,trans-2,4-decadienal	++	
2-undecenal	++	
hexadecanoic acid	+++	

[a] From Takeoka et al. (1996). Soybean and cottonseed oils used to fry meat products.
Analyzed by capillary gas chromatography of fractions isolated by simultaneous steam distillation-extraction.
[b] From Perkins (1996).
[c] ++ = 0.5-2%, +++ = >2%.

decomposition of hydroperoxides and polyunsaturated aldehydes. These volatile decomposition products are found in relatively small amounts because a large portion is removed from the oil by steam distillation and the sweeping action of steam generated during frying. The remaining volatile compounds are of concern because they are partially absorbed by the fried foods and contribute to their flavor, and to the odor of the room where frying is carried out, also referred to as *room odor*. Volatile compounds produced by thermal oxidation include aldehydes, ketones, alcohols, acids, esters, hydrocarbons, lactones, substituted furans, and aromatic compounds. Gas chromatographic analyses of fat samples after different frying treatments represent mainly the more stable volatile compounds remaining in the fats (Table 11-1). The major volatile constituents identified in frying oils are derived from the decomposition of lipid oxidation products and include 1-pentanol, hexanal, furfuryl alcohol, *trans*-2-heptenal, 5-methylfurfural, 1-octen-3-ol, octanal, 2-pentylfuran, *trans*-2-nonenal and hexadecanoic acid.

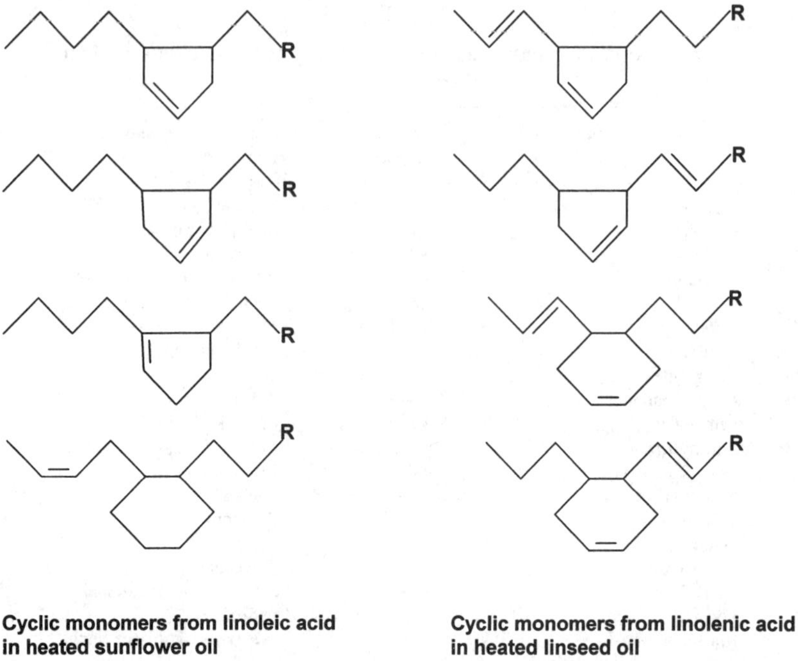

Cyclic monomers from linoleic acid in heated sunflower oil

Cyclic monomers from linolenic acid in heated linseed oil

Figure 11-1. Structures of some cyclic monomers from linoleic and linolenic acids in heated oils. Adapted from Le Quéré and Sébédio (1996).

Additional minor volatile compounds are found in both frying oils and in fried foods, including cis,trans- and trans,trans-2,4-decadienal derived from oxidized linoleate and 2,4-heptadienal derived from linolenate. The isomers of 2,4-decadienal impart a desirable "fried flavor" in fried potatoes when present in small amounts, but excessive amounts of this aldehyde would be expected to cause undesirable rancid flavors. Furfural compounds may be derived from interactions between food sugars and proteins. Minor amounts of sulfur compounds and nitrogen-containing heterocyclic compounds (methyl pyrazine and 2,5-dimethylpyrazine in potatoes) may originate from Maillard reactions (Chapter 10).

2. Monomeric Compounds

Thermal oxidation of unsaturated fats is accompanied by considerable isomerization of double bonds leading to products containing *trans* double bonds and conjugated double bond systems. Non-volatile decomposition products are formed by thermal oxidation, hydrolysis and cyclization. A large number of hydroxy, alkoxy-substituted, epoxy and keto compounds are produced by oxidation, free fatty acids and mono- and diacylglycerols by hydrolysis, and cyclic monomers by cyclization of polyunsaturated fatty acids in heated fats. Aldehydo glycerides and keto glycerides are also produced when triacylglycerol hydroperoxides undergo cleavage on the glyceride end of the

TABLE 11-2.
Gas chromatographic analysis of heated soybean oil.[a]

Oils	Frying at 195°C [b]	Cyclic monomers	Polar compounds	Noneluted material
soybean oil	52 hr. intermittent	0.45	0.41	18.4
hydrogenated soy. [c]	52 hr. intermittent	0.33	0.41	18.3
hydrogenated soy. [c]	104 hr. continuous	0.57	0.91	20.7

[a] Meltzer et al. (1981). Gas chromatographic analyses carried out on completely hydrogenated samples after concentration by low-temperature crystallization.
[b] Oils were heated intermittently or continuously and potato slices were fried at different intervals.
[c] Hydrogenated soybean oil to an iodine value of 107.

chain. Free fatty acids are particularly damaging to the stability of the frying medium and the fried foods because they are rapidly oxidized; they catalyse further oxidation of polyunsaturated fats by solubilization and activation of metal catalysts, and by their surface activity, they may increase their contact with oxidizable components of foods and promote lipid-protein interactions.

Cyclic monomers have received special attention because they are considered nutritionally harmful if they are present in sufficient quantities. However, the potential harmful effects of heated fats have been controversial because many nutritional studies have been carried out under exaggerated conditions that do not represent actual home or commercial applications (Section F). A large number of cyclic fatty acids have been identified in heated vegetable oils containing linoleic and linolenic acids. Main structures include disubstituted five-membered (cyclopentyl) and six-membered (cyclohexyl) compounds with unsaturation inside and outside the rings (Figure 11-1). About 0.3 to 0.6% cyclic monomers were found in partially hydrogenated soybean oil after prolonged frying of potato slices (Table 11-2). Polar materials varying from 0.4 to 0.9% included oxygenated monomeric products. Non-eluted products ranging from 18 to 21% represented high molecular-weight polymeric compounds. Surveys of commercial frying oils in the U.S. showed a range of 0.1 to 0.5% cyclic monomers and 1 to 8% polar and non-eluted materials. Although the structures of cyclic monomeric compounds in frying oils have been well defined, there is evidence for a large number of oxygenated monomeric compounds that have not been completely characterized because of their complex composition.

3. Polymeric Compounds

The predominant group of non-volatile compounds formed during frying of unsaturated fats includes dimers and oligomers. These high-molecular weight compounds are mostly formed in the termination stages of free radical oxidation (Chapter 1). Non-polar dimers are formed by addition of alkyl radicals and linked by carbon-carbon bonds. According to the type of fatty

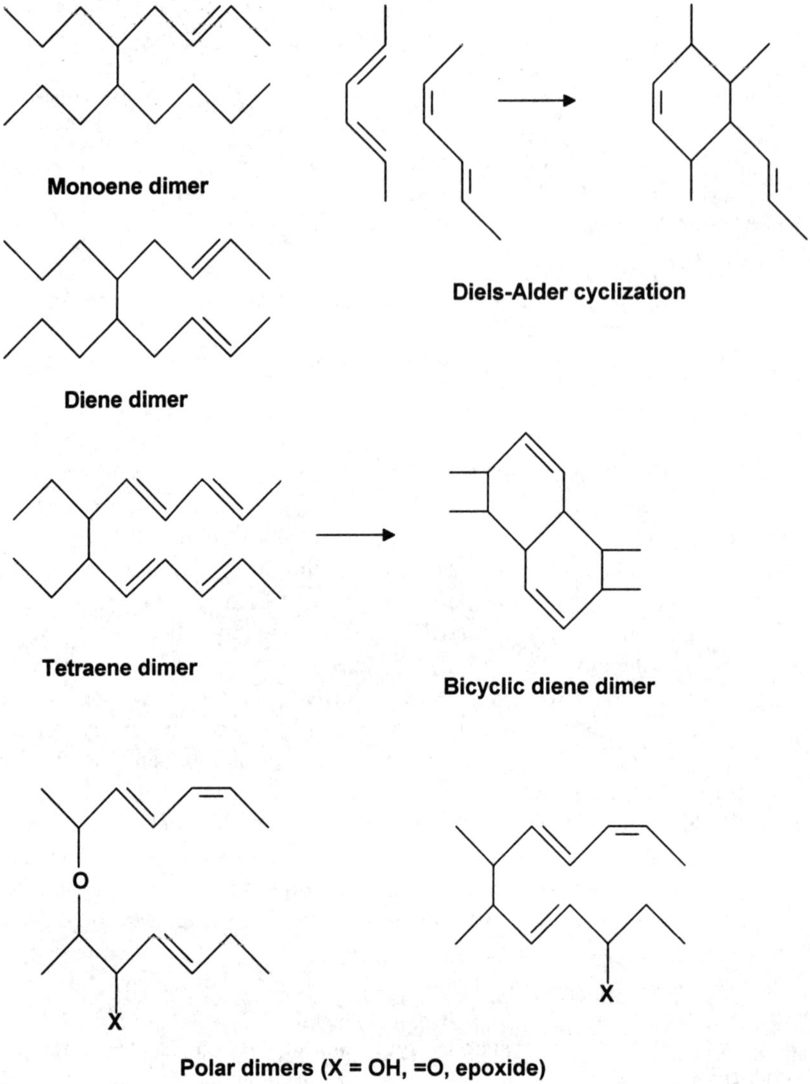

Monoene dimer

Diels-Alder cyclization

Diene dimer

Tetraene dimer

Bicyclic diene dimer

Polar dimers (X = OH, =O, epoxide)

Figure 11-2. Structures of different dimers formed in heated oils.

acid precursors these carbon-carbon dimers include monoene, diene and tetraene structures (Figure 11-2). Polar dimers are oxygenated and form either by combining alkyl and alkoxyl radicals and linked by ether bonds (C-O-C), or by combining radicals containing oxygenated functions (hydroxy, keto, epoxy). Polar dimers linked by peroxy bonds (C-OO-C) are only formed at low temperatures and decompose at elevated temperatures (above 100°C). At these temperatures, peroxy-linked dimers may be formed as intermediates and produce either ether-linked or other oxygenated polar dimers.

$$R-CH=CH^{\cdot} + {}^{\cdot}R \longrightarrow R-CH=CH-R \quad \text{non-polar dimers}$$

$$R-CH=CH^{\cdot} + {}^{\cdot}OR \longrightarrow R-CH=CH-O-R \quad \text{polar ether dimers}$$

$$R-\underset{\underset{OH}{|}}{CH}-HC^{\cdot} + {}^{\cdot}R \longrightarrow R-\underset{\underset{OH}{|}}{CH}-HC-R \quad \text{polar oxygenated dimers}$$

Cyclic dimers are also formed by intramolecular addition of an intermediate dimeric radical to the double bond in the same molecule.

$$\begin{array}{c} -CH-CH=CH- \\ | \\ -CH_2-\overset{\cdot}{C}H-CH- \end{array} \longrightarrow \begin{array}{c} -CH-\overset{\cdot}{C}H-CH-CH \\ | \quad \quad \diagup \\ -CH_2-CH-CH- \end{array}$$

Non-radical polymerization can also take place at elevated temperatures by Diels-Alder reactions between either two conjugated diene fatty acids or between a conjugated and a non-conjugated diene fatty acid, to form non-polar substituted cyclohexene unsaturated dimers (Figure 11-2). These types of dimers are found in only small quantities because the conjugated diene precursors are formed in minor concentrations. More complex structures have been identified in oils used after frying, including unsaturated bicyclic compounds with conjugated and non-conjugated double bonds, trimers, acyclic polar dimers, and tetrahydrofuran tetrasubstituted dimers. These polar and non-polar dimeric and oligomeric compounds are not completely characterized because of their complex composition.

B. FATS USED FOR FRYING

A wide range of fats is used commercially, varying from unhydrogenated refined fats to blends of partially and "fully" hydrogenated fats. The selection is based on the fatty acid composition, the commercial application, and the ultimate performance in various fried foods. The main parameters for the selection of frying fats include the turnover rate of the fat (removal and replenishment during frying), the amount of fat absorbed by the food, and the shelf-life of the fried foods. The fats used for foods consumed immediately or shortly after frying (*e.g.* restaurants) have different formulations ("light-duty") from fats used for foods consumed after storage (*e.g.* potato chips). Generally, the frying performance and stability of oils are greatly dependent on the amounts of total oxidizable polyunsaturated fatty acids. For repeated frying, to minimize the formation of undesirable rancid and fishy flavors as well as cyclic monomers, the level of linolenate must be below 2 to 3%. For this reason, soybean and canola oils are generally hydrogenated to reduce their linolenate content. If hydrogenation is used extensively to reduce the linolenate content below 1%, the oils acquire undesirable "hydrogenation" off-flavors

TABLE 11-3.
Properties of commercial fats used for frying.[a]

Characteristics	RBD soybean oil	All purpose shortening	Opaque shortening	Heavy duty shortening
Appearance	clear, liquid	plastic, solid	plastic, pourable	plastic, solid
m.p. °C	liquid	43-48	33-37	40-43
AOM, hr	10-25	40+	35+	200+
Polyunsaturates	34-61	15-20	35-40	4-8
Saturates	13-27	20-30	15-20	22-25
Iodine value	130-135	80-85	102-108	65-75

[a] From Erickson (1996). RBD = refined, bleached and deodorized; AOM = active oxygen method.

(see Chapter 4), and high *trans* unsaturation. Although the nutritional hazards of *trans* isomers in edible fats are controversial, current efforts in industry are aimed at minimizing the amounts of these isomers in our food supply.

Frying fats can be formulated by blending partially hydrogenated vegetable oils (soybean, corn, cottonseed and canola) of varying iodine values (64 to 107) with almost completely hydrogenated fats (of iodine value of 1-8) referred to as *hardstock*. A fluid light-duty shortening may contain 98% partially hydrogenated vegetable oils (of iodine value 104-106), and 2% hardstock. A heavy-duty frying fat may be formulated to contain 97% partially hydrogenated vegetable oils (of iodine value 80-82), and 3% hardstock. Standard specifications for various commercial frying fats are based on melting points, AOM stability, polyunsaturated and saturated fat contents, and iodine values (Table 11-3). The fluid light-duty shortenings contain appreciably more polyunsaturated fatty acids (35-40%) than the heavy-duty shortenings (4-8%).

Plant breeders have developed new genetically modified varieties of soybeans and canola seeds that produce oils with reduced linolenate contents. The new oils containing less than 3% linolenate have improved flavor and oxidative stability and can be used successfully for frying without hydrogenation. A high-oleic sunflower oil was also recently developed that is particularly suitable for frying. This oil can be used either directly or after blending with the more readily available polyunsaturated soybean and canola oils to reduce the linolenate content below 2 or 3% and obviate hydrogenation (Chapter 7, Section D).

C. METHODS TO ASSESS FRYING DETERIORATION

Several methods are used to determine compounds from thermal oxidation that cause significant changes in the physical, chemical and nutritional properties of frying fats. Gross changes resulting from frying include increase in viscosity and density, dark color development, tendency to foam and decrease in unsaturation. Fats used for frying show an increase in free fatty acids, a decrease in iodine value, small increases in peroxide value and conjugated diene absorption at 231 nm. The formation of polymers is one of the most important changes in fats during frying. Polymeric materials constitute more than 50% of the compounds formed by thermal deterioration due to frying.

A number of routine methods have been used to evaluate the extent of oxidative and thermal damage to frying fats. These methods can be divided into those that can be used later on oil samples collected sometime after frying, those employed immediately after frying for quality control, and those applied to fried foods.

1. Post-Frying Methods

The degrees of alteration caused by frying foods are commonly estimated by determining *total polar materials* by simple column chromatography, and polymers by gel-permeation chromatography, or size-exclusion chromatography. The amounts of *free fatty acids* produced during frying are generally too small to use as a basis to monitor the quality of food. More sensitive and quantitative techniques include column adsorption and partition chromatography and size exclusion chromatography to determine directly polar and non-polar monomeric and dimeric compounds. Gas-liquid chromatography is used to estimate indirectly the *polymeric polar materials* that do not elute ("non-eluted" materials) under standard conditions. These chromatographic techniques provide only semi-quantitative information owing to the complexity of the mixture of oxidized polar and non-polar fatty acid components formed under frying conditions.

The routine and well-recognized determination of total polar materials in heated fats is carried out by eluting the sample through a silica column (adsorption chromatography) and collecting first the non-polar or unchanged triacylglycerols with a mixture of 87% petroleum ether:13% diethyl ether (by volume), followed by the polar materials with pure diethyl ether. The weight of the polar material fraction, isolated after evaporating the solvents, is a coarse measure of thermal deterioration. In practice, the first and larger fraction of unchanged triacylglycerols is isolated and weighed and the polar material fraction is calculated by difference. On the basis of analyses of a large number of used frying fats, heated fats are acceptable until they reach a level of 25% polar materials. The formation of 25-30% polar materials in frying fats is recognized as an objectionable level of deterioration. The determination of total polar materials has been adopted as a standard reference method in several Europeans countries to test for excessive use of fats for frying. However, the complex polar materials in frying fats have not been well characterized. This determination also does not measure non-polar materials that are known to be formed in frying fats.

Different frying operations produced varying amounts of polar material (Figure 11-3). *Pan frying* with a shallow layer of fat forms the smallest amount of polar material (about 10%), because it proceeds quickly and usually without

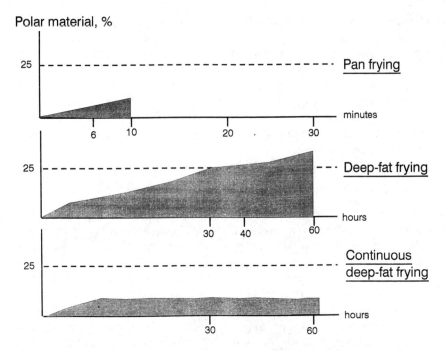

Figure 11-3. Effect of different treatments on development of polar materials during frying. Adapted from Billek (1983).

repeated use of the fat. *Deep-fat frying* in a household or restaurant produces an objectionable level of 30% polar materials after 40 hours, because the fat is used for intermittent frying and with slow turnover. *Commercial deep-fat frying* is carried out with a fast turnover by continuously replacing the fat used by food frying with fresh fat. By this operation, the fat approaches a steady and constant level of polar material that reached about 10% after a few hours of continuous frying.

2. Rapid Control Methods

These methods are employed during or immediately after frying for either quality control or for an inspection purpose. Change in the *dielectric constant* has been used to monitor frying by a rapid instrument called the "food oil sensor." The dielectric constant increased with frying time much more in soybean oil than in hydrogenated vegetable oil, and a hydrogenated animal-vegetable oil blend (Table 11-4). The changes in dielectric constant correlated with other commonly used tests to assess frying fats, including free fatty acids, peroxide values, conjugated dienes, polar material, foam heights and decrease in iodine value. The highest correlation with an increase in polar material indicated that the changes in dielectric constant were largely caused by oxidation products. Peroxide values and conjugated dienes are not reliable tests

TABLE 11-4.
Analytical tests of various oils during frying of potatoes.[a]

Frying at 190°C, hr	Instrument reading [b]	FFA %	Peroxide Value	Conjug. dienes	Polar material,%	Foam height, mm	Iodine value
			Soybean oil [c]				
0	1.0	0.02	1.3	0.40	2.8	40	122.8
16	4.1	0.18	21.9	0.67	14.1	50	121.0
24	5.3	0.27	21.3	0.67	17.1	80	118.3
32	7.0	0.30	22.9	0.69	21.2	80	—
Change	6.0	0.28	21.6	0.29	18.4	40	—
			Hydrogenated vegetable oil [d]				
0	1.0	0.02	0.9	0.11	3.7	30	70.2
16	1.6	0.17	2.4	0.18	5.0	35	69.3
24	1.8	0.28	3.4	0.19	5.5	40	69.7
32	2.0	0.38	3.9	0.22	7.6	25	—
Change	1.0	0.36	3.0	0.11	3.9	-5	—
			Hydrogenated animal-vegetable oil blend [e]				
0	1.0	0.05	0.9	0.39	3.0	45	49.3
16	2.8	0.32	11.5	0.49	11.0	30	46.4
24	3.7	0.50	12.5	0.48	13.4	40	46.3
32	4.8	0.78	14.5	0.46	16.5	30	—
Change	3.8	0.73	13.6	0.07	13.5	-15	—

[a] From Fritsch et al. (1979). FFA = free fatty acids as C18:1; Polar material was determined on a nylon chromatography column packed with neutral alumina, eluting neutral oil with toluene and hexane, cutting top section with razor and eluting polar material with chloroform/ methanol mixture; Foam height was read on the surface of a graduated cylinder cutoff at the bottom and inserted into the hot oil until it almost touched the bottom of the fryer.
[b] Food oil sensor for detecting changes in dielectric constants.
[c] Refined, bleached, deodorized soybean oil: iodine value: 123, initial peroxide value: 1.3, AOM: 10 hr, saturates: 17.1%, 18:1: 23.1%, 18:2: 52.2%, 18:3: 6.9%.
[d] Commercial frying shortening containing dimethyl silicone: iodine value: 70.2, initial peroxide value: 0.9, AOM: over 200 hr, saturates: 26.2%, 18:1: 70.6%, 18:2: 2.5%, 18:3: 0.6%.
[e] Commercial animal-vegetable shortening containing 10 ppm BHA and 30 ppm BHT: iodine value: 49.3, initial peroxide value: 0.9, AOM: 65 hr, saturates: 50.7%, 18:1: 17.6%, 18:2: 4.0%.

because the hydroperoxide precursors are largely decomposed during frying, and these measurements reflect mainly further oxidation taking place after sampling and cooling for analyses. Determinations of free fatty acids, foam heights and iodine value are the least sensitive and reliable measures of frying deterioration.

The procedure for dielectric constant has several limitations. The initial dielectric constant varies among different fats, and changes with different food

TABLE 11-5.
Effect of hydrogenation, methyl silicone and antioxidants on flavor and volatile formation in fried bread cubes after storage.[a,b]

Oil samples + citric acid [c]	Flavor quality [a]		Volatiles peak area [b]
	0-time	Stored	Stored
soybean oil	5.5	2.7	246.6
soybean oil + MS	4.8	3.8	6.2
soybean oil + TBHQ	5.4	2.0	146.8
soybean oil + TBHQ + MS	4.8	4.3	3.5
hydrogenated (4.6)	6.5	3.8	127.0
hydrogenated (4.6) + MS	6.6	4.8	4.8
hydrogenated (4.6) + TBHQ	6.8	2.0	76.1
hydrogenated (4.6) + TBHQ + MS	6.2	5.1	1.3
hydrogenated (2.7)	6.6	3.0	130.5
hydrogenated (2.7) + MS	7.0	6.4	1.9
hydrogenated (2.7) + TBHQ	6.6	1.9	56.3
hydrogenated (2.7) + TBHQ + MS	6.5	5.0	0.4
hydrogenated (0.4)	5.9	2.8	31.7
hydrogenated (0.4) + MS	5.9	4.5	1.7
hydrogenated (0.4) + TBHQ	6.4	2.0	23.2
hydrogenated (0.4) + TBHQ + MS	5.5	5.2	0.2

[a] From Frankel et al. (1985).
Treatment: 2.5 hr frying bread cubes (1 inch) + 14 hr heating at 190°C. Storage at 60°C for 4 days.
Flavor quality scale: 1-10, 10 = excellent, 1 = extremely poor.
[b] From Snyder et al. (1986). By static headspace gas chromatography, peak area x 10^{-3}
[c] citric acid, 100 ppm, MS = methyl silicone, 5 ppm, TBHQ = tertbutyl hydroquinone, 200 ppm; hydrogenated oils (%18:3).

fats extracted into the frying fat. The fresh oil must be available for comparison with the heated sample. Caution is necessary in interpreting dielectric constant changes in heated fats because these measurements represent the net balance between polar and non-polar compounds, both of which are produced during frying. Although the polar compounds predominate, the relationship between both types of products and oil quality is confounded by several factors including moisture, mono- and diglycerides, and free fatty acids. Several other commercial colorimetric kits have been developed to rapidly test the heat abuse of frying fats. These tests are based on redox indicators, carbonyl compounds, free fatty acids, and alkaline materials formed by interaction of fried foods and deteriorated oils.

3. Methods Applied to Fried Foods

To establish more reliable quality criteria for consumer acceptability the fried foods must be tested rather than the frying fats. More sensitive and

TABLE 11-6.
Analyses of frying cottonseed oil (CSO) and high-oleic sunflower oils (HOSUN) and their mixtures, and sensory evaluations of potato chips after frying and storage.[a]

Analyses	CSO	2:1 CSO/HOSUN	1:2 CSO/HOSUN	HOSUN
Fatty acids, %				
saturates	26.4	19.2	11.7	8.0
18:1	16.4	42.9	67.5	78.0
18:2	54.9	35.7	18.7	12.1
Polar compounds, %				
0-time	5.6	4.5	3.1	2.9
12 hours	8.9	7.1	6.5	4.8
Flavor quality				
0-time	7.7	7.6	7.0	6.6
6 months	5.0	5.7	5.4	5.1
Fried flavor intensity				
0-time	5.9	5.6	5.0	4.8
6 months	3.4	4.5	3.9	3.2
Rancid flavor intensity				
6 months	1.3	0.9	0.9	0.9
Volatiles (6 months)				
hexanal	38.7	21.8	15.6	2.3
pentanal	17.5	10.9	9.4	1.3
octanal	1.2	0.6	0.5	0.0
2,4-decadienal	62.5	36.5	20.2	6.7

[a] From Warner *et al.* (1997). Oils used to fry potato chips for 12 hours at 187-192°C.
Fatty acid composition on fresh oils; polar compounds on oils after frying
Flavor quality scores: 1 = bad, 10 = excellent; fried and rancid flavor intensity scores: 0 = none, 10 = strong.
Potato chips were tasted fresh at 0 time and after storage for 6 month at 25°C.
Potato chips were analyzed directly for volatiles by static headspace gas chromatography after storage for 6 month at 25°C.

reliable measures of frying performance of fats are based on sensory evaluations and gas chromatographic analyses of volatiles in fried foods carried out initially and after storage. Bread cubes proved to be a suitable model food because they can be prepared in uniform size and moisture content. After frying under standard conditions, bread cubes could be stored and evaluated by sensory techniques and by static headspace gas chromatography for volatiles. Although the flavor quality of bread cubes after initial frying was not different between hydrogenated oils and unhydrogenated oils containing citric acid, the flavor quality of bread cubes used repeatedly for frying decreased significantly after accelerated storage at 60°C (Table 11-5). Initially, the bread cubes fried in hydrogenated oils had better flavor quality than those fried in unhydrogenated soybean oil. However, after storage, the flavor quality scores decreased significantly with bread cubes fried in all oils containing citric acid and tert-butyl hydroquinone (TBHQ). However, those oils containing citric acid and methyl silicone or methyl silicone and TBHQ improved the stability of the fried bread cubes after storage. These results showed that if bread cubes are fried in repeatedly used oils, they can also be consumed immediately after frying. However, these bread cubes cannot be stored without marked deterioration of quality. The use of bread cubes as a food model to test frying fats was later confirmed with fried potato chips.

Gas chromatographic analyses of volatiles showed that bread cubes fried in all hydrogenated oils produced less volatiles than non-hydrogenated control oil. Although TBHQ reduced volatile formation in all oils, methyl silicone had a much greater effect. The mixture of TBHQ and methyl silicone further decreased the total volatile in bread cubes after storage. Therefore, gas chromatographic analysis of volatile showed more significant improvement by hydrogenation and antioxidants than sensory evaluations. Volatile analyses need to be complemented therefore by sensory evaluations.

When potato chips were fried in cottonseed oil, high-oleic sunflower oil, and their mixtures, the total polar compounds in the oils after frying decreased with increasing content of oleate (Table 11-6). However, the potato chips had better flavor quality and "fried flavor" intensity scores when evaluated fresh after frying in cottonseed oil, than chips fried in either high-oleic sunflower oil or in mixtures of these oils. After storage for 6 months at 25°C, potato chips fried in the oil blend containing 36% linoleate had the best flavor quality and fried flavor. However, potato chips fried in cottonseed oil had a higher intensity score for rancid flavor than in high oleic sunflower oil and all the mixtures with cottonseed oil. As expected, gas chromatographic analyses showed decreases in volatile formation with decreasing content of linoleate. Analyses of volatile compounds agreed with the total polar compounds in showing greater frying oil stability and oxidative stability of the fried chips. The higher flavor quality and fried flavor scores in the cottonseed oil and blends containing higher levels of linoleate were attributed to increased levels of 2,4-decadienal, a decomposition product of oxidized linoleate. However, 2,4-decadienal at high levels also contributes to rancidity. The relative concentrations of 2,4-decadienal necessary to optimize a desirable fried food flavor, without also producing rancidity, are not known and are difficult to control.

D. CONTROL MEASURES

The degradation of fats during frying varies depending on the properties of fats, the conditions and wide variations possible in the handling and processing of fats. The extent of fat deterioration reflects complex interactions among many factors: degree of water volatilization and resulting air agitation, extent of protection by steam blanketing, level of hydrolysis depending on the type of food and temperature, the surface to volume ratio of the fat, the uniformity of heating in the frying equipment. The nature and oxidative stability of fats used for frying were discussed in Section B, as one of the critical factors controlling the quality of fried foods. Other important factors are discussed below.

1. Temperature

The temperature of the oil is controlled to allow for moisture bubbles to evolve after immersing the food until it is removed. Depending on the foods, for satisfactory results, frying temperatures range from 160 to 190°C. Continuous evolution of steam from the food is essential to minimize the penetration of oil into the surface of the food. During frying fat, deterioration is delayed by the removal of volatile decomposition products by the stripping action of the steam. Fats decompose rapidly if they are held at frying temperatures without foods. This deterioration can be delayed by reducing the temperature of the fat when it is not used for frying.

2. Turnover Rate

This refers to the amount of fat added to the kettle to make up for the amount removed by the food during frying. The turnover rate is an important factor for the control of frying oil condition. Daily replacement of the oil absorbed by food is estimated as the proportion of the total oil capacity of the fryer. The turnover rate is a function of the absorption of fat by the frying food. The absorption of fat by the food during frying increases with the increase in viscosity. For any given rate of frying the rate of fat turnover will increase as the viscosity increases, and the fat is thus protected by this effect. The fat reaches a steady state condition, through the removal and replenishment processes, and does not deteriorate further with continuous frying. Addition of fresh fat to used fat in good condition maintains the ability of the fat to contribute desirable sensory attributes to fried foods. A rapid turnover rate is important to minimize thermal deterioration of fats and to maintain the quality of fried foods. If a low turnover is necessary when an insufficient amount of food is fried daily, then a certain amount of fat must be discarded to increase the amount of make-up fat. The used oil must be replaced as needed to maintain good quality of fried foods.

3. Filtration

Daily filtration of frying fat to remove accumulated food particles, charred batter and breading is important to reduce deterioration, excessive color formation and development of undesirable bitter flavors and odors. Systems used for filtration include metal screens, paper filters and plastic cloths. The use of diatomaceous earth or filter aid is effective to reduce free fatty acids and color compounds.

TABLE 11-7.
Loss of tocopherols in heated fats and fats treated under simulated frying.[a,b,c]

Oils	Oxidation at 100°C (AOM)[a]	Simulated frying[b]
	initial rate ppm/hr	loss %
Soybean oil	4.9	19.8
Soybean oil + CA[c]	3.6	—
Hydrogenated soybean oil + CA[c]	12.1	—
Safflower oil	3.9	8.8
Corn oil	21	46.5
Cottonseed oil	67	25.3
Lard + α-tocopherol[d]	180	—
Palm oil	—	67.4
Coconut oil	—	100

[a] From Frankel et al. (1959).
[b] From Yuki and Ishikawa (1976).
[c] From Evans et al. (1959); CA = citric acid, 0.01%; hydrogenated to iodine value of 101.7.
[d] 1500 ppm

4. Antioxidants

Synthetic antioxidants such as BHA, BHT, propyl gallate and TBHQ (Chapter 8) are sometimes included in commercial shortenings even though they are volatilized and decomposed at the temperatures of frying. The depletion rates in soybean oil heated under frying conditions was about the same for TBHQ as for BHA and BHT. These antioxidants are partially retained during frying and stabilize the fried foods. There is evidence that shelf-life of potato chips is extended by the use of shortenings containing synthetic antioxidants such as TBHQ and BHT. BHA has the disadvantage of producing an objectionable phenolic odor.

The effect of natural tocopherols and of added tocopherols to frying fats is controversial. Natural occurring tocopherols in vegetable oils are generally not effective as antioxidants at high temperatures and may even have prooxidant activity at high concentrations (Chapter 8. A.1). Tocopherols are lost during frying in varying degrees depending on the fats without apparently affecting the deterioration rate of the oils. During oxidation at 100°C under AOM conditions, tocopherols were depleted more slowly in polyunsaturated than more saturated fats and after partial hydrogenation (Table 11-7). This trend was explained by the more rapid decomposition of polyunsaturated hydroperoxides before they react with the tocopherols. During simulated frying conditions by heating at 180°C with continuous water spraying, natural tocopherols were more slowly decomposed in polyunsaturated soybean, cottonseed and safflower oils than in the more saturated palm and coconut oils (Table 11-7).

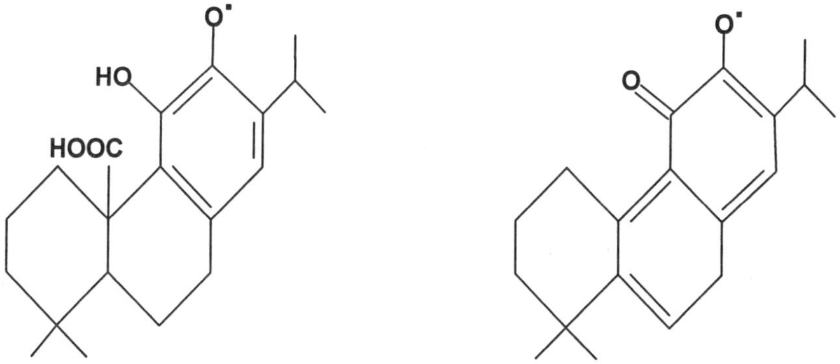

Figure 11-4. Structures of quinone radicals suggested by electron spin resonance studies of carnosic acid heated with oxidized methyl oleate. From Löliger et al. (1996).

Natural tocopherols, which are present at relatively high concentrations in soybean oil (Table 8-4), can produce undesirable red colored tocoquinones during prolonged frying. Thus, the observation that the oxidative stability of heated olive oil and high-oleic safflower oil is significantly increased by tocopherol addition may not necessarily apply under conditions of frying foods. The frying life of soybean oil can actually be improved on the basis of reduced dark color formation after deep-frying of various foods, by removing a significant portion of the tocopherols, by vacuum steam distillation or deodorization at elevated temperatures for short times. The fat produced by this process is suitable for foods that are consumed shortly after frying such as in restaurants. However, the decreased level of tocopherols would be expected to reduce the shelf-life of fried foods that are consumed after storage, such as packaged potato chips. Therefore, certain levels of tocopherols have a stabilizing effect on fried foods after storage, but would not be expected to increase the oxidative stability of fats during frying. For this reason, it is important to test the effectiveness of antioxidants on the oxidative stability of the fried foods and not that of the frying fats.

Rosemary extracts provide an important source of natural antioxidants used commercially in foods. These extracts are particularly active as antioxidants at the high temperatures in frying fats. They protect the oils (peanut and palm oils) during frying and their antioxidant activity is carried over into the fried foods (oriental noodles). The active components of rosemary, carnosic acid and carnosol, are readily decomposed during thermal oxidation into products that remain active as antioxidants in heated fats. Electron spin resonance studies of carnosic acid and carnosol in oxidized methyl oleate heated at 130-160°C suggested the formation of o-semi-quinone radicals (Figure 11-4), but the stable oxidation products of rosemary extracts that remain active as antioxidants in frying fats have not been identified.

5. Polymerization inhibitors

Silicones, such as methyl silicone or poly(dimethylsiloxane), are very effective additives for retarding thermal oxidation and deterioration by polymerization in frying fats. Silicones are only effective at frying temperatures and at very low concentrations (0.5-5 ppm). Methyl silicone improved significantly the flavor quality and reduced the formation of volatile compounds in the bread cubes fried repeatedly in soybean oil and hydrogenated soybean oils (Table 11-5). The use of mixtures of antioxidants with silicones shows synergistic effects in improving the shelf life of fried foods. The frying performance of partially hydrogenated soybean oil was improved by adding TBHQ with methyl silicone (Table 11-5). This result suggests a synergistic effect. Other synergistic mixtures were used to improve the quality of oils used for frying, including mixtures of dimethyl silicone, BHA and ascorbyl palmitate, methyl silicone, citric acid and TBHQ, phenolic antioxidants, silicone and esters of EDTA. The effectiveness of these mixtures may be explained by the stabilizing effect of silicone in retarding the depletion of phenolic antioxidants during frying.

Silicones increase the smoke point of frying oils apparently by providing a surface to air barrier. The mechanism for protection provided by silicones is not well understood. Silicones form a mono-molecular protective film at the air-oil interface that acts as an oxygen barrier. The addition of silicones to an oil inhibits foam formation by increasing the interfacial surface tension. There is also evidence that silicones slow down the degree of convection in fats during frying. Silicones cannot be used indiscriminately. They have disadvantages by causing failure in cake baking, poor performance in doughnut frying by defoaming the batter, and loss of desirable crispness in fried potato chips. The level of silicone is slowly depleted during frying by transfer to the surface of foods. The use of silicones is not recommended for potato chips because of their adverse effect on texture.

Plant sterols isolated from the unsaponifiable fraction of certain vegetable oils, and oat oil have been shown to retard oxidative polymerization of vegetable oils heated at frying temperature (180°C), on the basis of decrease in iodine value and polymer formation. The active sterols contain an ethylidene side chain (vernosterol, Δ^5-and Δ^7-avenasterol and fucosterol). However, no data are available on the effect of these sterols during actual frying of foods.

E. CRITERIA FOR USED FRYING FATS

One of the most difficult quality control problems in frying operations is to decide when a fat is so thermally abused that it should be discarded and replaced. Many reasons are used for discarding used frying fats: culinary criteria based on appearance, color, odor, and quality of fried foods, excessive foam and smoke, and routine schedules for fat replacement. When fats are heated under practical frying conditions, the extent of deterioration is generally

TABLE 11-8.
Selected animal studies testing the toxicity of heated fats.

Heat treatment	Polar material,%	Observations	References
Heat abused, without frying foods	> 50%	gastrointestinal irritations, diarrhea, growth retardation	Crampton et al. (1951) Chalmers et al. (1954)
Frying fats	10-20	no effect	Billek et al. (1983)
DNUA [a] from frying fats	—	growth retardation, diarrhea, toxic effects	Artman et al. (1972) Michael (1966) Nolen et al. (1967)
Polar fraction from frying fats	90%	growth retardation, slight liver damage	Billek et al. (1983)

[a] DNUA = distilled non-urea adducts.

small and limited by changes that make the fat unsuitable for culinary reasons. Samples of used fats from commercial frying operations showed levels of free fatty acids of 1-3%, polar triglycerides ranging from 5 to 30%, and were judged to have rancid or "heated" or "burnt-type" flavors.

Different countries have established different guidelines for frying deterioration or rejection points (also referred to as maximum allowable or "cut-off" level) above which thermally abused fats should be discarded. One German standard is based on 1% oxidized fatty acids, measured as the petroleum ether-insoluble fatty acid fraction. This endpoint is equivalent to 15% polymeric triglycerides and 27-28% total polar compounds. Other rejection points include polymeric and dimeric triglyceride contents (not exceeding 10%) or dimeric triglycerides (not exceeding 6%), acid values (2.0-2.5) and smoke point. (170-180°C). Only a few countries have established formal regulations to control the quality of frying fats. However, the methods for approving or rejecting frying oils do not work in practice because they are almost impossible to enforce or exercise.

F. HEALTH EFFECTS OF FRYING FATS

The biological and nutritional effects of frying fats have been a subject of extensive investigations over the last five decades. Early studies with animals fed heat-abused fats, containing sometimes more than 50% polar materials, showed significant growth retardation and other severe consequences (irritation of gastrointestinal tract, diarrhea) (Table 11-8). When fats that are used for normal frying of foods (containing 10-20% polar material) were fed to animals, no effects were observed even at high levels. Many studies were conducted with individual fractions separated from fats used for normal frying. The distilled non-urea adduct fraction, containing a high amount of cyclic monomers, was shown to be somewhat toxic in animal tests when used at concentrations much higher than the animal could consume in the whole fat.

The polar material separated chromatographically from fats used for normal frying caused small but significant reduction in growth and slight liver damage when the diets of animals contained 20% polar materials. The polymers present in frying fats as major components were regarded to be innocuous because they are not absorbed.

Most of the adverse effects observed in animal feeding experiments appear to be due to the use of fats abused by heating at high temperature without frying foods. Fats obtained under conditions of actual frying foods are subject to less heat damage, because of the moderating effect of steam generation from the food, and are considered to be not toxic. There is convincing evidence that frying fats prepared under these normal conditions are safe when consumed at reasonable levels unless they are severely abused. However, some compounds produced from polyunsaturated fats (*e.g.* cyclic monomers) under frying conditions can impair the nutritional value of frying fats, but the level and degree of toxicity are generally considered to be too low to be of any significance. The control of frying fats to prevent abusive conditions is, therefore, important to assure their safety.

The presence of sterol oxidation products (see Figure 12.8) in foods fried in animal, and various mixed animal/vegetable fats and vegetable oils have received attention recently because of their possible effects on human health. Sterol oxidation products were reported to be mutagenic, cytotoxic and carcinogenic, to inhibit cholesterol biosynthesis and to cause arterial damage. Analyses of commercial french fries showed levels of sterol epoxides and diols ranging from 20 to 200 $\mu g/g$ of lipids. However, no information is available on the level of toxicity of these compounds.

Hydrogenation of polyunsaturated vegetable oils has become less appealing because of recent evidence that *trans* isomers may have adverse nutritional effects. The linolenate content of vegetable oils such as soybean and canola oils can be easily reduced to less than 3% without hydrogenation, by blending with high-oleic sunflower or safflower oils (see Chapter 7, Section D.). These blends of all *cis* vegetable oils are more thermally stable and have improved frying performance. Increasing the level of thermally stable monounsaturated fats used for frying is also recommended in view of accumulating evidence that these fats provide other nutritional benefits, a subject that will be discussed in the next chapter.

BIBLIOGRAPHY

Artman,N.R. The chemical and biological properties of heated and oxidized fats. *Adv. Lipid Chem.* **7**, 245-330 (1969).

Artman,N.R. and Smith,D.E. Systematic isolation and identification of minor components in heated and unheated fats. *J. Am. Oil Chem. Soc.* **49**, 318-326 (1972).

Billek,G., Guhr,G. and Waibel,J. Quality assessment of used frying fats: A comparison of four methods. *J. Am. Oil Chem. Soc.* **55**, 728-733 (1978).

Billek,G. Lipid stability and deterioration, in *Dietary Fats and Health*, pp. 70-89 (1983) (edited by E.G. Perkins and W.J. Visek), American Oil Chemists' Society, Champaign, Illinois.

Boskou,D. Stability of frying oils, in *Frying of Food*, pp. 174-182 (1988) (edited by G. Varela, A.E. Bender and I.D. Morton), Ellis Horwood, Ltd., Chichester, England.

Chalmers,J.G. Heat transformation products of cottonseed oil. *Biochem. J.* **56**, 487-492 (1954).
Christopoulou,C.N. and Perkins,E.G. Isolation and characterization of dimers formed in used soybean oil. *J. Am. Oil Chem. Soc.* **66**, 1360-1370 (1989).
Crampton,E.W., Farmer,F.A. and Berryhill,F.M. The effect of heat treatment on the nutritional value of some vegetable oils. *J. Nutr.* **43**, 431-440 (1951).
Crampton,E.W., Common,R.H., Farmer,F.A., Berryhill,F.M. and Wiseblatt,L. Studies to determine the nature of the damage to the nutritive value of some vegetable oils from heat polymerization. I. The relation of autoxidation to decrease in the nutritive value of heated linseed oil. *J. Nutr.* **43**, 533-539 (1951).
Dobarganes,M.C. and Márquez-Ruiz,G. Dimeric and higher oligomeric triglycerides, in *Deep Frying*, pp. 89-111 (1996) (edited by E.G. Perkins and M.D. Erickson), American Oil Chemists' Society Press, Champaign, IL.
Dutta,P.C., Przybylski,R., Appelqvist,L-Å. and Eskin,N.A.M. Formation and analysis of oxidized sterols in frying fats, in *Deep Frying*, pp. 112-150 (1996) (edited by E.G. Perkins and M.D. Erickson), American Oil Chemists' Society Press, Champaign, IL.
Erickson,D.R. Production and composition of frying fats, in *Deep Frying*, pp. 4-28 (1996) (edited by E.G. Perkins and M.D. Erickson), American Oil Chemists' Society Press, Champaign, IL.
Evans,C.D., Frankel,E.N. and Cooney,P.M. Tocopherol oxidation in fats. Hydrogenated soybean oil. *J. Am. Oil Chem. Soc.* **36**,73-77 (1959).
Figge,K. Dimeric fatty acid[1-^{14}C]methyl esters. I. Mechanisms and products of thermal- oxidative reactions of unsaturated fatty acid esters - Literature review. *Chem. Phys. Lipids.* **6**, 159-177 (1971).
Firestone,D. Worlwide regulation of frying fats and oils. *Inform* **4**, 1366-1371 (1993).
Frankel,E.N., Evans,C.D. and Cooney,P. Tocopherol oxidation in fats. Natural Fats. *J. Agric. Food Chem.* **7**, 438-441 (1959).
Frankel,E.N., Smith,L.M., Hamblin,C.L., Creveling,R.K. and Clifford,A.J. Occurrence of cyclic fatty acid monomers in frying oils used for fast foods. *J. Am. Oil Chem. Soc.* **61**, 87-90 (1984).
Frankel,E.N., Warner,K. and Moulton,K.J. Effects of hydrogenation and additives on cooking oil performance of soybean oil. *J. Am. Oil Chem. Soc.* **62**, 1354-1358 (1985).
Frankel,E.N. and Huang,S-W. Improving the oxidative stability of polyunsaturated vegetable oils by blending with high-oleic sunflower oil. *J. Am. Oil Chem. Soc.* **71**, 255-259 (1994).
Freeman,I.P., Padley,F.B. and Sheppard,W.L. Use of silicones in frying oils. *J. Am. Oil Chem. Soc.* **50**, 101-103 (1973).
Fritsch,C.W., Egberg,D.C. and Magnuson,J.S. Changes in dielectric constant as a measure of frying oil deterioration. *J. Am. Oil Chem. Soc.* **56**, 746-750 (1979).
Fritsch,C.W. Measurements of frying fat deterioration: A brief review. *J. Am. Oil Chem. Soc.* **58**, 272-274 (1981).
Geoffroy,M., Lambelet,P. and Richert,P. Radical intermediates and antioxidants: An ESR study of radicals formed on carnosic acid in the present of oxidized lipids. *Free Rad. Res.* **21**, 247-258 (1994).
Gordon,M.H. and Magos, P. The effect of sterols on the oxidation of edible oils. *Food Chem.* **10**, 141-147 (1983).
Le Quéré, J-L. and Sébédio,J-L. Cyclic monomers of fatty acids, in *Deep Frying*, pp. 49-88 (1996) (edited by E.G. Perkins and M.D. Erickson), American Oil Chemists' Society Press, Champaign, IL.
Löliger,J., Lambelet,P., Aeschbach,R. and Prior,E.M. Natural antioxidants: From radical mechanisms to food stabilization, in *Food Lipids and Health*, pp. 315-344 (1996) (edited by R. E. McDonald and D.B. Min), Marcel Dekker, Inc., New York.
Maerker,G. Cholesterol autoxidation. *J. Am. Oil Chem. Soc.* **64**, 388-392 (1987).
Márquez-Ruiz,G. and Dobarganes,M.C. Nutritional and physiological effects of used frying fats, in *Deep Frying*, pp. 160-182 (1996) (edited by E.G. Perkins and M.D. Erickson), American Oil Chemists' Society Press, Champaign, IL.
Meltzer,J.B., Frankel,E.N., Bessler,T.R. and Perkins, E.G. Analysis of thermally abused soybean oils for cyclic monomers. *J. Am. Oil Chem. Soc.* **58**, 779-784 (1981).
Michael,W.R. Characterization of C18 cyclic esters. *Lipids* **1**, 365-368(1966)
Michael,W.R., Alexander,J.C. and Artman,N.R. Thermal reactions of methyl linoleate. I. Heating conditions, isolation techniques, biological studies and chemical changes. *Lipids* **1**, 353-358 (1966).

Nolen,G.A., Alexander,J.C. and Artman,N.R. Long-term rat feeding study with used frying fats. *J. Nutr.* **93**, 337-348 (1967).
Paquot,C. and Hautfenne,A. Determination of polar compounds in frying fats, in *Standard Methods for the Analysis of Oils, Fats and Derivatives*. pp. 216-219 (1987), Blackwell Scientific Publ., Oxford.
Paradis,A.J. and Nawar,W.W. A gas chromatographic method for the assessment of used frying oils: Comparison with other methods. *J. Am. Oil Chem. Soc.* **58**, 635-638 (1981).
Perkins,E.G. Volatile odor and flavor components formed in deep frying, in *Deep Frying*, pp. 43-48 (1996) (edited by E.G. Perkins and M.D. Erickson), American Oil Chemists' Society Press, Champaign, IL.
Scavone,T.A. and Braun,J.L. High temperature vacuum steam distillation process to purify and increase the frylife of edible oils. U.S. patent No. 4,789,554 (December 6, 1988).
Schwarz,K. and Ternes,W. Antioxidative constituents of *Rosemarinus officinalis* and *Salvia officinalis*. II. Isolation of carnosic acid and formation of other phenolic diterpenes. *Z. Lebensm. Unters. Forsch.* **195**, 99-103 (1992).
Sébédio,J.L., Garrido,A. and López, A. (editors) *Utilization of Sunflower Oils in Industrial Frying Operations*. Grasas y Aceites **47**, 1-99 (1996).
Sims,R.J., Fioriti,J.A. and Kanuk,M.J. Sterol additives as polymerization inhibitors for frying oils. *J. Am. Oil Chem. Soc.* **49**, 298-301 (1972).
Smith,L.M., Clifford,A.J., Hamblin,C.L. and Creveling,R.K. Changes in physical and chemical properties of shortenings used for commercial frying. *J. Am. Oil Chem. Soc.* **63**, 1017-1023 (1986).
Snyder,J.M., Frankel,E.N. and Warner,K. Headspace volatile analysis to evaluate oxidative and thermal stability of soybean oil. Effect of hydrogenation and additives. *J. Am. Oil Chem. Soc.* **63**, 1055-1058 (1986).
Stevenson,S.G., Vaisey-Genser,M. and Eskin,N.A.M. Quality control in the use of deep frying oils. *J. Am. Oil Chem. Soc.* **61**, 1102-1108 (1984).
Takeoka,G., Perrino,C. and Buttery,R. Volatile constituents of used frying oils. *J. Agric. Food Chem.* **44**, 654-660 (1996).
Waltking,A.E., Seery,W.E. and Bleffert,G.W. Chemical analysis of polymerization products in abused fats and oils. *J. Am. Oil Chem. Soc.* **52**, 96-100 (1975).
Waltking,A.E. and Wessels,H. Chromatographic separation of polar and nonpolar components of frying fats. *J. Assoc. Off. Anal. Chem.* **64**, 1329-1330 (1981).
Warner,K., Orr,P. and Glynn,M. Effect of fatty acid composition of oils on flavor and stability of fried foods. *J. Am. Oil Chem. Soc.* **74**, 347-356 (1997).
Weiss,T.J. *Food Oils and their Uses*. Second Edition. pp. 157-180 (1983) Avi Publ. Co., Inc., Westport, CT.
White,P.J. and Armstrong,L.S. Effect of selected oat sterols on the deterioration of heated soybean oil. *J. Am. Oil Chem. Soc.* **63**, 525-529 (1986).
Yuki,E. and Ishikawa,Y. Tocopherol contents of nine vegetable frying oils, and their changes under simulated deep-fat frying conditions. *J. Am. Oil Chem. Soc.* **53**, 673-676 (1975).

CHAPTER 12

BIOLOGICAL SYSTEMS

The role of polyunsaturated lipids in the diet and their effects on health have received much attention in the last three decades. Concerns about the problems of lipid oxidation stem not only from the deterioration of foods but also from the possible biological damage resulting from changes in diet and environment. Intense interest in this field has been generated by the compelling evidence that increased oxidative damage is implicated in many human diseases. In the early 1960's, vegetable oils became the main source of edible fats. They have higher contents of polyunsaturated fatty acids and lower contents of saturated fatty acids and lack cholesterol. When saturated fats and dietary cholesterol were shown to be associated with heart diseases in humans, diets high in polyunsaturated vegetable oils were recommended for a long time because they lower serum cholesterol. However, more recently too much linoleic acid in the diet was associated with a predisposition toward coronary and inflammatory diseases. Because n-3 polyunsaturated fatty acids moderated the adverse effects of n-6 polyunsaturated fatty acids, an increase in consumption of fish and fish oils has also been advocated. The oxidation of polyunsaturated fatty acids components in lipoproteins in blood is now implicated in the development of atherosclerosis (see Sections D and E). Therefore, in recommending increased intakes of polyunsaturated fats in the diet, the potential problems of oxidation of polyunsaturated lipids should not be overlooked.

This chapter considers the nutritional consequences of unsaturated lipids in the diet on free radical reactions in biological systems and diseases. The term lipid "peroxidation," is now used broadly to include both non-enzymatic and enzymatic oxidative reactions of free fatty acids, phospholipids, triacylglycerols, cholesterol and cholesteryl esters, and lipoproteins (see Glossary at the end of this Chapter). An ultimate goal of nutrition is to improve our diet by reducing or minimizing the risk factors from oxidative deterioration of polyunsaturated lipids, proteins and DNA that are implicated in the developments of certain human diseases, including atherosclerosis, cancer, rheumatoid arthritis and the process of aging.

A. BIOLOGICAL PEROXIDATION

The biological significance of the free radical process known as lipid peroxidation has moved this field beyond our concerns with the problems of rancidity in foods. Lipid peroxidation is induced under certain conditions by the formation of more readily oxidizable substances, referred to collectively as *reactive oxygen species*. These oxidant species are responsible for the toxic effects of oxygen in the body, and together with the free radicals generated from unsaturated lipids (Chapter 1) may cause tissue damage accompanying disease states. Reactive oxygen species affecting lipid peroxidation include:

a. Singlet oxygen ($^1O_2$). By photosensitization reactions, energy produced by exposure to light and an excited sensitizer, is transferred to oxygen, which becomes excited from the triplet state to the singlet state (Chapter 3). Many biological sensitizers are known to be effective, including chlorophylls, riboflavin and derivatives (flavin mononucleosides and flavin adenine dinucleotide), bilirubin, retinal, and porphyrins. The resulting $^1O_2$ reacts with linoleic acid at least 1500 times faster than normal triplet oxygen to form hydroperoxides causing damage to the membrane lipids in the retina. Visual cells can be damaged by exposure to light due to sensitization of retinal causing the formation of $^1O_2$. Exposure of cells to light causes damage to the mitochondria which are rich in heme and flavin-containing proteins. In addition to polyunsaturated lipids, a large number of biologically important compounds are damaged by $^1O_2$, such as α-tocopherol, DNA, cholesterol, β-carotene, and proteins, by reactions with several amino acids (tryptophan, methionine, cysteine and histidine).

b. Superoxide radical ($O_2^{-\cdot}$). This reactive species is produced in aerobic cells by one-electron reduction of oxygen through the cellular electron transport chain of mitochondria, chloroplasts, and phagocyte cells. An important antibacterial mechanism proceeds by neutrophils, monocytes, and macrophages, which kill bacteria by producing $O_2^{-\cdot}$.

In aqueous solutions, superoxide may act as a weak base by accepting protons to form a hydroperoxyl radical ($HO_2^{-\cdot}$) that can dissociate to regenerate protons, by the equilibrium (1). However, not much $HO_2^{\cdot}$ is present at physiological pH.

$$O_2^{-\cdot} + H^+ \rightleftharpoons HO_2^{-\cdot} \qquad (1)$$

Superoxide can act either as a reducing agent or as a weak oxidizing agent. Superoxide reduces the iron in cytochrome c from Fe^{3+} to Fe^{2+}, and oxidizes ascorbic acid. In the oxidation of haemoglobin in erythrocytes to oxyhemoglobin, superoxide can be produced in its conversion to methemoglobin through an intermediate complex with oxygen (2).

$$\text{hemoglobin} + O_2 \longrightarrow [\text{hem-Fe}^{2+}\cdots O_2] \longrightarrow \text{hem-Fe}^{3+} + O_2^{-\cdot} \qquad (2)$$
$$\text{oxyhemoglobin} \qquad \text{methemoglobin}$$

The cells are protected against the damage from excess superoxide by various enzymes in a sequence of reactions involving superoxide dismutase (SOD) producing hydrogen peroxide, which is removed by catalase, as discussed in Section B. More recently, superoxide has been suggested to regulate vascular tone as a vasoconstrictor, by interacting with nitric oxide (NO˙), which acts on smooth muscle cells to produce relaxation.

c. **Hydrogen peroxide.** The dismutation of O_2^- (3) in the presence of SOD produces hydrogen peroxide, which occurs in whole bacteria, phagocytic cells, spermatozoa, mitochondria, microsomes, and chloroplasts.

$$2\,O_2^{-\cdot} + 2\,H^+ \longrightarrow H_2O_2 + O_2 \quad (3)$$

Hydrogen peroxide crosses biological membranes such as red blood cell membranes and is also found in human eye lens and human breath (from pulmonary macrophages). Although hydrogen peroxide *per se* is a weak oxidizing agent, in the presence of Fe^{2+} in the cell, it forms hydroxyl radicals, which are extremely reactive and responsible for the toxicity of hydrogen peroxide in killing bacteria. Catalase provides biological protection by removing hydrogen peroxide from the cell.

d. **Hydroxyl radicals (˙OH).** In the presence of transition metals, H_2O_2 and O_2^- produce hydroxyl radicals according to equation (4).

$$M^{n+} + H_2O_2 \longrightarrow M^{(n+1)+} + {}^\cdot OH + OH^- \quad (4)$$

Hydroxyl radicals are the most reactive and damaging oxygen species generated by a wide range of systems, including activated phagocytes, autoxidizing compounds, mixtures of hypoxanthine and xanthine oxidase (an important enzyme that synthesizes uric acid from xanthine). Hydroxyl radicals are so highly reactive that when produced *in vivo* they can react at or in close proximity to their site of formation.

The decomposition of hydrogen peroxide by Fe^{2+} and some Fe^{3+} complexes, is known as the *Fenton reaction* and is represented by equations (5) and (6) (see chapter 10. A. 4).

$$Fe^{2+} + H_2O_2 \longrightarrow Fe^{3+} + {}^\cdot OH + OH^- \quad (5)$$

$$Fe^{3+} + H_2O_2 \longrightarrow \text{ferryl} + H_2O_2 \longrightarrow \text{perferryl} + H_2O_2$$

$$\downarrow \quad (6)$$

$${}^\cdot OH$$

The occurrence of the Fenton reaction *in vivo* has been debated for a long time. If this reaction occurs, it is limited by the availability of iron ions. The hydroxyl radicals formed are so highly reactive that they combine at or near the location of the metal ions. If catalytic iron is bound to membrane lipids, lipid peroxidation could be initiated by hydroxyl radicals formed in the presence of hydrogen peroxide.

e. Hypochlorous acid (HOCl). This powerful oxidant is generated at sites of inflammation by the enzyme *myeloperoxidase* in activated phagocytic cells. The resulting process known as *phagocytosis*, takes place by engulfing foreign particles such as bacteria with cells known as neutrophils. The myeloperoxidase can kill bacteria by oxidizing chloride ions in the presence of hydrogen peroxide to produce HOCl. Bacteria are killed by the strong oxidizing action of HOCl on proteins, especially the amino acids components containing sulfhydryl (-SH) groups, and other biological molecules such as ascorbic acid (vitamin C).

B. BIOLOGICAL ANTIOXIDANTS

Organisms have evolved antioxidant defense systems for protection against reactive oxygen species and repair systems to prevent the accumulation of products from oxidatively damaged lipids, proteins and DNA. A number of *extracellular* and *intracellular antioxidant systems* are known to effectively inactivate reactive oxygen species and oxidants, discussed in Section A.

a. Singlet oxygen quenchers. Carotenoids such as lycopene and β-carotene from fruits and vegetables protect lipids and other biological systems by interfering with the activation of oxygen. The mechanisms recognized for this effect include: quenching singlet oxygen and sensitizers, and scavenging peroxyl radicals at low oxygen pressures (Chapter 3). Relatively low concentrations of β-carotene are effective in preventing lipid peroxidation by singlet oxygen. Other compounds such as α-tocopherol, amino acids and glutathione behave as weak quenchers by oxidation with singlet oxygen.

b. Metal binders. Living organisms are well protected against the cytotoxic effects of iron by using *transferrin* as a transport protein and *ferritin* as a storage protein to minimize the intracellular pool of free iron. Transferrin is a glycoprotein carrier molecule that under normal conditions is about 30% loaded with iron. No free iron salts are thus expected in blood plasma. The availability of iron and other transition metals necessary to generate reactive oxygen species *in vivo* has thus been much debated in the literature. However, under *oxidative stress*, induced by superoxide or by hydrogen peroxide, small amounts of free iron may be released and mobilized from ferritin and other protein binders to catalyse free radical reactions. Cells may also contain small amounts of free iron that may catalyse the formation of hydroxyl radicals. *Albumin* can inhibit lipid peroxidation by binding copper ions. In addition to its metal binding capacity, albumin is also a good scavenger of peroxyl radicals (by reacting with the sulfhydryl groups of its amino acids components) and of hypochlorous acid. Metal chelation by plasma proteins is generally regarded as one of the most important antioxidant defense mechanisms.

c. Superoxide scavengers. *Superoxide dismutases* (SOD) catalyse the dismutation of superoxide by reaction (3) to produce hydrogen peroxide, which is removed by catalase and glutathione peroxidase. Several SOD enzymes have been characterized, including the copper-zinc-SOD in mammalian tissues, fish and plant tissues, the manganese-SOD and the iron-SOD in higher organisms and bacteria.

d. Enzymatic peroxide destroyers. *Glutathione peroxidase* removes hydrogen peroxide within cells by selectively using glutathione (GSH), a tripeptide of glutamic acid, cysteine and glycine, as a hydrogen donor. Two GSH molecules condense by reaction (7), through the -SH groups of cysteine to form a disulfide bridge, -S-S- .

$$2\,H_2O_2 + 2\,GSH \longrightarrow GSSG + H_2O \qquad (7)$$

Oxidized glutathione, GSSG, is reduced back to GSH by *glutathione reductase* to maintain a ratio of GSH/GSSG greater than 10 in normal cells.

Glutathione peroxidase accepts other hydroperoxides and reduces lipid and steroid hydroperoxides, to form stable hydroxy lipids that do not decompose to form alkoxyl radicals and cell damaging aldehydes. There is ample evidence indicating that glutathione peroxidase provides an important protective mechanism against lipid peroxidation *in vivo*.

Two types of glutathione peroxidases are known, selenium and non-selenium containing, and vary widely in different animals and organs. The former enzyme consists of four protein units, each containing one atom of selenium. The non-selenium glutathione peroxidase reacts with artificial hydroperoxides (*e.g.* cumene and tert-butylhydroperoxides) but not with hydrogen peroxide, and its activity may be due to glutathione-*S*-transferases. The selenium containing glutathione peroxidase is most important as indicated by the requirement of selenium in trace amounts as a cofactor in animal and human diets.

Catalases remove hydrogen peroxide by catalysing reaction (8).

$$2\,H_2O_2 \longrightarrow 2\,H_2O + O_2 \qquad (8)$$

Catalase activity is particularly high in the liver and erythrocytes of animals. These enzymes consist of proteins containing a heme group (Fe^{3+}-protoporphyrin), and can catalyse peroxidase-type reactions.

e. Non-enzymatic peroxide destroyers. *Ascorbic acid* has multiple antioxidant effects in biological systems. It is an active reducing agent that can scavenge singlet oxygen, superoxide, hydrogen peroxide, hydroperoxyl radicals, hydroxyl radicals and hypochlorous acid, to give semidehydroascorbate. One of the multiple action of ascorbic acid is to reduce lipid hydroperoxides to stable hydroxy lipids. It is regarded as the most effective water-soluble antioxidant in plasma. However, it can also reduce Fe^{3+} to Fe^{2+} which can participate in the Fenton reaction by producing damaging hydroxy radicals in the presence of hydrogen peroxide, and stimulate iron-catalysed lipid peroxidation (Section A.d). This prooxidant effect is counterbalanced under certain conditions by the scavenging properties of ascorbic acid toward hydroxyl radicals. The main protective role of ascorbic acid may be the regeneration of vitamin E after its oxidation in plasma proteins (Section B.f.).

Uric acid is a water-soluble antioxidant found in relatively high concentrations in plasma. It inhibits lipid peroxidation by tightly binding iron and copper ions into inactive forms, and by scavenging various oxidants such as hydroxyl radicals, peroxyl radicals, singlet oxygen and hypochlorous acid. By complexing with iron, uric acid stabilizes ascorbic acid in human serum.

Ubiquinol-10 (reduced form of ubiquinone-10, or coenzyme Q10) is a reactive lipid-soluble antioxidant that may act like ascorbic acid by regenerating vitamin E by reducing the vitamin E radicals produced during lipid peroxidation.

f. Radical chain breakers. α-*Tocopherol* (vitamin E) is considered the major lipid-soluble antioxidant in membranes, acting by several mechanisms, including: by scavenging free radicals (superoxide, hydroxyl radicals), which can initiate and propagate lipid peroxidation, by reacting with nitric oxide, and by deactivating singlet oxygen. Vitamin E reacts with lipid peroxyl and alkoxyl radicals by donating a labile hydrogen and terminating the lipid peroxidation sequence (Chapter 8). The resulting α-tocopherol radical is stabilized by electron delocalization around the phenol ring structure. We shall see later that α-tocopherol is important in protecting lipoproteins against oxidation. To be effective *in vivo*, the α-tocopherol radicals are reduced back to α-tocopherol by ascorbic acid or other reducing agents, such as cysteine, or glutathione. This synergistic effect is assumed to take place at the membrane interface because the hydrophilic reducing agent, in solution in the aqueous phase, would have an affinity for the membrane phospholipids at the surface.

C. OXIDANT-ANTIOXIDANT BALANCE

The field of lipid peroxidation has generated considerable attention by compelling evidence that increased oxidative damage due to reactive oxygen species is implicated in many human diseases. Lipid peroxidation and degradation products may be involved in a number of biological effects and may be relevant to certain types of cancer. A multitude of biological macromolecules, such as lipids, proteins and DNA, are susceptible to oxidative damage by free radicals and reactive oxygen species. A number of decomposition products of lipid hydroperoxides, such as malonaldehyde and hydroxynonenal, are known to be cytotoxic. The integrity of cell membranes

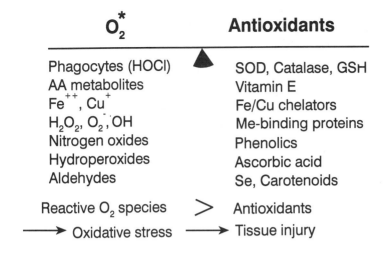

Figure 12-1. Oxidant - Antioxidant balance. $O_2^{\cdot}$, reactive oxygen species; AA, arachidonic acid; SOD, superoxide dismutase; Me, metals; Se, selenium.

provides an important protective mechanism by separating unsaturated lipids from biological oxidants and activated oxygen species and by maintaining a low intracellular concentration of oxygen. Any factors that affect this structural separation will cause damaging lipid peroxidation. Lipid peroxidation occurs thus more rapidly in disrupted than in healthy tissues. However, it is often difficult to establish unequivocally if free radical lipid peroxidation is the cause of biological damage leading to diseases, or if it takes place as a consequence of this damage.

Protection against peroxidation comes from antioxidant nutrients and intracellular defenses from the enzymes, antioxidants, proteins acting as metal binders, reducing agents and cofactors discussed in section B. *In vivo*, there is a constant competition between these oxidation and protective processes that depend on the polyunsaturated fatty acid composition of tissue lipids and their antioxidant levels (Figure 12-1). For optimum health, a balance is required between the oxidizing species and the antioxidants. An imbalance resulting from excess oxidants or oxygen species will lead to a disturbance, defined as *oxidative stress*, causing tissue damage and susceptibility to diseases.

Reactive oxygen species produced by several stimuli are now recognized as important in the mechanism of toxicological responses of various compounds referred to as *toxins*. The relative toxicity of toxins could be influenced by the concentrations of antioxidants in tissues to ensure the removal of excess oxidants. Many kinds of tissue injuries may initiate free radical reactions such as lipid peroxidation by destroying or inactivating cellular antioxidants or by liberating metal catalysts from chelating proteins in cells. Oxygen radicals are recognized mediators of various degenerative disease and inflammatory diseases, including rheumatoid arthritis, diabetes, cancer, cataract formation, immune and brain dysfunctions, and the universal problem of aging.

TABLE 12-1.
Composition of lipoprotein particles in human blood plasma, expressed as % particle mass.[a]

Components and characteristics	Chylomicrons	VLDL[b]	LDL[b]	HDL[b]
Triacylglycerols	83	50	10	8
Total Cholesterol	8	22	48	20
Phospholipids	7	20	22	22
Protein	2	7	20	50
Apoproteins	A, B-48, C	B-100, E	B-100	A
Mass (daltons)	0.4-30 x 10^6	10-100 x 10^6	2.35 x 10^6	1.75-3.6 x 10^5
Density (g/mL)	>70	30-90	18-22	5-12

[a] From Gurr and Harwood (1991).
[b] VLDL, very-low-density lipoproteins; LDL, low-density lipoproteins; HDL, high-density lipoproteins.

Extensive studies are now carried out worldwide to test this oxidant-antioxidant balance hypothesis. The health of an individual may be influenced by the efficiency of various protection systems against repetitive and accumulated oxidant damage. The stress and tissue injuries resulting from oxidant damage may range in intensity from viral infections, trauma, inflammation, cigarette smoke or environmental pollution. Antioxidant requirements will vary in relation to the oxidative cellular damage or oxidant stress, and would be therefore very difficult to establish for a general population. The evidence is accumulating that food antioxidants may inhibit biologically harmful oxidation reactions in the body. Recently, the occurrence of relatively high levels of polyphenolic antioxidants in fruits, vegetables and certain beverages (especially red wine and tea) has received special attention for their potential protective effects against the damage from biological oxidants. The problems of oxidative stress could thus be alleviated by including in the diet more phenolic antioxidants found in fruit and vegetables. However, the present knowledge is incomplete on the great diversity of plant phenolic compounds and their biological effects. The extent that these plant antioxidants are absorbed in the body and their molecular mechanisms of protection *in vivo* and of disease intervention are not well established. This subject is now under intensive investigation around the world.

D. PLASMA LIPOPROTEINS

Since triacylglycerols and cholesterol are not soluble in blood they are carried through by lipoproteins serving as special transport systems. To understand the role of oxidized lipoproteins in the development of vascular heart diseases, it is necessary to review briefly some of the basic metabolic pathways of lipoproteins. Lipids are transported in the blood stream through a series of lipoprotein particles varying in sizes and phospholipid and protein

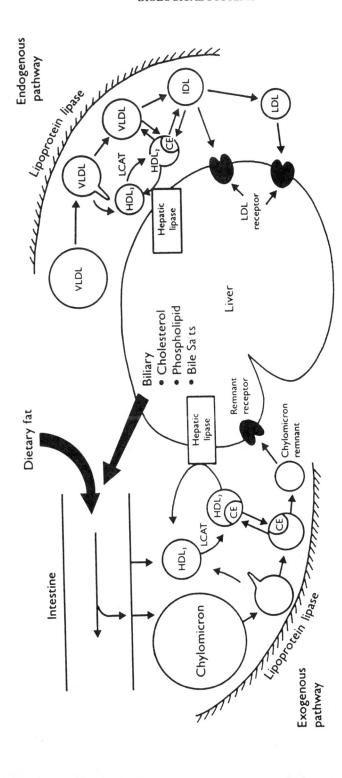

Figure 12-2. Metabolism of lipoproteins. From Gurr (and Marel Dekker Inc.) (1997) with permission. CE, cholesteryl esters; HDL, high density lipoprotein which is interconverted into two forms, HDL_2 and HDL_3; VLDL, very low density lipoprotein; IDL, intermediate density lipoprotein; LDL, low density lipoprotein; LCAT, lecithin cholesterol acyltransferase (see Figure 12-3).

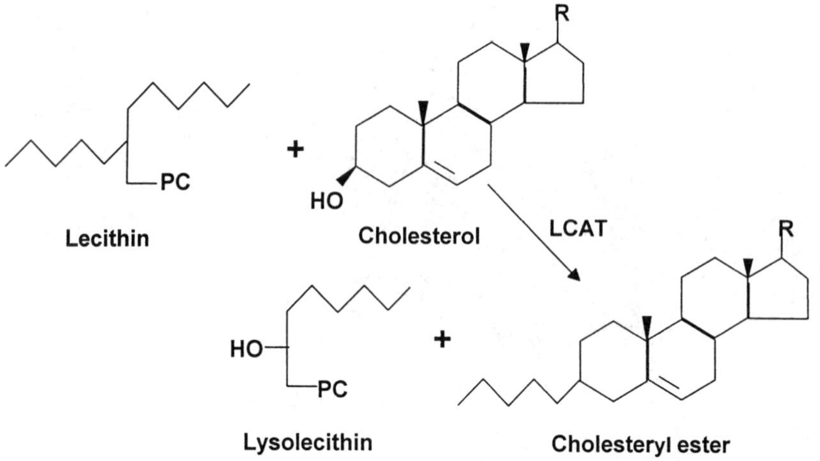

Figure 12-3. Lecithin cholesterol acyltransferase (LCAT). This enzyme esterifies cholesterol by transferring unsaturated fatty acids (mainly linoleic acid) from lecithin to form cholesteryl esters and lysolecithin.

constituents (Table 12-1). These particles have a lipid core, containing mainly triacylglycerols and cholesteryl esters, and surrounded by a shell of phospholipids and a protein, called *apoprotein*. These lipoproteins are separated by ultracentrifugation into different classes, from the lowest to the highest density: *chylomicrons, very low density lipoproteins* (VLDL), *low density lipoproteins* (LDL) and *high density lipoproteins* (HDL). These particles are described in terms of particle mass because they are structural aggregates of individual lipid and protein molecules. The apoprotein moieties of lipoproteins provide structural specificity to the particles by directing their metabolic pathways to various cell surface receptors throughout the body. The following functions are recognized for each lipoprotein: a) chylomicrons are the principal carriers of dietary lipids composed mainly of triacylglycerols in the blood stream, b) VLDL transport endogenous lipids synthesized in the liver and intestine, c) LDL are the main carriers of plasma cholesterol, and d) HDL carry excess cholesterol to the liver.

Lipoproteins are transported and metabolized through two pathways. The exogenous pathway concerns the fat from the diet; the endogenous pathway concerns the fats synthesized in the body from carbohydrates (Figure 12-2). After absorption, triacylglycerols, phospholipids and cholesteryl esters are packaged as chylomicrons.These large particles are removed from the plasma by the hydrolytic action of *lipoprotein lipase* forming fatty acids for either storage in adipose tissue or for fuel in the blood. The chylomicrons are thus partially degraded to smaller particles called *chylomicron remnants*, and removed by the liver where they are taken up by remnant receptors and metabolized to LDL particles which are enriched in cholesterol.

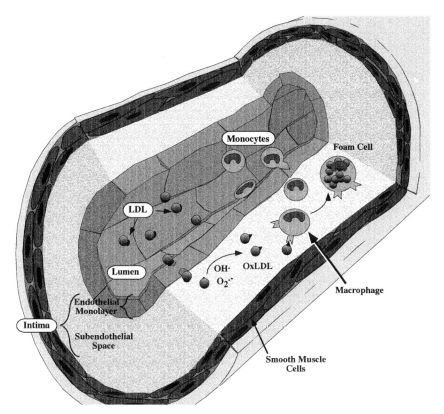

Figure 12-4. Oxidative modification of low density lipoprotein. Formation of atherosclerotic lesions by oxidation of LDL in the intima, the inner layer concentric to the endothelial layer of the artery. From Pearson (1998).

The production of cholesterol in the blood is controlled by reacting with specific LDL receptors on the cell surface, forming a lipoprotein-receptor complex in the cell, where the lipoprotein is broken down. Portions of chylomicrons are converted to HDL subfractions (HDL_1 and HDL_2), which remove excess cholesterol by converting it to cholesteryl esters by the action of another enzyme, *lecithin-cholesterol acytransferase* (LCAT) (Figure 12-3). By the endogenous pathway, the liver exports fat molecules as VLDL in the blood. These particles circulate in the blood and metabolize cholesterol by similar pathways as the chylomicrons, by forming smaller VLDL particles called intermediate density lipoproteins (IDL), which are metabolized into LDL and taken up by the LDL receptor.

The level of cholesterol in the body is delicately controlled by the number of LDL receptors which varies according to the intracellular production of cholesterol. When cholesterol is present in blood in excessive amounts, it may lead to metabolic defects with undesirable nutritional consequences. The HDL particles play a vital part in removing excessive free cholesterol, converting it to cholesteryl esters by the action of LCAT and transporting it to the liver.

E. CORONARY HEART DISEASE

This disease is characterized by the thickening of the inner linings of arteries. Atherosclerotic lesions are initiated by an increase of LDL in plasma which triggers an increase in circulating monocytes that adhere to endothelial cells in the artery (Figure 12-4). This process leads to more LDL particles entering into the intima (the next layer concentric to the endothelial layer) where they become oxidized by cells found in arterial lesions. The strong evidence that oxidation of LDL is a critically important event in the initiation and development of atherosclerosis has generated considerable interest in the biomedical significance of lipid peroxidation. Medical interventions against this disease are now based not only on lowering the serum level of LDL cholesterol, but also on reducing the susceptibility of LDL to oxidation.

1. Development of the Disease

Atherosclerosis is believed to develop by a response to *injury* mechanism, also referred to as the *endothelial injury* hypothesis, involving a chain of complex cellular interactions leading to the formation of fatty streaks.

i. *Arterial injury.* An injury occurs initially to the endothelial lining of the artery in response to a range of factors, including aging, toxins, viral infection, immunological reactions, smoking, hypertension and lipid oxidation products in the diet and in oxidized LDL (oxidized cholesterol, phospholipids, triacylglycerols and free fatty acids). Although the origin of this arterial injury is unknown, it may be initiated by LDL particles which become more susceptible to oxidation in circulating plasma when present at high concentrations.

ii. *Monocyte adherence and macrophage production.* In response to injury, monocytes migrate from the blood into the intima and become macrophages that ingest particle cells, within the vessel wall. Monocytes attach themselves to the endothelium by the presence of *adhesion molecules* expressed in response to an injury. Activated monocytes and macrophages may cause further injury to endothelial and smooth muscle cells by secreting reactive oxygen species (superoxide, hydrogen peroxide).

iii. *Oxidative modification of LDL.* LDL in the arterial intima is oxidized by endothelial cells, smooth muscle cells and macrophages, and becomes recognized by the abnormal *scavenger receptors*. This oxidation of LDL may be the result of a local deficiency of antioxidants.

iv. *Foam cell formation.* Oxidized LDL particles passing through into the lining of the blood vessel become engulfed and ingested by the macrophage and other cells which are converted into lipid-laden distorted *foam cells*.

v. *Fatty streak formation.* Fatty streaks in the sub-endothelial area are formed by the prolific production of foam cells from macrophages. Platelets adhere to denuded areas in endothelium, recruit and proliferate smooth muscle cells.

vi. *Intimal thickening.* Smooth muscle cells were assumed to produce foam

TABLE 12-2.
Chemical composition of human LDL.[a]

Constituents	mean mol / mol LDL
Phospholipids	700
Phosphatidylcholine	450
Lysophosphatidylcholine	80
Sphingomyelin	185
Free cholesterol	600
Cholesteryl esters	1600
Triacylglycerols	170
Free fatty acids	26
Apoprotein B	1
Fatty acid composition	
Saturated acids	880
Oleic acid	454
Linoleic acid	1100
Arachidonic acid	153
Docosahexaenoic acid	29
Antioxidants	
α-Tocopherol	6.4
γ-Tocopherol	0.5
β-Carotene	0.3
α-Carotene	0.1
Lycopene	0.2
Ubiquinol-10	0.1

[a] From Esterbauer *et al.* (1992).

cells. These cells are now believed to be derived from circulating macrophages which adhere to the endothelium and get into the sub-endothelial space (Figure 12-4) where they assimilate lipoproteins laden with cholesteryl esters.

The injury cycle repeats with ultimate production of extensively raised plaques, which narrow and begin to obstruct the lumen of the artery. The process of atherosclerosis is very complex and takes several years to develop. Key steps that may promote this development include oxidative stress from excessive dietary cholesterol, elevated blood lipid concentrations, hypertension, cigarette smoking, or other genetic factors associated with this disease. Plaques cause disease by limiting blood flow to the heart or brain, causing eventually a stroke or heart attack when the artery becomes completely occluded.

Oxidized LDL may have several other functions in the development of atherosclerosis, including the expression of chemotactic factors that attract macrophages in the endothelial space, to induce endothelial cells to secrete a chemotactic protein for monocytes activating inflammatory and immune responses, and altering the production of nitric oxide. The endothelial injury produces platelet aggregation, which is followed by the release of growth factors that initiate smooth muscle cell proliferation. Certain growth factors,

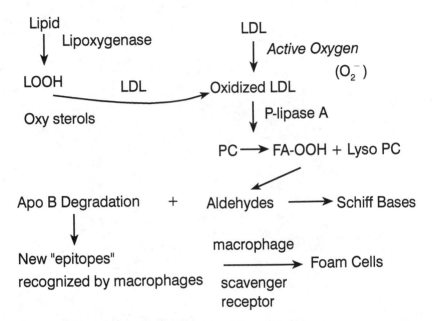

Figure 12-5. Pathways for the oxidative modification of low density lipoprotein. The lipoxygenase pathway produces hydroperoxides that catalyse the oxidation of LDL. The active oxygen pathway involves the direct oxidation of LDL with reactive oxygen species such as superoxide. The oxidized fatty acids (FA-OOH) of phosphatidylcholine (PC) in LDL are preferentially hydrolysed by phospholipase A and decomposed into aldehydes that interact with ApoB of LDL, leading to the formation of foam cells via macrophage and scavenger receptors.

eicosanoids (oxygenated products of arachidonic acid) and nitric oxide, can also affect the activities of the cells in atherosclerotic lesions. Macrophages release growth factors that affect the transport of cholesterol within the body. Endothelial cells release potent substances such as *prostacyclin*, through the prostaglandin cascade, and nitric oxide, that inhibit blood clot formation. The effect of dietary lipids through the conversion of polyunsaturated fatty acids to physiologically important molecules (eicosanoids, prostaglandins and leukotrienes) will be discussed further in Section F. 2.b.

2. Oxidative Modification of LDL

The average LDL particle consists of a very complex mixture consisting mainly of cholesteryl esters, phospholipids, free cholesterol, triacylglycerols, and apolipoprotein B (apoB) containing 4536 amino acid residues (Table 12-2). Minor antioxidant components include tocopherols, carotenoids and ubiquinol. Both lipid and apoB components become oxidized and the products of oxidation may interact to cause damage.

BIOLOGICAL SYSTEMS 263

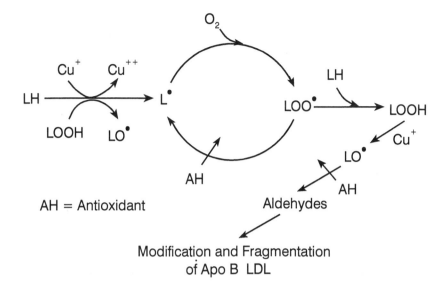

Figure 12-6. Free radical oxidation of polyunsaturated fatty acids (LH) components of low density lipoprotein (LDL). Formation of aldehydes that causes modification and fragmentation of amino acid residues in the Apo B of LDL. Antioxidants (AH) can inhibit LDL oxidation by reacting with either peroxyl radicals (LOO·) or alkoxyl radicals (LO·).

The type and sources of biological oxidants initiating LDL oxidation are unknown and controversial. Several biological pathways have been advanced for the oxidative modification of LDL, which can take place either in the arterial wall or at inflammatory sites. In one pathway, cell lipids are first

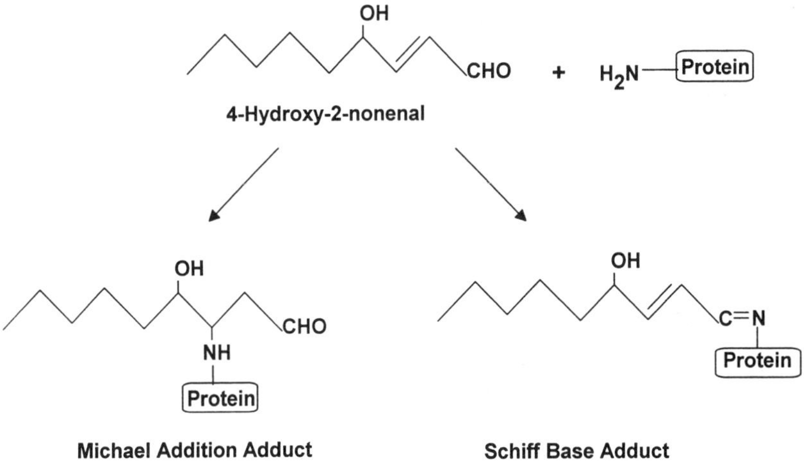

Figure 12-7. Formation of Michael addition and Schiff base adducts by interaction of 4-hydroxy-2-nonenal with amino residue of proteins.

oxidized by enzymes such as the 12 and 15-lipoxygenases, and the oxidized lipids transfer into the LDL in the medium and initiate chain reactions with extensive modification of the lipids in LDL (Figure 12-5). A second pathway postulates the generation of active oxygen species such as superoxide within the cell, followed by release into the medium, where LDL lipids become oxidized. During oxidative modification, LDL phospholipids undergo extensive conversion to lysolecithin. Phospholipid hydroperoxides are selectively hydrolysed from the 2-position by phospholipase-A2, which is intrinsic to apoB of LDL. The resulting lysolecithin may be chemotactic to monocytes and produce lesions.

Metal catalysts appear to be necessary to initiate lipid peroxidation in cells since LDL oxidatively modified in the presence of cells has the same physicochemical and biological properties as LDL oxidized by metals in the absence of cells. The polyunsaturated lipids (LH) in cholesteryl esters, phospholipids and triacylglycerols in LDL undergo oxidation by the same free radical chain mechanism as that established for food lipids with formation and decomposition of hydroperoxides (Chapters 2 and 4) (Figure 12-6). In the presence of metals, the hydroperoxides undergo extensive fragmentation into aldehydes that react covalently with the apoB moiety of LDL by the formation of Schiff bases with the lysine ε-amino groups of apoB, and form cross-links between lipids and proteins. In addition to Schiff base formation, 4-hydroxy-2-nonenal and other 2,4-dienals can form protein adducts by Michael addition, leaving a free aldehyde moiety to undergo further condensation by intra- or intermolecular cross-linking changing the charge and configuration of the protein (Figure 12-7). The positive charge of lysine ε-amino residues is neutralized and the LDL particles acquire a net negative charge. When sufficient lysine amino acid residues are altered, a new epitope is formed on apoB that causes oxidized LDL to be recognized by the macrophage scavenger receptor and to be internalized and undergo endocytosis to become converted into lipid-engorged foam cells.

Other mechanisms postulated for the oxidation of human LDL involve the direct oxidation of apoB amino acid residues. The copper-mediated oxidation of tryptophan residues in apoB was suggested as an initiation step in the oxidation of LDL particles. Hypochlorous/hypochlorite (HOCl) produced by myeloperoxidase in microphages caused the preferential oxidation of lysine, tryptophan and cysteine residues of apoB. Although HOCl is found in atherosclerotic lesions, the direct intervention of myeloperoxidase in the *in vivo* oxidative modification of LDL is disputed. Several mechanisms have been advanced for the atherogenic properties of oxidized LDL. In addition to macrophage uptake causing foam cell formation, oxidized LDL induce chemotactic proteins and growth factors in endothelial cells that cause cell adhesiveness. Oxidized LDL is chemotactic for monocytes and facilitates their migration into the arterial wall.

Several lines of evidence support the *in vivo* oxidation of LDL. The LDL extracted from atherosclerotic lesions manifest the same physicochemical and biological properties as LDL oxidized under *in vitro* conditions. The susceptibility of plasma LDL to copper-induced oxidation *in vitro* correlates with the extent of coronary atherosoclerosis in patients with previous experience of myocardial infraction. As we shall see below, antioxidants inhibit atherosclerosis in some animal models (Section E. 5.). Much of the current research in this area is directed at accumulating evidence to support the potential antiatherogenic effects of antioxidants as preventive measures to slow atherosclerosis by inhibiting LDL oxidation *in vivo*.

3. Methodology

The results of *in vitro*, *ex vivo* and *in vivo* studies of LDL oxidation are extremely difficult to interpret in view of the great complexity of the LDL structure, which varies widely in particle heterogeneity. The degree of oxidation is greatly affected by the methods used to prepare the lipoprotein samples and by the wide range of conditions used to induce LDL oxidation. The determinations of aldehydes, oxysterols, hydroxy-enals, linoleate and cholesteryl linoleate hydroperoxides, fluorescence, and oxidized proteins or protein carbonyls provide useful information regarding oxidative damage most relevant to the modification of LDL. Oxysterols are found in oxidized LDL, in the blood of fasting humans, and in animals on a vitamin E-deficient diet (see Section E. 4.). Linoleate hydroperoxides are specific oxidation products formed initially from LDL (Figure 12-6), but they only act as precursors of aldehydes which actually cause the damage to apoB. Fluorescence formation is largely due to modification of apoB. Measurements of oxidized proteins, loss of amino acids (tryptophan and cysteine), or protein carbonyls are most relevant to the oxidative modification of apoB, but they are not sensitive and require significant oxidative changes for detection. Measures of protein carbonyls are also not specific and are influenced by factors other than peroxidation. Modification of LDL can be measured by macrophage uptake due to a process leading to aggregation resulting from the covalent cross-linking of apoB of LDL particles.

Other methods used to measure LDL oxidation include the loss of linoleic acid (18:2), arachidonic acid (20:4) and cholesteryl esters, formation of conjugated dienes and TBARS, and electrophoretic mobility. The loss of 18:2, 20:4 and cholesteryl esters is an insensitive method measuring a small change compared to a very large amount of starting material. Oxidative damage of lipids can in fact be observed before significant changes in 18:2 and 20:4 contents can be measured (above 5%). Determinations of conjugated dienes derived from polyunsaturated fatty acids hydroperoxides, and several variations of the TBA test have been used most commonly to determine oxidation in LDL. Electrophoretic mobility measures the net negative charge of LDL resulting from the cross-linking of amino acid residues in apoB.

The rapid determination of conjugated dienes spectrophotometrically has been used extensively to make continuous measurements of LDL oxidation and

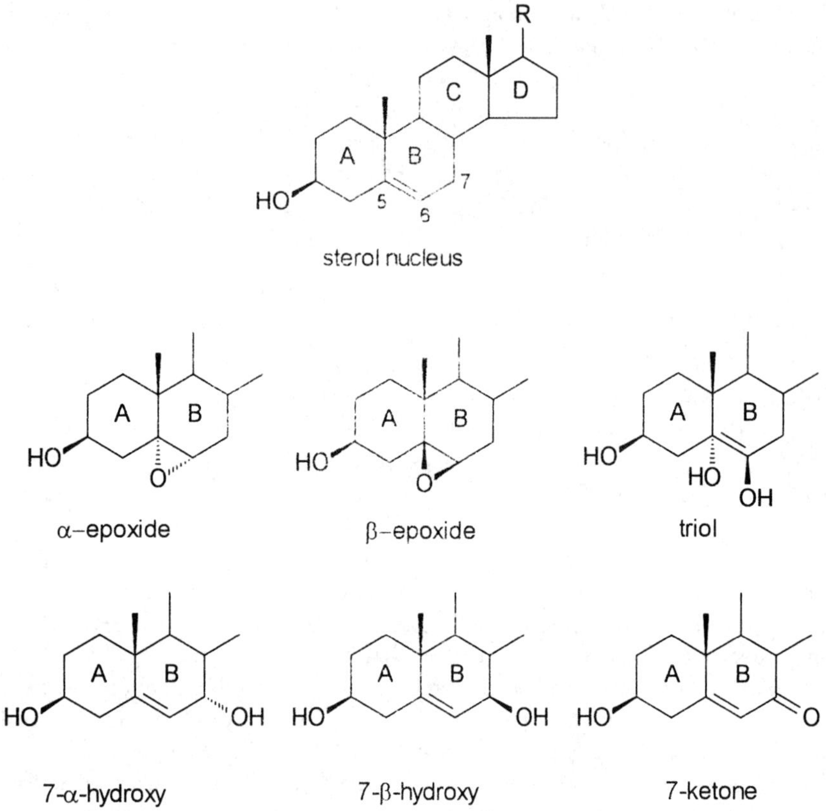

Figure 12-8. Common oxidation products of cholesterol.

to determine duration of the lag phase as a sensitive method for oxidative susceptibility. Although diene conjugation is a useful method to measure lipid peroxidation in pure lipids, it reflects hydroperoxide formation but not the decomposition aldehyde products that actually cause oxidative damage to apoB.

Because of its convenience, the TBA method has become a common assay to determine the degree of peroxidation and oxidative susceptibility of a wide range of biological materials, including LDL. However, the validity of the TBA determination as an index of lipid peroxidation in biological samples has been a matter of considerable debate in the literature. The determination of TBARS inherently lacks specificity, and is subject to interference by many compounds including materials that are not due to lipid peroxidation. This method is also flawed by analytical artifacts, and is affected by the same factors as lipid peroxidation.

The potential to be misled by inappropriate methodology is particularly critical when varying dietary unsaturated fats in animal studies. Measurements of TBA values and conjugated dienes are not appropriate if the effect of oleic acid is investigated since these methods do not measure oleic acid oxidation. Similarly, if the effect of n-3 PUFA is investigated, the degree of oxidation measured by TBARS may be exaggerated, because the results of this method are markedly affected by the oxidation of polyunsaturated fatty acids containing more than two double bonds (Chapter 5. E). Fish oil diets were reported to increases LDL oxidation on the basis of formation of TBARS. Because TBARS are formed by oxidative decomposition of polyunsaturated fatty acids containing more than two double bonds, the oxidation of any lipids containing n-3 polyunsaturated fatty acids would be expected to produce high levels of TBARS. Therefore, TBARS determinations cannot be used as the only determination of oxidative susceptibility to compare the effects of dietary n-6 versus n-3 polyunsaturated fatty acids on LDL oxidation (see Section F. 2.).

The technique of static headspace gas chromatography (Chapter 5. F) proved to be useful in our laboratory for the quantitative analysis of volatile aldehydes in oxidized human LDL. This method was used to distinguish volatile oxidation products of n-6 polyunsaturated fatty acids (pentane and hexanal) and n-3 polyunsaturated fatty acids (propanal) in LDL samples from hypertriglyceridemic human subjects fed fish oil supplements. This headspace assay of LDL oxidation thus permits the specific analysis of multiple products of lipid peroxidation (see Section F. 2.b.).

More specific and sophisticated methodology should be used to elucidate how lipid oxidation products act in the complex mechanism of LDL oxidation. Chemical analyses of specific oxidation products should be complemented by quantitative measures of biological changes that create an atherogenic LDL particle, such as increase macrophage intake or electrophoretic mobility.

4. Autoxidation Products of Cholesterol

Oxidized cholesterol oxides have been suggested to initiate atherosclerotic lesions by stimulating LDL interactions with arterial smooth muscle cells and macrophages thus producing foam cells. Cholesterol is oxidized especially at the allylic 7-position and a wide range of products have been identified and detected in LDL and other biological systems, including a mixture of epoxy, keto and hydroxy compounds (Figure 12-8). These products also referred to either as *oxysterols* or as *cholesterol oxides* are formed enzymatically or by non-enzymatic processes involving active oxygen species. In biological systems, evidence for oxysterols has been reported by oxidation induced with iron and carbon tetrachloride, and in tissues of animals on vitamin E-deficient diets. Although this type of evidence supports the involvement of oxysterols in human health, the physiological significance of cholesterol oxidation in normal tissues is not well understood. Cholesterol oxides have also been reported to be angiotoxic, mutagenic and to inhibit cholesterol biosynthesis.

Cholesterol oxides have been detected at ppm levels in dehydrated foods, potatoes fried in tallow (Chapter 11) and cured animal products. However, there is no adequate evidence of any harmful biological effects on the amounts

TABLE 12-3.
Inhibition of *in vitro* LDL oxidation following *in vitro* or *in vivo* supplementation of vitamin C and vitamin E.[a]

In vitro LDL oxidant	*In vitro* supplementation		*In vivo* supplementation	
	Vitamin C	Vitamin E	Vitamin C	Vitamin E
Endothelial cells	++	++, partial	–	++
Macrophages	++	++	n.d.	++
Cu^{2+}	++	+, partial	–	++
AAPH [b]	++	prooxidant	n.d.	Prooxidant

[a] From Frei (1995). ++ = yes; + = slight. n.d. = not determined.
[b] AAPH = 2,2′-azobis(2-amidinopropane) hydrochloride

ingested in a normal diet and on the degree of absorption and metabolism of oxysterols found in foods.

5. Inhibition of LDL Oxidation by Antioxidants

Although LDL is well protected against oxidation in blood plasma by an ample supply of endogenous antioxidants, this protection is apparently not adequate in the arterial wall where the oxidative modification of LDL is induced by endothelial cells. One hypothesis proposed is that natural antioxidants may be depleted within the arterial sub-endothelial space where oxidation takes place. The oxidative modification of LDL by endothelial cells can be completely inhibited in the presence of sufficient vitamin E or other antioxidants such as butylated hydroxytoluene. These antioxidants can inhibit two steps in the free radical oxidation pathway, by reacting with the peroxyl radicals to stop the propagation chain, and with the alkoxyl radicals to inhibit the breakdown of the hydroperoxides, and the formation of aldehydes (Figure 12-6). Therefore, antioxidants can inhibit the formation of hydroperoxides and aldehydes, which in turn will prevent the oxidative modification of LDL.

a. Endogenous antioxidants. In addition to the lipid-soluble antioxidants associated with LDL (Table 12-2), human plasma contains antioxidant proteins and enzymes and water-soluble enzymatic and non-enzymatic peroxide destroyers (see Section B). These antioxidants in plasma are also found in extracellular fluids and may inhibit *in vivo* oxidation of LDL in the intimal subendothelial space (Figure 12-4). The non-enzymatic antioxidants (metal-binding proteins, uric acid and ascorbic acid) are more important in extracellular fluids than the enzymatic antioxidants (superoxide dismutase and glutathione peroxidase). Plasma protein thiol groups are also active by trapping any aqueous peroxyl radicals found in plasma. Vitamin E is one of the most abundant lipid-soluble antioxidants in plasma and LDL that protects the lipids of LDL particles.

b. Antioxidant supplementation. Several studies showed that antioxidants can reduce the oxidative modification of LDL *in vitro*, and can inhibit experimental atherosclerosis in animals. Various antioxidants react differently in LDL and their effectiveness varies greatly according to their concentration and the types of oxidant used (Table 12-3). When added to isolated LDL, vitamin C was an effective antioxidant against oxidation induced by endothelial cells, macrophages, copper and the aqueous azo initiator, AAPH (2'-azobis[2-amidinopropane] hydrochloride). Vitamin C may inhibit LDL oxidation by scavenging water-soluble peroxyl radicals and by regenerating α-tocopherol present in LDL. Vitamin E added *in vitro* inhibited LDL oxidation by endothelial cells and macrophages more effectively than oxidation induced by copper. In contrast, vitamin E showed a prooxidant effect when LDL oxidation was initiated by AAPH. With this artificial initiator, α-tocopherol appears to promote radical chain transfer.

Supplementation of vitamin C *in vivo* showed no effect on LDL oxidation tested *in vitro* (Table 12-3). This result is expected with hydrophilic antioxidants such as Vitamin C, which cannot be tested with isolated LDL because they are removed from LDL when separated from plasma. Similar negative effects were observed in studies of the hydrophilic flavonoid antioxidants when tested in LDL after isolation from plasma (Section F.5.). Supplementation with vitamin E resulted in a significant improvement of LDL resistance against oxidation by cells, macrophages and copper, but a prooxidant effect was found with AAPH. The relevance of using AAPH as an oxidant to test antioxidants for their biological effects is, therefore, questionable.

The administration of antioxidants in the diets of several experimental animals can inhibit LDL oxidation and retard the progression of atherosclerosis (see Section F. 5.). LDL oxidation is effectively inhibited *in vitro* by a large number of natural antioxidants, including flavonoids (Chapter 8). Flavonoids inhibit the oxidation of LDL by a multiplicity of mechanisms, including scavenging reactive oxygen species (superoxide, oxidized lipids, oxysterols), binding iron or copper catalysts, and protecting α–tocopherol present in LDL. Flavonoids are also known to inhibit lipoxygenase and cyclooxygenase enzymes that play key roles in eicosanoid synthesis (see Section F. 2.). The evidence supports the possible use of antioxidants to ameliorate the stability of LDL oxidation, but the precise mechanism of how atherosclerosis is initiated *in vivo* needs to be clarified before antioxidants can be used as a mode of intervention to prevent this disease.

F. NUTRITIONAL CONSEQUENCES

Several hypotheses have been advanced to explain the development of cardiovascular disease and the *lipid hypothesis* has been given the most attention. According to this hypothesis, the primary cause of atherosclerosis is the consumption of too much fat and particularly saturated fats in the diet. Hyperlipidemia and hypercholesterolemia in plasma are recognized as the most important symptoms of this disease. Modification of lipoproteins is another factor affecting the atherogenic potential, and high plasma LDL accelerates atherogenesis. Epidemiological and clinical studies showed that the effect of diet on the level of plasma LDL cholesterol has the strongest relationship to heart diseases. Plasma HDL is inversely related to heart disease risk. Thus, the main biochemical risk factors for cardiovascular diseases are high plasma LDL cholesterol and low plasma HDL cholesterol. On this basis, the desired nutritional goals are a decrease of LDL cholesterol and an increase in HDL cholesterol. Reduction of LDL cholesterol would limit the uptake of cholesterol by cells in fatty streaks. An increase in HDL cholesterol would increase the rate of cholesterol removal from the arteries.

The nature and amount of lipids in the diet may not be the only causative factors, however. Heart disease is multifactorial and other risk factors of environment and life style as well as genetically inherited traits, may exert an even greater influence than diet, and these include stress, hypertension, obesity, and lack of exercise. Factors that interact to cause high blood pressure, hypertension, and blood toxins are all implicated in the development of heart diseases.

1. Effect of Saturated Fatty Acids

Early work established a link between the amount and quality of dietary fats and plasma cholesterol concentration. The general consensus among nutritionists and epidemiologists is that saturated fats in the diet are the most important contributors to cardiovascular disease by influencing blood cholesterol and LDL cholesterol. Although the mechanism is not clear, saturated fats may interfere with the ability of LDL receptors to clear LDL from the blood stream (Figure 12-2). Another hypothesis is based on evidence suggesting that abnormalities in lipoprotein metabolism predispose certain subjects to increased risk of cardiovascular disease. A poor ability to cope with postprandial lipemia may increase this risk because cholesteryl esters can be shunted from LDL and HDL into chylomicron remnants that become atherogenic as they are slow to clear from plasma. According to this hypothesis, the remnants of chylomicrons and VLDL (Figure 12-2) are more atherogenic than LDL. Subjects predisposed to this abnormal lipoprotein metabolism have higher blood triacylglycerol levels, low concentrations of HDL and a disturbance in lipoprotein lipase activity.

Certain saturated fatty acids, such as lauric (C12:0), myristic (C14:0) and palmitic (C16:0), are known to raise plasma LDL cholesterol, especially at low levels of linoleic acid. Stearic acid (C18:0) does not increase LDL cholesterol, apparently because it is rapidly converted to oleic acid by desaturation (Figure 12-9). Palmitic and stearic acids make up about 36-40% of the fatty acids of animal fats and 11-16% of the fatty acids of vegetable fats. Although myristic acid is present at higher levels in butter fat (12%) and in palmkernel and coconut oils (45-48%), it is usually a rather minor component of the diet. It is generally agreed that high levels of saturated fatty acids in dietary fats constitute a risk factor for cardiovascular disease.

2. Effect of Polyunsaturated Fatty Acids

The replacement of saturated fatty acids with polyunsaturated fatty acids is generally accompanied by a lowering of plasma total cholesterol and LDL cholesterol. Linoleic acid reverses the effects of saturated fatty acids in raising blood triacylglycerols and LDL. Although the results of many studies are inconsistent, most studies show that plasma HDL cholesterol levels tend to decrease when the saturated fatty acids in the diet are replaced with polyunsaturated fatty acids. However, the level of HDL cholesterol may actually be determined by the saturated fatty acid content of the diet.

Although polyunsaturated fats in the diet lower LDL cholesterol, they increase the susceptibility of LDL to oxidation. Higher levels of polyunsaturated fats in the diet also raise the requirements for vitamin E and other antioxidants. In view of the important role that oxidized LDL plays in the development of atherosclerotic lesions, the unsettled question remains of whether the effect of dietary polyunsaturated fatty acids in lowering plasma LDL is greater than the risk factor of increasing the oxidative susceptibility of LDL.

Linoleic acid (18:2 n-6) and linolenic acid (18:3 n-3) are *essential fatty acids* and are required for good health because they cannot be synthesized by animals which lack the required desaturases (Figure 12-9). These polyunsaturated fatty acids must be supplied by plant foods because they are necessary for growth and reproduction and serve as precursors of various oxygenated fatty acids, called *eicosanoids*. These compounds are physiologically active at extremely low concentrations. As an essential fatty acid, linoleic acid is required in the diet at a level of about 1% of dietary energy and linolenic acid at even lower levels of about 0.2%.

a. n-6 Polyunsaturated fatty acids. In membranes, linoleic acid in phospholipids is the precursor of arachidonic acid, which is released by hydrolysis with phospholipase (Figure 12-10). The conversion of arachidonic acid (20:4 n-6) to biologically active eicosanoids occurs by a complex cascade of reactions catalysed by cyclooxygenase producing prostaglandins, and by lipoxygenase producing hydroperoxides and leukotrienes (Figures 12-10 and 12-11). The resulting mixtures of eicosanoids play a significant role in cellular signal communication. They are important mediators of leucocytes (monocyte, macrophage, neutrophil) which play a role in atherogenesis. They regulate immune function by altering activity of monocytes, macrophages and neutrophils, and are implicated in atherosclerosis, cancer, and autoimmune disorders (Table 12-4).

Lipid hydroperoxides play an important role in the biosynthesis of prostaglandins from arachidonic acid. The control of the concentration of

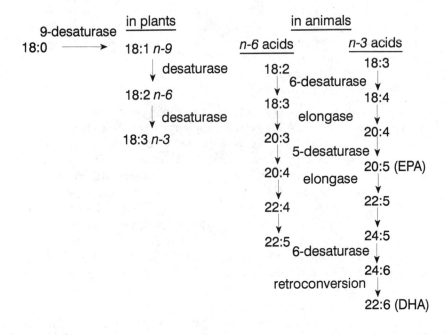

Figure 12-9. Biosynthesis of fatty acids. Oleate (18:1) formed by desaturation of stearate (18:0) can undergo further desaturation to linoleate (18:2) and linolenate (18:3) in plants. In animals, the n-6 and n-3 polyunsaturated fatty acids undergo sequential desaturation and elongation to produce arachidonic acid (20:4 n-6), eicosapentaenoic acid (20:5 n-3) and docosahexaenoic acid (22:6 n-3).

hydroperoxides, also referred to as *peroxide tone*, is important to maintain normal cell functions. Hydroperoxides can either stimulate cyclooxygenase activity at low concentrations (0.01 μM) or inhibit this enzyme at high concentrations (1 μM). The enzymes glutathione peroxidase and glutathione transferase control prostaglandin synthesis in the cell by suppressing hydroperoxides and protecting cells against the accumulations of undesirable levels of hydroperoxides.

When the dietary level of linoleic acid is excessive, its conversion to arachidonic acid may cause serious pathophysiological actions by overproducing eicosanoids that generate abnormal signals to the cells (Figure 12-12). The enzyme *prostacyclin synthetase* is extremely sensitive to hydroperoxides which control the activation and inhibition of cyclooxygenase. The basal rate of converting arachidonic acid to eicosanoids requires relatively low levels of linoleic acid (less than about 1% of energy). However, the Western diet which includes about 8 to 10% of total calories as linoleic acid greatly favors the biosynthesis of arachidonic acid. In the United States a significant dietary shift from animal to vegetable fats has resulted in a substantial and continuing increase in consumption of linoleic acid from about

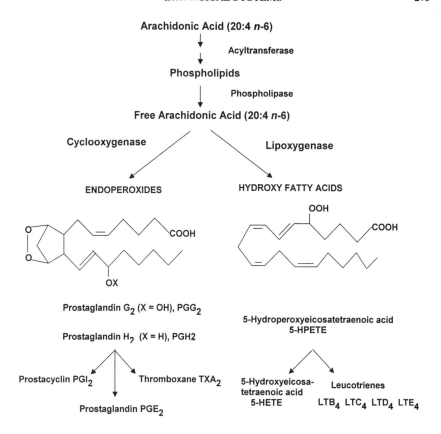

Figure 12-10. Eicosanoid synthesis by cyclooxygenase and lipoxygenase pathways. (See Table 12-4 for biological functions of eicosanoids).

8 g per person per day in 1910 to about 27 g per person per day in 1985 (National Research Council, 1989).

The blood level of arachidonic acid formed from dietary linoleic acid critically affects the response of vessels to stress. This response is delicately balanced by two prostaglandin systems (TXA_2 and PGI_2.) (Figure 12-10, Table 12-4). The production of *thromboxane* TXA_2 causes vasoconstriction resulting from adhesion and excessive platelet aggregation that promotes blood clotting, a critical event in preventing bleeding. This effect is counteracted by the production in the walls of blood vessels of *prostacyclin* PGI_2 causing vasodilation and limiting the aggregation process away from an injury. The flow of blood is thus delicately controlled by the steady state formation of TXA_2 and PGI_2. Since they are both formed from arachidonic acid, the optimum level of dietary linoleic acid to minimize the risk of thrombosis is uncertain.

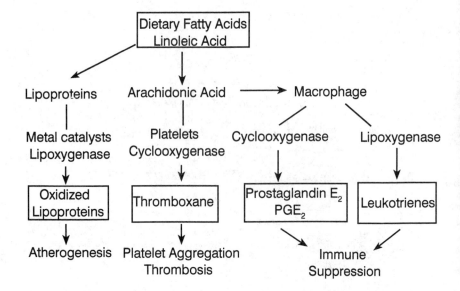

Figure 12-11. Metabolic pathways for the formation of oxidized lipoproteins and eicosanoids. These oxidation products contribute to atherogenesis, thrombosis and immune suppression.

b. n-3 Polyunsaturated fatty acids. Eicosapentaenoic (20:5 n-3) (EPA) and docosahexaenoic (22:6 n-3) (DHA) acids from seafood and fish oils and linolenic acid (18:3 n-3) from soybean and rapeseed oils are known to antagonize and compete with the metabolism of linoleic acid by lowering the

TABLE 12-4.
Biological functions of eicosanoids.[a]

System	Effects	Active species
Prostanoids and thromboxanes (cyclooxygenase products)		
Blood vessels	vasodilation	$PGI_2 > PGI_3 > PGE_1$
	vasoconstriction	TXA_2
Platelets	adhesion	
aggregation		TXA_2
	antiaggregation	$PGI_2 > PGI_3 > PGE$
Tissue	Pain	PGI_2
	Cytoprotection	PGI_2, diMe PGE
Leukotrienes (lipoxygenase products)		
Vascular	constriction	LTC_4, LTD_4
	permeability	LTC_4, LTD_4
Neutophils, monocytes	adhesion	LTB_4
	chemotaxis	LTB_4, HETE (5,9,11)
	lysozyme secretion	LTB_4, HETE (5,12)

[a] From Kinsella (1987). See Figures 12-10, 12-11, 12-12.

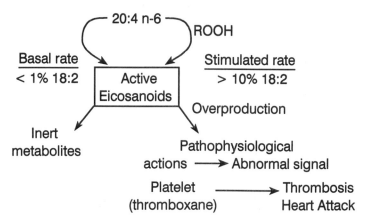

Figure 12-12. Effect of dietary linoleic acid on synthesis of active eicosanoids and metabolism. Pathophysiological actions result from faster eicosanoid formation by increased hydroperoxide levels. From Lands et al. (1986).

biosynthesis of arachidonic acid. n-3 Polyunsaturated fatty acids compete with n-6 polyunsaturated fatty acids for the enzymes involved in the formation of arachidonic acid from linoleic acid. The eicosanoids derived from EPA and

TABLE 12-5.
Effect of fish oil supplementation on *in vitro* LDL oxidation
determined by static headspace gas chromatography.[a]

Analyses	Unsupplemented	Supplemented
Volatiles (nmol/mL LDL)		
Pentane	9.6	7.5
Propanal	3.7	13.4
Pentanal	14.7	11.4
Hexanal	138	108
Total volatiles [b]	189	189
Fatty acids (%)		
n-6 PUFA	43.7	35.0
n-3 PUFA	3.2	14.6
Total PUFA	46.8	49.6
Tocopherol (nmol/mL LDL)	14.0	18.0

[a] Frankel et al. (1994). LDL obtained from blood of 9 subjects after consuming fish oil each day for 6 weeks (supplemented) or after consumer control diet without fish oil (unsupplemented). LDL samples oxidized with $CuSO_4$ were analyzed by static headspace gas chromatography (Frankel et al., 1992). See Figure 12-13.

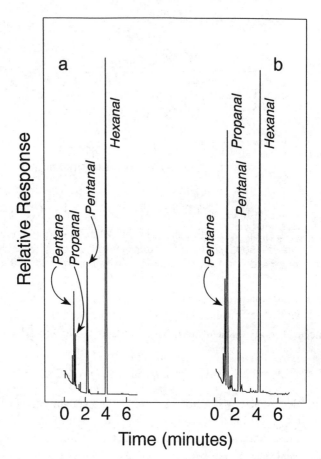

Figure 12-13. Headspace gas chromatograms of copper-oxidized low density lipoprotein isolated from a, unsupplemented subject, and b, fish oil supplemented subject. From: Frankel et al. (1994), with permission from the AOCS Press.

DHA in fish produce weakly inflammatory eicosanoids, in contrast to the strongly pro-inflammatory eicosanoids from the arachidonic acid produced by excessive linoleic acid in the diet. n-3 Polyunsaturated fatty acids are known for their antithrombotic activity by inhibiting platelet aggregation and converting arachidonic acid to thromboxane. Because n-3 polyunsaturated fatty acids require a much higher level of hydroperoxides than n-6 polyunsaturated fatty acids to be converted to eicosanoids, the n-3 acids act as competitive inhibitors of prostaglandin biosynthesis from the n-6 acids. An increase in the n-3/n-6 PUFA ratio in cell membranes alters the composition of eicosanoids by the actions of the enzymes cyclooxygenase and lipoxygenase (Figures 12-10 and 12-11). These enzymes need more hydroperoxides in the presence of EPA than with arachidonic acid. The leukotriene LTA5 produced by EPA is less

BIOLOGICAL SYSTEMS 277

TABLE 12-6.
Reported effects of dietary n-3 polyunsaturated fatty acids on manifestations associated with cardiovascular disease.[a]

Plasma lipids
    Reduce hepatic fatty acid synthetase, triacylglycerol synthesis in liver,
    VLDL triacylglycerol production, ApoB synthesis
    Increase tissue deposition of cholesterol, HDL cholesterol
Arachidonated metabolism
    Inhibit 6-desaturase and arachidonic acid synthesis
    Reduce peroxide tone
    Compete with arachidonic acid for cyclooxygenase and lipoxygenase
Eicosanoids
    Reduce TXA_2 synthesis (by platelets)
    Convert EPA to TXA_3, a weak agonist
    Form bioactive PGI_3 and LTB_5 from EPA
    Decrease ratio TXA to PGI_2 and reduce platelet aggregation and vasoconstriction
    Reduce LTB4 (chemotaxis and adhesion)
Platelets
    Decrease arachidonic acid and increase EPA and DHA contents, TXA_2 synthesis
    Decrease aggregation, adhesion and release reactions
Endothelium
    Reduce PGI_2 synthesis (slightly)
    Enhance PGI_3 synthesis
Monocytes and macrophages
    Reduce adhesion, synthesis of LTD_4 (chemotaxls), inflammatory reactions
Other effects
    Enhance bleeding tendency
    Cause vasorelaxation

[a] From Kinsella et al. (1990). See Figures 12-9, 12-10, 12-11, 12-12.

effective as chemotactic agent. Fatty acids of the n-3 PUFA series from fish also have a triacylglycerol lowering effect in people with hypertriglyceridemia, and play a vital role in vision and brain function.

The fatty acid composition of LDL is influenced by supplementation of fish oil, but the effect on LDL oxidation is not clear because the increased level of EPA and DHA is accompanied by a corresponding decreased level of n-6 polyunsaturated fatty acids. The assessment of susceptibility of LDL oxidation varies also according to the methods used to follow lipid peroxidation. When hyperglyceridemic subjects were supplemented with fish oil, LDL oxidation resulted in a significant increase in propanal formation. During *in vitro* oxidation of LDL with copper, propanal formation increased significantly in subjects supplemented with fish oil (Figure 12-13). This increase correlated directly with increases in n-3 polyunsaturated fatty acids in LDL, but the oxidative susceptibility of LDL did not change, based on total volatile oxidation products (Table 12-5). However, other studies reported significant increases in formation of TBARS resulting from fish oil diets. Because TBARS are formed by oxidative decomposition of polyunsaturated fatty acids containing more than two double bonds (Chapter 5.E), the oxidation of any lipids containing n-3 polyunsaturated fatty acids would be expected to produce high levels of TBARS. Therefore, TBARS determinations cannot be used as

TABLE 12-7.
Oxidation of LDL from animals and humans fed diets containing sunflower oil or high-oleic sunflower oil.[a,b]

Feeding periods - oxidation tests	Sunflower oil	High-oleic sunflower oil
Rabbits on diets of 10% oil [a]		
6 weeks - Conj. dienes, 8 hr	0.41	0.5
10 weeks - Conj. dienes, 8 hr	0.48	0.7
6 weeks, TBARS, 8 hr	32	10
6 weeks - Macrophage degradation, 8 hr	4.5	0.8
Humans on diets containing 39.2 % fat [b]		
5 weeks - Conj. dienes, 8 hr	1.2-1.3	0.75-0.90
5 weeks - TBARS, 8 hr	43.6	37.6
5 weeks - Macrophage degardation, 8 hr	11.9	6.8

[a] From Parthasarathy et al. (1990). [b] From Reaven et al. (1991). Rabbits or humans fed either sunflower oil (67% 18:2, 20% 18:1) or high-oleic sunflower oil (83% 18:1, 8% 18:2). Conjugated dienes (absorbance at 234 nm) and TBARS (as nmol MDA/ml LDL protein) measured on LDL oxidized with 5 μM copper, and macrophage degradation of ^{125}I-LDL (μg/mg cell protein / 5hr). MDA, malonaldehyde.

the sole determination of oxidative susceptibility to compare the effects of dietary n-6 versus n-3 polyunsaturated fatty acids on LDL oxidation. Better methodology is needed to obtain the basic chemical information necessary to understand the effects of dietary polyunsaturated fatty acids on the oxidation of LDL and other biological systems implicated in many diseases.

Dietary n-3 polyunsaturated fatty acids contribute additional effects on the cellular prooxidant activity of monocytes. Dietary supplementation of humans with n-3 polyunsaturated fatty acids increased the DHA and EPA of monocyte membrane and decreased arachidonic acid. These monocytes produce less superoxide and other reactive oxygen species. The mechanism for this effect of n-3 polyunsaturated fatty acids is related to the decrease in arachidonic acid and the resulting bioactive metabolites of arachidonic acid.

Increased intakes of n-3 polyunsaturated fatty acids in the diet have been recommended because of their multitude of important biological functions and their known effects in competing with excessive linoleic acid (Table 12-6). Fish oils are presumed to reduce the risk of coronary heart disease by decreasing plasma lipids and by balancing eicosanoid metabolism. The cardiovascular system is closely regulated by a balance between the production of thromboxane (causing vasoconstriction) and prostacyclin (causing vasodilation). Although dietary α-linolenic acid is less effective than EPA and DHA in depressing eicosanoid synthesis, the more consistent consumption of linolenate-containing leafy vegetables and vegetable oils (soybean and canola oils) may be also desirable in low-fat diets. However, the great susceptibility of n-3 polyunsaturated fatty acids to oxidation may constitute a potential risk factor in the development of cardiovascular disease. Studies with animals fed high amounts of menhaden or linseed oil for prolonged periods showed evidence of yellow discoloration of depot fat, referred to as *yellow fat disease*, which is attributed to the formation of *lipofuscin* by the polymerization of end-products of lipid peroxidation. This condition may also reflect a deficiency of vitamin E.

The higher requirement for vitamin E in diets containing higher level of linoleic acid is generally satisfied by the consumption of vegetable oils rich in tocopherols. However, fish oils contain very little vitamin E (40-200 $\mu g/g$), and increased consumption of n-3 polyunsaturated fatty acids leads to an even higher requirement for vitamin E. Furthermore, vitamin E does not appear to provide sufficient protection to animals on diets high in n-3 polyunsaturated fatty acids. Practically, the production of oxidatively stable fish and fish oils is extremely difficult because of their high susceptibility to oxidation. This problem underscores the need to limit the amount of n-3 polyunsaturated fatty acids to the minimum required to antagonize the effects of linoleic acid, estimated at about 1% of calories. More effective methods are also needed to control the oxidation of fish oils and seafood products if their consumption is to be increased.

3. Effect of Monounsaturated Fatty Acids

Diets high in oleic acid are now recognized to be as effective in lowering serum cholesterol as diets rich in linoleic acid. However, in contrast to polyunsaturated fats, monounsaturated fats have the important advantages of not only lowering LDL, but also of being much less susceptible to oxidation. These effects are both considered to be important risk factors. The desirable effects of monounsaturated fats are maintained even when present in the diet at relatively high levels. Experiments with rabbits and humans fed diets rich in high-oleic sunflower oil compared to standard sunflower oil, showed that LDL was much less susceptible to copper-catalysed oxidation and underwent less degradation by macrophages after incubation with endothelial cells (Table 12-7). The increased resistance of LDL to oxidation was directly related to its increased oleate content. Several other studies showed that the linoleate content of LDL strongly correlated with the extent of oxidation and the reverse trend was observed with the oleate content of LDL. In studies with animal fed an atherogenic diet high in cholesterol, diets rich in oleic acid produced LDL more resistant to oxidation and significantly decreased fatty streak formation. Oleate-enriched diets may, therefore, reduce LDL oxidation and slow progress of atherosclerosis. Diets with lower linoleic acid and higher oleic acid contents have a clear advantage by lowering plasma cholesterol by the same degree as polyunsaturated fatty acids and at the same time decreasing the risk of LDL oxidation.

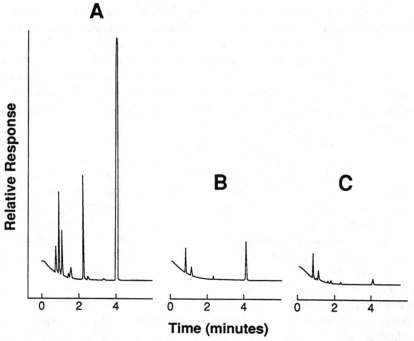

Figure 12-14. Headspace gas chromatograms of samples of oxidized human low density lipoprotein. A, control LDL, B, plus 3.8 µM wine phenolics, C, plus 10 µM wine phenolics. From Frankel et al. (1993), with permission from The Lancet Ltd.

4. Effect of Trans Unsaturated Fatty Acids

Recent evidence showed that *trans* mono-unsaturated fatty acids are hypercholesterolemic in human studies (11% *trans* of energy intake) compared with oleic acid by raising the level of atherogenic LDL in serum and lowering the level of HDL. Further work confirmed that *trans* fatty acids from hydrogenation in margarines and shortening have similar effects as saturated fatty acids in the diet of humans (7.7% of energy) by lowering HDL cholesterol and increasing LDL cholesterol compared to linoleic acid. Epidemiological studies showed no consistent association between *trans* fatty acid intake and coronary heart disease, however. In addition to the effect of *trans* fatty acids in elevating LDL cholesterol, its effect on the oxidative susceptiblity of human LDL as another risk factor for cardiovascular disease remains to be determined. Since the average consumption of *trans* fatty acids is estimated to be 3-4% of daily energy intake in the United States, there is little concern that *trans* fatty acids may increase the risk of cardiovascular disease.

5. Effect of Antioxidants

Many studies tested directly the effects of antioxidants on the inhibition of atherosclerosis in animals and humans. With cholesterol-fed rabbits positive effects were reported with probucol (a dithio-butylated diphenol), butylated hydroxyanisole and diphenylenediamine. With humans, epidemiological evidence indicated that high intakes of vitamin E and quercetin (Chapter 8) correlate with reduced coronary disease. Several clinical trials with human subjects demonstrated that dietary supplementation with vitamin E produced significant enrichment of LDL vitamin E content and *ex vivo* protection of LDL against oxidation induced by copper ions, endothelial cells and smoking. The effect of vitamin C supplementation cannot be tested because it is water-soluble and removed from LDL when isolated from plasma (Table 12-3). A large number of studies showed that natural phenolic antioxidants reduce the oxidative modification of LDL *in vitro*, and a few animal studies showed that they can inhibit experimental atherosclerosis. LDL particles from hypercholesterolemic individuals are believed to undergo more oxidant stress by prolonging their exposure to intravascular oxidation. Therefore, these individuals may require higher antioxidant intakes for protection against the damaging effects of higher circulating levels of oxidatively susceptible LDL.

Considerable epidemiological evidence has shown an association between diets rich in fruits and vegetables and decreased risk of cardiovascular disease and certain types of cancer. The protective effects of fruits and vegetables have been usually associated with the level of vitamins E and C and β-carotene. However, significant amounts of phenolic flavonoid compounds (Figure 8-10) are also present in fruits and vegetables and their potential nutritional effects as antioxidants must also be considered for their cardioprotective actions. Similarly, flavonoid antioxidants from fruits and vegetables may be important in regulating eicosanoids. Because of the ability of flavonoids to inhibit cyclooxygenase and lipoxygenase activity, these antioxidants may exert anti-inflammatory effects (Figure 12-11), and may have broad metabolic and physiological effects. Our daily intake of plant flavonoids is estimated to range from 20 mg to 1 g. However, little is known about the absorption, metabolism and *in vivo* antioxidant activity of flavonoid compounds in humans. The bioavailability of plant flavonoids may vary in different foods. How the food matrix and various flavonoid-nutrient interactions affect absorption and metabolism of flavonoid antioxidants is largely unknown.

The results of various methods used to measure the effectiveness of dietary antioxidants *ex vivo* or *in vivo* do not always follow predictable patterns. Antioxidants have been commonly evaluated by determining their relative concentrations or the so-called "antioxidant status" in blood or other tissues after dietary supplementation. These analyses are confounded by variations in the relative stability of antioxidants and by the up-regulation of a wide array of antioxidant enzymes and scavenging systems in response to oxidative stress (Section B). The effectiveness of antioxidant intake may be more usefully evaluated by specific analyses of products formed at different stages of lipid oxidation than by the determination of their relative concentrations. Antioxidants have also been evaluated on the basis of their reactivity toward peroxyl radicals and their inhibition of hydroperoxide formation. However, their effects in scavenging alkoxyl radicals and in inhibiting aldehyde formation, have been overlooked. Yet, the production of aldehydes is likely to be more relevant to oxidative damage due to their interaction with the apoB of LDL (Figures 12-5 and 12-6).

Several nonspecific assays known as "radical trap methods" have been used to test the "antioxidant capacity" of biological tissues and fluids by measuring the relative reactivity of antioxidants toward artificial radicals (see Chapter 8.A.1.c). However, these methods provide no quantitative information on what biological targets are protected by the antioxidants. Several studies have shown marked variations in antioxidant activity with different lipid systems and oxidizing conditions. Various flavonoids showed widely different antioxidant activities and even prooxidant activities according to their abilities to scavenge hydroxyl or superoxide radicals or lipid peroxidation. Because the properties of various phenolic compounds are very system-dependent, caution is required in interpreting the antioxidant activities determined with non-specific assays in the absence of suitable substrate targets. The validity of the antioxidant capacity or radical trap methods may be questionable without product analyses to indicate what specific biological substrate is protected.

The effects of phenolic compounds in wine on the oxidative susceptibility of human LDL were investigated by measuring the amounts of hexanal and conjugated dienes formed by Cu^{2+} catalysed-oxidation of human LDL. Hexanal formation was inhibited by 60 and 98% by the addition of diluted de-alcoholized red wine containing 3.8 and 10 μM phenolic compounds (Figure 12-14). The relative inhibition of LDL oxidation varied from 46 to 100% with red wines and from 3 to 6% with white wines. The antioxidant activity of wines toward LDL oxidation was distributed widely among the principal phenolic compounds, including gallic acid, catechin, myricetin, quercetin, caffeic acid, rutin, epicatechin, cyanidin and malvidin-3-glucoside (see Chapter 8). The antioxidant activities of extracts of different table and wine grapes in human LDL *in vitro* were comparable to those for wines. The extracts of various grape varieties showed different trends in antioxidant activity between liposome and LDL systems. Rosemary extracts and components, tea catechins and commercial green teas had different antioxidant activities when compared in different lipid systems (bulk oil, lecithin liposomes, and human LDL). Therefore, the relative activity of natural phenolic antioxidants is greatly affected by the test system used and the biological target to be protected.

In addition to preventing LDL oxidation, phenolic flavonoids are known to inhibit phospholipase, cyclooxygenase and lipoxygenase, which play a key role in the arachidonate cascade (Figure 12-11). By down-regulating cyclooxygenase and lipoxygenase, flavonoids and other phenolic compounds in fruits and vegetables may reduce thrombotic tendencies and inflammatory reactions in the body. By preventing or reducing monocytes from collecting in vessel walls, these phenolic compounds interfere with the immune response and with platelet aggregation involved in the blood clotting process and prevent thrombosis.

Several recent studies have produced mixed results in indicating *in vivo* antioxidant activity of phenolic compounds in red wine. However, the results of these studies are unconvincing because non specific assays were used for antioxidant action that may be inadequate due to the confounding and indirect effects of many serum and plasma components. The effect of *in vivo* supplementation of polyphenols in red wine on the oxidizability of LDL cannot be tested *ex vivo* because, like vitamin C, these hydrophilic compounds are removed from LDL during isolation from plasma (Table 12-3). Not surprisingly red wine consumption showed no effect on LDL oxidation *ex vivo*. The total antioxidant activity of polyphenols of grapes, wines and green teas was evaluated by determining their scavenging ability toward an artificial radical cation system. However, this approach for the evaluation of natural antioxidants by an artificial radical model system provides no information on what lipid or protein is protected.

To determine the real effects of natural antioxidants, it is important to obtain specific chemical information about what products of lipid peroxidation are inhibited. Several specific assays are needed to elucidate how lipid oxidation products act in the complex multi-step process of lipid peroxidation (Figure 12-6) and oxidative damage in biological tissues. The results of many complementary methods are required to determine oxidation products formed at different stages of the free radical chain. Since antioxidants show different activities toward hydroperoxide formation and decomposition, it is important that more than one method be used to monitor the oxidation process. The great difficulty of conducting *in vivo* experimental work to demonstrate the activity of plant antioxidants has added to the uncertainty of their nutritional benefits.

6. Effect of Iron Supplementation

Healthy subjects are well protected against oxidative damage from free iron by the strong metal binding properties of plasma proteins (Section B). However, metal ions may become available to promote oxidation as a result of tissue injury from bacterial infections, inflammations and human disease. Oxidative stress by tissue injury in unhealthy subjects can release enough free iron ions to promote the generation of active oxygen species causing tissue damage. Lowering the amount of storage *in vivo* may, therefore, decrease injury due to oxygen radicals. In excess, dietary iron and iron supplementation may contribute to oxidative damage and may overload the defense system. The interaction of iron with superoxide, hydrogen peroxide and lipid hydroperoxides in producing more reactive oxygen species is now recognized as one of the most serious sources of free radical injury.

A major form of antioxidant defense in human plasma is the prevention of iron and other transition metal ions from promoting the generation of reactive oxygen species. Although many therapeutic measures to inactivate metals by chelation therapy have been advocated, the avoidance of iron supplementation in the diet may provide a simpler and more economical approach to this

TABLE 12-8.
Human consumption of dietary fats (percent calories).

Fats	Current intakes [a]	Recommended Intakes				
		NATO [a]	Kinsella [b]	Dupin et al [c]	COMA [d]	SCF [e]
Saturated	15	6-7	8	—	—	—
Monounsated	18	12-14	14	—	—	—
Polyunsaturated	7	6-7	8	—	6-10	max 15
n-6	7	4.8	6	3-6	min 1	2 (min 0.5)
n-3	0.8	1.3 [f]	2	0.5-1	min 0.2	0.5 (0.1-5)
n-6/n-3	9	4	3	4-10	—	—
Total	40	24-28	30	—	—	—

[a] Galli and Simopoulos (1988).
[b] Kinsella (1991).
[c] Dupin et al. (1992). Centre national d'études et de recommandations sur la nutrition et l'alimentation (CNERNA) (France) (1992).
[d] Committee on medical aspects of food policy (UK) (1991).
[e] Scientific Committee for Food (SCF) (European Union) (1993).
[f] Includes 1.0 % 18:3, 0.27% EPA +DHA, and 18-3 n-3 / n-3 as EPA + DHA = 4/1.

potential health problem. The current practice of iron supplementation of flour and cereal products may be imprudent in view of the possible risk of excessive iron in generating oxidative stress in certain segments of the adult and older population. Iron deficiency was even advocated as a more effective and practical antioxidant treatment than supplementation with either synthetic or natural antioxidants.

G. OPTIMUM DIETARY INTAKES

Nutritional advice on optimum fat intakes has been generally based on epidemiological, clinical and animal studies showing an association between dietary fat, serum cholesterol levels and coronary heart disease. These studies have concluded that an elevated level of serum cholesterol is a major risk factor for a heart attack. Keys et al. and Hegsted et al. (1965) have formulated equations based on a large number of human feeding trials indicating that saturated fatty acids raise serum cholesterol twice as much as polyunsaturated fatty acids decrease it, and that dietary cholesterol has a less important effect in raising serum cholesterol. These simplified relationships have been based on the assumption that the ratio of polyunsaturated to saturated fatty acids (p/s ratio) is fundamental to the control of coronary heart disease by reducing elevated serum cholesterol. This assumption has led to the unfortunate tendency in many Western countries to markedly increase the polyunsaturated fats in the diet.

We have seen in Section F that the dietary factors linked with the incidence of coronary heart disease reflect complex relations that cannot be simply based on ratios of polyunsaturated to saturated fats. It is now evident that not all saturated, polyunsaturated and monounsaturated fats play the same role in nutrition. n-3 Polyunsaturated fatty acids are not particularly effective in lowering serum cholesterol, but markedly reduce serum triacylglycerols in hypertriglyceridic subjects. The tendency has been recently to make nutritional recommendations on the basis of total fat intake and on the ratio of n-6 to n-3 polyunsaturated fatty acids. Reducing n-6 polyunsaturated fatty acids may be an effective way of utilizing relatively low and more practical amounts of fish oils.

Current fat intakes are considered to have high contents of saturated and polyunsaturated fats and a high n-6 to n-3 PUFA ratio (Table 12-8). Recommendations have been made to reduce total fat, saturated fats, and decrease the n-6 to n-3 PUFA ratio from 4/1 to 3/1. These recommendations reflect much guesswork, however. The traditional p/s ratio and now the n-6/n-3 PUFA ratio may be of little value. These recommended ratios fail to recognize the multiple effects of dietary n-6 and n-3 polyunsaturated fatty acids and their metabolic interrelationships. In a multivariate system, attempts to describe a ratio of two variables have little meaning and is an inappropriate description of a mixture of fatty acids having a multiplicity of biosynthetic effects. These values should be considered only as possible targets. It should be no longer valid to base dietary recommendations on the attainment of plasma polyunsaturated lipid concentrations in view of the great multiplicity of functions mediated by eicosanoid metabolites.

The unduly increased level of dietary linoleate may be excessive in promoting a variety of chronic diseases including coronary artery disease by the excessive release of eicosanoids in blood and by increasing the oxidative susceptibility of n-6 polyunsaturated fatty acids in lipid components of LDL. n-3 Polyunsaturated fatty acids have anti-aggregatory and anti-inflammatory functions by inhibiting eicosanoid synthesis. On this basis, an increased intake of n-3 polyunsaturated fatty acids (18:3 from vegetable oils, EPA and DHA from fish oils) has been advocated. However, the levels necessary for these beneficial effects of n-3 polyunsaturated fatty acids are not well defined. Levels of total n-3 polyunsaturated fatty acids above 1% of calories in the diet may become a risk factor by increasing the oxidative susceptibility of LDL and other lipoproteins and oxidative stress. One important question that concerns nutritionists today is whether we need additional or more effective natural antioxidants in our diet to reduce oxidative stress from either excessive dietary polyunsaturated lipids or from other environmental factors.

H. CONCLUSIONS AND FUTURE PERSPECTIVES

Knowledge of the effects of dietary fatty acids in the development of coronary artery disease is still incomplete. Atherosclerotic lesions develop slowly by a complex series of cellular reactions in which lipid peroxidation and various forms of oxidative stress may play an important part at each stage of lesion progression. It is now firmly established that coronary artery disease is reduced by diets that lower LDL cholesterol by replacing saturated fatty acids with polyunsaturated fatty acids. However, Western populations have tended to consume high amounts of polyunsaturated fats that are susceptible to oxidation. Excessive levels of polyunsaturated fatty acids in the diet, without providing sufficient protection from antioxidants such as vitamin E, can promote LDL oxidation and upset the oxidant-antioxidant balance (Section C) in favor of cellular prooxidants and promote coronary heart disease. On the other hand, diets rich in oleic acid have clear advantages in not only lowering serum cholesterol but also in decreasing the oxidative susceptibility of LDL.

The prevention of LDL oxidation by antioxidants has been viewed as an important means of decreasing the risk of heart disease. Plant phenolic antioxidants in fruits and beverages (wine, grapes, grape juice and tea) may have a protective effect against coronary heart disease and cancer, but the molecular mechanism of protection is not understood. There is now intensive interest in developing plant sources of polyphenolic antioxidants for foods and biological systems, and for their possible health benefits. There is, however, a significant gap in our knowledge on the relative absorption of these natural antioxidants in the body and their bioavailability, and a lack of reliable methods to test their activity *in vivo*.

Polyunsaturated fatty acids from fish (EPA and DHA) may be useful to alleviate inflammation, reduce blood pressure and thrombosis. The recommended ratio of n-6 to n-3 PUFA in the diet is estimated to be around 3:1. Essential fatty acid requirements are less than 1% of calories. Therefore, current consumption in Western countries of more than 8 to 10% polyunsaturated fatty acids may be excessive and conducive to increased susceptibility to oxidative stress. An increased intake of vitamin E and other natural plants antioxidants may be required. Alternatively, increased consumption of monounsaturated fats may be highly recommended. Several population studies suggested that diets enriched in oleic acid and n-3 polyunsaturated fatty acids decrease coronary artery disease. The nutritional benefits from the so-called Mediterranean diet are generally attributed to olive oil rich in oleic acid, α-linolenic acid and natural antioxidants from fruits and vegetables. However, the minimum requirements for the n-3 as well as the n-6 polyunsaturated fatty acids need to be more closely defined because these fatty acids are highly susceptible to oxidation, especially during cooking and frying (see Chapter 11). Diets rich in EPA and DHA from fish should, therefore, be used prudently and in moderate amounts. An adequate supply of antioxidants from fruits and vegetables is also necessary to prevent peroxidation of polyunsaturated lipids.

In future research, we need to
   (a) define more precisely the role of lipid peroxidation in atherosclerosis and autoimmune disorders;
   (b) reduce oxidation products in foods;
   (c) optimize antioxidants from fruits and vegetables;

BIOLOGICAL SYSTEMS 287

(d) redefine the requirements of iron supplementation which may be excessive for some segments of the adult population;
(e) consider the oxidative susceptibility of polyunsaturated fats in the diet as an important *risk factor*. The dietary intake of polyunsaturated fats and minerals is clearly *not* a simple problem. Dietary recommendations of polyunsaturated fats must consider their susceptibility to oxidation as a possible risk factor. When the interrelationships between the metabolic effects of different polyunsaturated fatty acids are better understood an appropriate target may be to achieve a balanced eicosanoid metabolism. It would seem imprudent to make dietary recommendations to the public before the mechanisms of polyunsaturated lipid nutrition are better understood.

BIBLIOGRAPHY

Ames,B.A. Dietary carcinogens and anticarcinogens. Oxygen radicals and degenerative diseases. *Science* **221**,1256-1264 (1983)
Ames,B.A. Cancer and diet. *Science* **224**, 668-670, 757-760 (1984).
Ames,B.N., Shigenaga,M.K. and Hagen,T.M. Oxidants, antioxidants, and the degenerative disease of aging. *Proc. Natl. Acad. Sci. U.S.A.* **90**, 7915-7922 (1993).
Beare-Rogers,J. Challenges for lipid nutritionists, in *Dietary Fat Requirements in Health and Development.* pp. 201 206 (1988) (edited by J. Beare-Rogers) American Oil Chemists' Society, Champaign, IL.
Benedetto,C., McDonald-Gibson,R.G., Nigam,S. and Slater,T.F. (editors) *Prostaglandins and Related Substances. A Practical Approach.* (1987) IRL Press Ltd, Oxford.
Brown,M.S. and Goldstein,J.L. A receptor-mediated pathway for cholesterol homeostasis. *Science* **232**, 34-47 (1986).
Bruenner,B.A., Jones,A.D. and German,J.B. Direct characterization of protein adducts of the lipid peroxidation product 4-hydroxy-2-nonenal using electrospray mass spectrometry. *Chem. Res. Toxicol.* **8**, 552-559 (1995).
Commitee on Medical Aspects (COMA) of Food Policy, Department of Health. *Dietary reference values for food energy and nutrients for the United Kingdom.* (Report of the panel on dietary reference values of the committee on medical aspects of food policy. Report on health and social subjects No. 4. (1991) HMSO, London.
Dupin,H., Abraham,J. and Giachetti,I. *Apports nutritionels conseillés pour la population française.* 2nd Edition (1992) CNERNA-CNRS et Lavoisier, Paris.
Esterbauer,H. Effect of antioxidants on oxidative modification of LDL. *Annals Med.* **23**, 573-581 (1991) ,
Esterbauer,H. Cytotoxicity and genotoxicity of lipid oxidation products. *Am. J. Clin. Nutr.* **56**, 7796-7865 (1993).
Esterbauer,H., Gebicki,J., Puhl,H. and Jürgens,G. The role of lipid peroxidation and antioxidants in oxidative modification of LDL. *Free Radical Biol. Med.* **13**, 341-390 (1992).
Finocchiaro,E.T. and Richardson,T. Sterol oxides in foodstuffs: A review. *J. Food Prot.* **46**, 917-925 (1983).
Frankel,E.N. Oxidation of polyunsaturated lipids and its nutritional consequences, in *Oils-Fats-Lipids 1995* (edited by W.A.M. Castenmiller). Vol. 2, pp. 265-270 (1996) P.J. Barnes & Assoc., Bridgwater, UK.
Frankel,E.N., German, J.B. and Davis, P.A. Headspace gas chromatography to determine human LDL oxidation. *Lipids* **27**, 1047-1051 (1992).
Frankel,E.N., Kanner,J., German,J.B., Parks,E. and Kinsella,J.E. Inhibition of *in vitro* oxidation of human low-density lipoprotein with phenolic substances in red wine. *The Lancet* **341**, 454-457 (1993).
Frankel,E.N., Parks,E.J., Xu,R., Schneeman,B.O., Davis,P.A. and German,J.B. Effect of n-3 fatty acids rich-fish oil supplementation on oxidation of low density lipoproteins. *Lipids* **29**, 233-236 (1994).

Frei,B. Cardiovascular disease and nutrient antioxidants: Role of low-density lipoprotein oxidation. *Crit. Rev. Food Sci. Nutr.* **35**, 83-89 (1995).

Galli,C. and Simopoulos,A.P. (editors). *Dietary ω3 and ω6 Fatty Acids. Biological Effects and Nutritional Essentiality* (Proceedings of a NATO Advanced Research Workshop) pp. 403-404 (1989) Plenum Press, New York.

Gießauf,A., Steiner,E. and Esterbauer,H. Early detection of tryptophan residues of apolipoprotein B is a vitamin E-independent process during copper-mediated oxidation of LDL. *Biochim. Biophys. Acta* **1256**, 221-232 (1995).

Gottenbos,J.J. Nutritional evaluation of n-6 and n-3 polyunsaturated fatty acids, in *Dietary Fat Requirements in Health and Development.* pp. 107-119 (1988) (edited by J. Beare-Rogers) American Oil Chemists' Society, Champaign, IL.

Grundy,S.M. Monounsaturated fatty acids and cholesterol metabolism: Implications for dietary recommendations. *J. Nutr.* **119**, 529-533 (1989).

Gurr,M.I. and Harwood,J.L. *Lipid Biochemistry. An Introduction.* (1991) Chapman & Hall, London.

Gurr,M.I. Lipids and nutrition, in *Lipid Technologies and Applications.* pp. 79-112 (1997) (edited by F.D. Gunstone and F.B. Padley) Marcel Dekker, Inc., New York.

Hajjar,D.P. and Nicholson,A.C. Atherosclerosis. *American Scientist* **83**, 460-467 (1995).

Halliwell,B. Iron damage to biomolecules, in *Iron and Human Disease.*pp. 209-236 (1992) (edited by R. B. Lauffer) CRC Press, Boca Raton, FL.

Halliwell,B. and Gutteridge,J.M.C. *Free Radicals in Biology and Medicine.* 2nd Edition (1989) Clarendon Press. Oxford.

Halliwell,B. and Gutteridge,J.M.C. Role of free radicals and catalytic metals ions in human disease: An overview. *Methods Enzymology* **186**, 1-85 (1990).

Hazel,L.J. and Stocker,R. Oxidation of low-density lipoprotein with hypochlorite causes transformation of the lipoprotein into a high-uptake form for macrophages. *Biochem. J.* **290**, 165-172 (1994).

Hegsted,D.M., McGandy,R.B., Myers,M.L. and Stare,F.J. Quantitative effects of dietary fat on serum cholesterol in man. *Am. J. Clinical Nutr.* **17**, 281-295 (1965).

Jialal,I. and Devaraj,S. The role of oxidized low density lipoprotein in atherogenesis. *J. Nutr.* **126**, 1053S-1057S (1996).

Jessup,W., Rankin,S., De Whalley,C., Hoult,J. and Scott,J. α-Tocopherol consumption during low density lipoprotein oxidation. *Biochem. J.* **265**, 399-405 (1990).

Kearney,J.F. and Frei,B. Antioxidant protection of low-density lipoprotein and its role in the prevention of atherosclerotic vascular disease, in *Natural Antioxidants in Human Health and Disease.* pp. 303-350 (1994) (edited by B. Frei) Academic Press, San Diego, CA.

Keys,A., Anderson,J.T. and Grande,F. Serum cholesterol response to change in the diet. IV. Particular saturated fatty acids in the diet. *Metabolism* **14**, 776-787 (1965).

Kinsella,J.E. *Seafoods and Fish Oils in Human Health and Disease.* (1987) Marcel Dekker, Inc., New York.

Kinsella,J.E. Interactions among dietary polyunsaturated fatty acids. (S.S. Chang award lecture, American Oil Chemists' Society Annual meeting, Chicago, IL). *Inform* **2**, 578 (1991).

Kinsella,J.E., Frankel,E.N., German,B. and Kanner,J. Possible mechanisms for the protective role of antioxidants in wine and plant foods. *Food Technology* 85-89 (April 1993).

Kinsella,J.E. Lokesh,B. and Stone,R.A. Dietary n-3 polyunsaturated fatty acids and amelioration of cardiovascular disease: possible mechanisms. *Am. J. Clin. Nutr.* **52**, 1-28 (1990).

Lands,W.E.M. (1986) *Fish and Human Health.* Academic Press, Orlando, FL.

Lands,W.E.M., Kulmacz,R.J. and Marshall,P.J. Lipid peroxide actions in the regulation of prostaglandin biosynthesis, in *Free Radicals in Biology.* pp. 39-61 (1984) (edited by W.A. Pryor) Academic Press, Orlando, FL.

Lands,W.E.M. and Kulmac,R.J. The regulation of the biosynthesis of prostaglandins and leukotrienes. *Prog. Lipid Res.* **25**, 105-109 (1986).

Lynch,S.M. and Frei,B. Antioxidants as antiatherogens: Animal studies, in *Natural Antioxidants in Human Health and Disease.* pp. 353-385 (1994) (edited by B. Frei) Academic Press, San Diego, CA.

Maerker,G. Cholesterol autoxidation - Current status. *J. Am. Oil Chem. Soc.* **64**, 388-392 (1987).

Mattson,F.H. Effects of dietary fatty acids on plasma HDL and LDL cholesterol levels: A review, in *Dietary Fats and Health.* pp. 679-688 (1983) (edited by E.G. Perkins and W.J. Visek) American Oil Chemists' Society, Champaign, IL.

Mattson,F.H. and Grundy,S.M. Comparison of effects of dietary saturated, monounsaturated, and polyunsaturated fatty acids on plasma lipids and lipoproteins in man. *J. Lipid Res.* **26**, 194-202 (1985).
Mensink,R.P. and Katan,M.B. Effect of diet enriched with monounsaturated or polyunsaturated fatty acids on levels of low-density and high-density lipoprotein cholesterol in healthy women and men. *New Eng. J. Med.* **321**, 436-441 (1989).
Mensink,R.P. and Katan,M.B. Effect of dietary *trans* fatty acids on high-density and low-density lipoprotein cholesterol levels in healthy subjects. *New Eng. J. Med.* **323**, 439-445 (1990).
Mensink,R.P. and Katan,M.B. Effects of dietary fatty acids on serum lipids and lipoproteins. A meta analysis of 27 trials. *Atherosclerosis & Thrombosis* **12**, 911-919 (1992).
National Research Council. *Diet and Health.* (1989) National Academy of Science, Washington, DC.
Nettleton,J.A. *Omega-3 Fatty Acids and Health.* (1995) Chapman & Hall, New York.
Parthasarathy,S., Khoo,J.C., Miller,E., Barnett,J., Witztum,J.L. and Steinberg,D. Low density lipoprotein rich in oleic acid is protected against oxidative modification: Implications for dietary prevention of atherosclerosis. *Proc. Natl. Acad. Sci. U.S.A.* **87**, 3894-3898 (1990).
Pearson, D.A. *Vascular Targets and Actiaus of Plant Phenolic Compounds.* Ph. D. Thesis (1998) Universty of California, Davis, California.
Reaven,P.D., Parthasarathy,S., Grasse,B.J., Miller,E., Almazan,F., Mattson,F.H., Khoo,J.C., Steinberg,D. and Witztum,J.L. Feasibility of using an oleate-rich diet to reduce the susceptibility of low density lipoprotein to oxidative modification in humans. *Am. J. Clin. Nutr.* **54**, 701-706 (1991).
Reaven,P.D. and Witztum,J.L. Oxidized low density lipoproteins in atherogenesis: Role of dietary modification. *Ann. Rev. Nutrition,* **16**, 51-71 (1996).
Rice-Evans, C.A. and Miller, N.J. Total antioxidant status in plasma and body fluids. *Meth. Enzymol.* **234**. 279-293 (1994).
Rice-Evans,C.A., Miller,N.J. and Paganga,G. Structure-antioxidant activity relationships of flavonoids and phenolic acids. *Free Rad. Biol. Med.* **20**, 933-956 (1996).
Rice-Evans,C.A. and Packer,L. (editors). *Flavonoids in Health and Disease* (1997). Marcel Dekker, Inc. New York.
Ross,R. The Pathogenesis of atherosclerosis - An update, *New Engl. J. Med.* **314**, 488-500 (1986).
Salter,A.M. and White,D.A. Effects of dietary fat on cholesterol metabolism: Regulation of plasma LDL concentrations, *Nutr. Res. Rev.* **9**, 241-257 (1996).
Scientific Committee for Food (SCF). *Nutrient and Energy Intakes for the European Community.* (1993) Food Science and Techniques, 31th series. Commission of the European Communities. Directorate-General Industry, Luxenbourg.
Sethi,S., Gibney,M.J. and Williams,C.M. Postprandial lipoprotein metabolism. *Nutr. Res. Rev.* **6**, 161-183 (1993).
Sevanian,A. and Hochstein,P. Mechanisms and consequences of lipid peroxidation in biological systems. *Ann. Rev. Nutr.* **5**, 365-390 (1985).
Sies,H. *Oxidative Stress: Oxidants and Antioxidants.* (1991) Academic Press, New York.
Simic,M.G. and Karel,M. (editors). *Autoxidation in Foods and Biological Systems.* (1980) Plenum Press, New York.
Smith,L.L. Review of progress in sterol oxidations: 1987-1995. *Lipids* **31**, 453-487 (1996).
Smith,W.L., Borgeat,P. and Fitzpatrick,F. The eicosanoids: cyclooxygenase, lipoxygenase, and epoxygenase pathways, in *Biochemistry of Lipids, Lipoproteins and Membranes* (edited by D.E. Vance and J. Vance) pp. 297-325 (1991) Elsevier, Amsterdam.
Steinberg,D., Parthasarathy,S., Carew,T.E., Khoo,J.C. and Witztum,J.L. Beyond cholesterol. Modifications of low-density lipoproteins that increase its atherogenicity. *New Engl. J. Med.,* **320**, 915-924 (1989).
Steinberg,D. Low density lipoprotein oxidation and its pathological significance. *J. Biol. Chem.* **127**, 20963-20966 (1997).
Zock,P.L. and Katan,M.B. Hydrogenation alternatives: effects of *trans* fatty acids and stearic acid versus linoleic acid on serum lipids and lipoproteins in humans. *J. Lipid Res.* **33**, 399-410 (1992).

GLOSSARY

Agonist: compound which acts on a receptor to elicit a response
Aggregation: process of forming adhesions between particles such as cells
Antigens: foreign substance producing immune reaction or response
Atherogenesis: a long series of events that leads to development of atherosclerosis
Atherosclerosis: condition of decreased blood flow in vessels in which the inner space has been diminished by thickening of the vessel wall by deposition of cholesterol and fat.
Autoimmune: immune to self antigens
Cardiovascular: refers to the heart (cardio) and blood vessels *(vascular)* and the system for distributing blood, oxygen and nutrients to various tissues
Chemotactic: chemical agents able to attract or induce cells
Choloroplasts: plastid (photosynthetic cells) containing chlorophyll (site of photosynthesis)
Cytotoxic: toxic to cells
DNA: deoxyribonucleic acid
Dysfunction: impairment or abnormality in the function of an organ
Eicosanoids: oxygenated derivatives of arachidonic acid
Endocytosis: uptake of material into a cell by the formation of a membrane-bound vesicle
Epitope: group recognized by an antibody formed in presence of foreign substance
Erythrocytes: red blood cells
ex vivo: outside the living body
Hepatic: acting on or in the liver
Hypercholesterolemia: condition of elevated levels of serum cholesterol
Hyperlipidemia: condition of elevated levels of serum low-density lipoprotein
Hypertriglyceridic: condition of elevated levels of blood triglycerides
Immune: defensive network of protective response against a pathogenic agent or antigen
Immune suppression: reduction of immune response
Internalized: taken into the cell or a specific organelle
Intima: innermost layer of blood vessels or arteries
in vitro: in a test tube or artificial environment
in vivo: within the living body
Ischemia: localized tissue anemia due to obstruction of arterial blood
Leucocyte: white or colorless nucleated blood cell
Leucotriene: eicosanoid formed by the action of lipoxygenase on arachidonic acid (Figure 12-10)
Lipemia: presence of abnormal amount of lipid in circulating blood
Lipoprotein: complex of lipid and protein separated by ultracentrifugation
Lymphocyte: white blood cell, mediator of specific immunity
Macrophages: large mononuclear phagocytes (large scavenger cells that ingest dead tissue)
Microphages: small leucocytes which ingest chiefly bacteria
Microsome: one of the granules in protoplasm with minute cellular structure
Mitochondria: cellular organelles found outside nucleus
Monocyte: large mononuclear leucocyte in circulating blood
Myocardial infarction: condition of regional death of heart muscle due to impaired blood flow
Neutrophil: finely granular white cell (leucocyte) that is the chief phagocyte of blood
Organelle: a structurally discrete component of a cell
Peroxide tone: peroxide level required to activate or inhibit cyclooxygenase (Figure 12-12).
Phagocyte: cell that ingests bacteria, foreign particles, other cells and consumes debris
Plasma: fluid portion of blood in which blood cells are suspended
Platelets: small disk particles irregular in shape that adhere to blood vessel walls
Postprandial: occuring after a meal
Prostacyclin: important antithrombotic prostaglandin produced by endothelial cells of blood vessels (Figure 12-10).
Prostaglandin: oxygenated derivative of arachidonic acid that contain a five-membered ring (Figure 12-10).
Rheumatoid: inflammation on a joint
Serum: fluid portion of blood remaining after coagulation
Thrombi: clots of blood formed within blood vessel remaining attached to place of origin
Thrombosis: formation of blood clot of platelets and blood coagulation proteins within a blood vessel that restricts the flow of blood

Thromboxane: prostaglandin intermediate causing platelet aggregation and contraction of muscles (Figure 12-10)
Vasodilator: dilates blood vessels causing decrease in blood pressure
Vasoconstrictor: constricts blood vessels causing increase in blood pressure
Vesicle: a closed membrane shell, derived either physiologically (budding) or mechanically (by sonication).

Abbreviations

The following abbreviations are used at various points throughout this book.

AAPH / ABAP	2,2′-azobis-(2-amidino-propane) dihydrochloride
ADP	adenosine diphosphate
ATP	adenosine triphosphate
AIBN	α,α-azobisisobutyronitrile
AMVN	2,2′-azobis(2,4-dimethylvaleronitrile
ANOVA	analysis of variance
AOM	active oxygen method
ApoB	apolipoprotein B
BHA	butylated hydroxyanisole
BHT	butylated hydroxytoluene
DHA	docosahexaenoic acid
DLPC	dilinoleoyl-phosphatidylcholine
DLPG	dilinoleoyl-phosphatidylglycerol
DMVN	2,2′-azobis (2,4-dimethylvaleronitrile)
DNA	deoxyribonucleic acid
DNPH	2,4-dinitro-phenylhydrazine
DNUA	distilled non-urea adducts
EDTA	ethylenediamine tetraacetic acid
EPA	eicosapentaenoic acid
ESR	electron spin resonance
FFA	free fatty acids
GC	gas chromatography
GC-MS	gas chromatography-mass spectrometry
GC-SP	gas chromatography-sniffing port
GSH	glutathione
HPLC	high-performance liquid chromatography
HDL	high-density lipoproteins
IP	induction period
IDL	intermediate-density lipoproteins
LCAT	lecithin-cholesterol acyltransferase
LC-MS	liquid chromatography-mass spectrometry
LDL	low-density lipoproteins
LOOH	hydroperoxides
LOX	lipoxygenase
MDA	malonaldehyde
NADH	nicotinamide adenine dinucleotide
NADP	nicotinamide adenine dinucleotide phosphate
NMR	nuclear magnetic resonance spectroscopy
PC	phosphatidylcholine
PE	phosphatidylethanolamine
PG	propyl gallate
PI	phosphatidylinositol
ppm	parts per million
PS	phosphatidylserine
psi	pounds per square inches
PUFA	polyunsaturated fatty acids
PV	peroxide value
RBD	refined, bleached and deodorized
$^1O_2$	singlet oxygen
SDS	sodium dodecyl sulfate
SOD	superoxide dismutase
O_2^-	superoxide radicals
$^3O_2$	triplet oxygen
TBA	thiobarbituric acid

TBARS	thiobarbituric acid reactive substances
TBHQ	tert-butylhydroquinone
TLC	thin-layer chromatography
TMS	trimethylsilyl
UV	ultra-violet
VLDL	very-low-density lipoproteins

Index

Absorption 258, 268, 2816
  oil 227, 233, 241
  (see oxygen)
acetaldehyde 65, 66, 122
acids 229
  (see fatty acids, free)
activation energy 102, 110, 208, 217
active oxygen method 105
  (see AOM)
active oxygen species 262, 264, 269, 278
addition, reactions 9, 45
  cyclo 46, 48
  Michael 263, 264
aldehydes 9, 15, 62, 63, 66-75, 83, 87, 100, 103, 104, 135, 189, 206, 207, 211-213, 217, 218, 229, 253, 265, 282
  α-keto 70, 72
  core 63, 70
  di- 70, 75
aldehydo glycerides 63, 70, 85, 118, 230
alkadienals 80, 81, 87, 104, 212
alkanals 70, 80, 87, 104, 212
alkenals 80, 81, 87, 104, 212
allylic carbons 23, 24
allylic hydroperoxides 14, 23
amino acids 94, 170, 199, 200, 201, 205, 211, 252, 264
animal fats 3, 270
animal studies 91, 241
anisidine value/test 62, 84, 108, 119
anthocyanidins 155
antioxidant (s) 6, 11, 16, 19, 44, 86, 99-103, 105, 109, 110, 115, 129, 135, 142, 143, 162, 174, 176, 177, 181, 184, 202
  biological 252, 254, 267, 281,286
  capacity 134, 282
  cereal products 222
  effectiveness 133, 134
  flavonoid 269, 28
  frying fats 238-241, 244
  hydrophilic 101, 165, 176, 182, 184, 269
  LDL 261-265, 271
  lipophilic 101, 165, 176, 182
  milk products 205, 207
  meat/ fish products 209. 212, 216, 220
  natural 11, 129, 130, 139, 285, 286
  phenolic /poly- 19, 244, 256, 281, 286
  preventive 116, 135
  synthetic 115, 129, 130, 138, 205
  wine 280
AOM 155, 234
apo b 262, 264, 265, 282

apoprotein 258
arachidonic acid 94, 193, 217, 261, 265, 272
  biosynthesis 271, 272
  eicosanoid synthesis 273, 274, 275
arachidonate, autoxidation 33
Arrhenius, plot / equation 102, 103, 110, 111
aroma extract dilution 205, 207, 212, 213, 218
ascorbate 211, 213, 216, 220
ascorbic acid 61, 117, 135, 136, 140-142, 147, 149, 169, 170, 176-178, 181, 192, 199, 202 205, 208, 213, 220, 250, 252-254
ascorbyl palmitate 138, 139, 147, 149, 176, 181, 213, 240
atherogenesis 270, 271, 274
atherosclerosis 249, 259-261, 271, 280, 286
azo, initiators 17, 18, 21, 134, 175, 177, 268, 269
  cage effect/solvent 18, 174

β-scission 9, 28, 68
BHA 129, 138, 139, 177, 220, 222, 242, 244, 281
BHT 129, 138, 139, 142, 175, 177, 220, 242
bilayers 5, 39, 102, 177
  lamellar 140, 156, 166
bioavailability 281, 286
bleaching 51, 117, 119
  carotene 109, 155
blending fats 115, 123
bread cubes 238, 240
browning 195, 210, 215
  materials 139, 194
  non-enzymatic 194, 197, 198, 211, 219
  reaction 103, 148, 169, 220
butenal 66, 122
butter 164, 202, 205, 207
butylated hydroxyanisole 129, 281
  (see BHA)
butylated hydroxytoluene 129, 268
  (see BHT)

Caffeic acid 177, 184
cage effect / solvent 18, 174
cancer 249, 271
carbonyl (s) 83, 84, 118, 137
  compounds 62, 84, 106, 113, 205, 212, 218, 238
  protein 210, 265
  value 105, 108, 219
carnosic acid 148-151, 153, 243
carnosol 148-151, 153, 243

carotene 49-52, 75, 252, 281
  bleaching 109, 155, 252
carotenoids 6, 48, 187, 252
casein 167, 168, 205
catalase 156, 187, 209, 214, 251, 253
catalyst (s) 99, 187, 203
  see metal catalysts
catechin (s) 155, 177, 179, 182-184
  tea 55, 184, 282
  gallate (s) 155, 177, 179
cereal (s) 199, 220, 221, 223, 284
chemiluminescence 39
chlorophyll 6, 43, 44, 48, 52, 250
cholesterol 202, 257-261, 270, 271, 280, 284, 285
  LDL 259, 260, 271, 280, 281
  oxidation products 214, 266, 267
choline 5
  (see phosphatidylcholine)
  glycerophosphoryl 5
chromatography
  (see HPLC, GC)
  column 62, 85, 235
  gel permeation 235
  size-exclusion 62, 235
chylomicron 256-258
  remnants 258, 270
citric acid 115, 116, 135, 136, 138, 142, 143, 213, 244
cleavage 31, 66, 132, 189
  homolytic 63, 64, 66
  products 31, 65, 66, 67
colloidal systems 162, 185
color 221, 234, 241, 244
condensation 16
  aldol 199
conductivity 105, 106, 109
configuration(s) 26, 32
conjugatable products 95, 96, 121
conjugated
  dienes 58, 83, 105, 108, 113, 121, 133, 137, 151, 236, 265
  hydroperoxides 45, 83
  diene-trienes 61, 120
  trienes / tetraenes 58, 96
consumer 80, 108, 113
control 29, 99, 115, 236
  measures 215, 240
cooked meat 173, 195, 209
cooking 122, 198
copper 20, 59, 116, 172, 173, 175, 193, 203-205, 209, 269
  complexes 204, 253, 269
  oxidized LDL 276, 277
corn oil 3, 75, 101, 124, 133, 150-152, 182
  bulk 146-152
cyanidins 155
cyclic
  dimers 233
  monomers 230, 231, 245, 246

peroxide(s) 46-48, 86, 104
cyclization 9, 31, 32, 35, 37, 46-49, 55, 56, 105, 132, 230
cyclooxygenase 269, 271-274, 276, 281, 282
cysteine/cystine 199, 200, 210, 254
cytochrome c 172, 187

2,4-Decadienal 64, 65, 74, 92, 122, 189, 207, 213, 230, 240
decatrienal 65, 67, 71, 74, 218
2-decenal 64, 207, 211
decomposition 14, 20, 66, 68, 69, 73, 74, 132, 135, 142, 228
  hydroperoxide 16, 20, 55, 132, 134, 137, 147, 172, 208, 264, 283
  products 21, 104, 112, 133, 189, 227, 241
degumming 4, 117
dehydration 95
denaturation, protein 209, 219
deodorization 62, 117-119
DHA 3, 36, 216, 274, 278
dialdehydes 70, 75
dielectric constant 236, 237
diet 12, 258, 270, 286
diffusion 195, 197
diglycerides /diacylglycerols 4, 238
dimer 9, 59-62, 72, 104, 118, 119, 189, 200, 231, 233, 235
  carbon-carbon 16, 59, 60, 232
  ether- 16, 59, 60, 232
  oxidative 61, 76, 84, 119
  oxygenated 189, 233
  peroxyl 16, 59, 61, 72, 76, 232
  polar 232, 233
  non polar 231, 233
  tocopherol 140, 141
  triacylglycerol 62, 245
diseases
  autoimmune 271, 286
  chronic 11, 255, 270, 277
  coronary 255-260, 284, 285
docosahexaenoic acid 2, 36, 216, 274, 276
  (see DHA)

EDTA 117, 135, 173, 203, 216, 220
  esters 244
egg
  dried 39
  yolk 168
  lecithin 183
eicosanoids 3, 11, 262, 271, 272
  metabolism 278, 285, 287
  synthesis 273, 275
eicosapentaenoic acid 2, 36, 216, 274-276
  (see EPA)
elaidic acid 2
electron
  acceptor 140
  delocalization 131

donor 131, 140
transfer 20
electron spin resonance 147, 243 (ESR)
electrophilic 7, 44
electrostatic 164
   attraction 171, 179
emulsion 142, 151, 161, 162, 166-176, 180, 184
   fish oil 152
   food 104, 165
   oil-in-water 101, 133, 145, 148, 149, 162, 163, 168, 176, 182
   vegetable oil 152, 170
   water-in-oil 162, 163, 168
emulsifier (s) 101, 139, 162-171, 181, 184
endothelial 259, 260, 269
endoperoxides 86, 273
   bicyclo- 32, 34, 35, 48, 56, 58, 73, 75, 86
end-point (s) 99-102, 104, 107, 109, 111
enzyme (s) 188, 191, 202, 211, 217
   antioxidant 156, 281
   hydrolytic/oxidative 187, 221
   iso- 188, 189
EPA 2, 36, 38, 216, 278
epidioxides 48, 58
   hydroperoxy 31, 34, 37, 46, 56, 58, 61, 70, 74, 86
equilibrium 88, 195
   non- 89, 194
ethane 66, 73, 75, 91, 201
ethylene diamine tetraacetic acid (see EDTA)

Fat 233, 234, 241
   globule 163, 202, 203, 205
fats 1, 266, 279
   animal 3, 103, 211, 212, 272
   frying 227, 233, 234, 244, 245
   monounsaturated 115, 246, 286
   polyunsaturated 109, 287
   saturated 270, 285
fatty acids 1, 7, 18, 120, 164, 221, 279, 280
   essential 3, 271
   free 7, 70, 105, 109, 137, 164, 172, 173, 188, 211, 219, 221, 227, 230, 231, 235-238, 245
   furanoid 204, 207
   monounsaturated 120, 279, 280
   n-3 polyunsaturated 11, 84, 105, 249, 274-279, 285, 286
   n-6 polyunsaturated 249, 267, 271, 285
   polyunsaturated 36, 180, 233, 249, 267, 272, 284, 286
   saturated 269, 270, 284
Fenton reaction 192, 251, 253
ferric 61, 82, 223, 254
ferrous 223, 254
fish 39, 94, 173, 193, 219, 220, 275, 278, 286
   fish, oil 12, 36, 82, 104, 105, 115, 119, 120, 126, 139, 152, 267, 274, 275, 278

fishy 79, 122, 207, 218
flavan-3-ols 155
flavonoids/flavonols 12, 155, 178, 215, 220, 220, 282
flavor(s) 55, 61, 87, 108, 204, 205, 217-219
   cooked/ heated/burnt 209, 245
   compounds 113, 205, 206, 211, 212
   defects/off-flavors 79, 116
   deterioration 21, 62, 79, 111
   dilution factor 206, 207
   fried 230, 239, 240
   heated 209, 245
   hydrogenation 32, 120-123
   impact/significance 80, 92, 108
   meat 198, 201
   rancidity/oxidized 107, 202, 204, 208, 239, 240
   score 101, 190
   stability 101, 106
   quality 238, 240, 243
   warmed-over 210, 210, 214, 215
fluorescence 96, 260
foam formation 234-237, 245
foam cells 260, 262, 264, 267
food(s) 55, 138, 161, 169, 187, 191, 194, 197, 198, 241, 286
   emulsion 104, 166
   fried 230, 233, 238, 241, 244
   heated 169, 190
   moist 196, 197
   oil absorption 227, 233
   quality 187, 194
   stability 112, 194, 195, 197
   storage/shelf-life 99, 100
free radical reactions 8
   cyclization 47, 48
   decomposers 56
   initiators 56, 59
   injury 283
   oxidation 13, 44, 263
fruits 129, 198, 256, 281, 286
frying 61, 153, 198, 227, 234-236, 241, 244, 245
   deep-fat 122, 236
   fats / oils 227, 229, 230, 233, 234, 237, 244, 245
furans, substituted 80, 212, 229

Gallate 183
   catechin 155
   propyl 129, 155, 176, 182-184, 242
gallic acid 155, 176-178, 183, 184
gas chromatography (see GC)
gas chromatography-mass spectrometry (see GC-MS)
GC 57, 62, 85, 95, 106, 229
   direct injection 62
   static headspace 239, 267, 276, 280
   volatiles 106, 108
GC-MS 23, 31, 62, 67

genetic modification 115, 124, 234
glass transition theory 196-197
glassy state 196-197
glucosylamine 197, 198
grape extracts 154, 155
green tea 155, 156, 178, 282
  catechins 155
glutathione 158, 211, 219, 252, 254
  peroxidase 211, 213, 217, 253, 272
  reductase 253
  transferase 158, 211, 272
glycerophosphoric acid 5

HDL 256-259, 270, 271, 280
headspace 88, 106
  static method 87, 239, 267
  dynamic 89, 90
health 11, 223
frying fats 245
heme 172, 173, 187, 191, 195, 209, 210
  iron 191, 192, 193, 216
    non-heme 192, 216
  proteins 43, 44, 173
hemin 104, 135, 174, 187
hemoglobin 173, 187, 217
  met- 192, 217
2,4-heptadienal 66, 70, 73, 75, 92, 189, 218, 230
heterogeneous systems 20, 137
heterolytic, cleavage 7, 68, 69
2-/4-heptenal 65, 71, 74, 92, 189, 211, 218, 229
hexanal 64, 67, 73, 74, 92, 101, 112, 144, 145, 146, 154, 218, 229, 267, 276
high-density lipoproteins (see HDL)
high-oleic, oils 12, 123, 246
  soybeans 125
  sunflower oil 123, 234, 239, 240, 279
high performance liquid chromatography (see HPLC)
histidine 170, 192, 199, 213
homoallylic structure 31, 46, 47
homogenization 162, 163, 168, 209
homolytic, radical reactions 7, 15, 68, 69
  cleavage 7, 63, 66, 86
  decomposition 20, 132
HPLC 23, 27, 31, 36, 38, 39, 85, 106
hybrid, structure 8, 14
hydrocarbons 6, 62, 75, 80, 87, 91, 229
hydrogen 120, 131
  abstraction 8, 23, 28, 132
  donation 28, 129, 132, 135, 148
  peroxide 44, 156, 192, 216, 251-253, 283
hydrogenation 11, 84, 120, 233, 234, 238, 280
  flavor 32, 120
  partial 115, 120-123, 243
hydrolysis 219, 227, 228
hydroperoxide(s) 13, 14, 19, 39, 43, 44, 61, 64, 75, 100, 106, 108, 112, 201, 271, 272
  allylic 14, 23

analysis 137
conjugated 45, 83
decomposition 16, 137, 147, 172, 173, 208, 264
distribution 25, 28, 32, 45, 46
formation 16, 133, 134, 137, 283
inhibition 154
oleate 25, 65
linoleate 26-28, 38, 56, 59, 65, 68, 69, 265
linolenate 29-32, 67, 69
polymerized 118
polyunsaturated/PUFA 95, 108
precursors 91
reduction 149
steroids 253, 265
hydroperoxy
  bicylcloendoperoxides 32, 33, 48, 58, 73, 75, 86
  epidioxides 31, 34, 37, 46, 48, 56, 58, 61, 73, 75, 86
  bis- 48, 58
  epoxides 71
hydroperoxyl radicals 250, 253
hydroxy, products 10, 23, 39, 135, 189, 221, 230, 273
  enals/nonenal 254, 264, 265
  hydroperoxides 39, 56
  octadecanoate/stearate 23, 31
hypochlorous acid 44, 252, 254

Induction period 18, 19, 100, 102, 103, 112, 134, 143
inhibitor (s) 99, 190, 244, 276
initiation 13, 21
initiator(s) 13, 29, 141, 174
  azo 17, 134, 175, 177, 178, 268, 269
  free radical 56, 59, 171
interaction(s) 38, 55, 94, 161, 169, 194, 198, 202, 228
  products 94, 199, 200, 231
interconversion 28, 63
interface 163, 166, 172, 176, 182, 185
  air-oil 101, 148, 152
  oil-water 148, 152, 162, 171, 172, 184, 185
  water-oil 176
interfacial 161, 166
  partitioning 150, 176
  phenomenon 150, 152, 161, 217
  properties 100, 147, 168
interference 109, 266
in vivo 91, 268, 286
iodine value 236, 237
ionic, reactions 7, 8, 66
  emulsifiers 164, 165, 171
iron 20, 116, 143, 188, 203, 210, 216, 219, 222, 223, 250
  binding 214, 269
  catalysts 193, 204, 210
  compounds/complexes 172, 173, 192,

210, 211, 216, 223, 252, 253
deficiency/dietary 284
elemental/free 223, 252, 283
fortification/supplementation 12, 115, 223, 283, 287
heme/non-heme 191-193, 216
meat 173, 193
isomer(s)(ic) 10, 120
hydroperoxides 46, 74
trans 120, 122, 234, 246, 280
isomerization
geometric 32
trans,trans 48

Ketones 10, 75, 81, 136, 212, 229
kinetic(s) 16, 134, 195
control 29
factors 24
stability 11

β-Lactoglobulin 167, 209
lactone(s) 75, 207, 229
LC-MS 38
LDL 12, 256-264, 270, 278, 280, 281
cholesterol 259, 260, 270, 271, 280, 286
oxidation 259-262, 275, 276, 278, 280, 286
oxidative modification 259-262, 281
oxidative susceptibility 259, 271, 277, 285
particles 258, 260
receptors 259, 270
lecithin 4, 38, 116, 139, 140, 163, 168-170
acetylcholesterolacyl-
transferase (LCAT) 259
liposomes 39, 155, 156, 175, 182, 183
lyso- 262, 264
light 14, 43, 53, 82, 100, 104, 115, 117, 204
UV catalysed 43, 135
linoleate 44, 45, 56, 182, 240
cholesteryl 265
emulsion 170
iso 121
hydroperoxides 26, 28, 38, 56, 59, 65, 68, 69, 75, 265
keto 56
oligomeric products 59
triacylglycerol 120
volatiles 101
linoleic acid 2, 146, 202, 271, 272, 272, 274
antioxidants 147, 155, 179
cooxidation 109
in LDL 271
micelles 180, 182
linolenate 29-36, 46, 56, 75, 82, 87, 233, 234
autoxidation 29-36
containing oils 122-125
methyl 29-36, 49, 56, 59
dimers 72
hydroperoxides 30-32, 67, 68, 68, 69, 84
triacylglycerols 120
volatiles 101

linolenic acid 2, 124, 271, 274, 278
lipase 3, 4, 187, 188, 190, 219, 221, 222, 257, 258, 270
lipolysis 187, 188, 202, 222
lipoprotein (s) 12, 39, 202, 203, 256-270, 274
LDL 12, 39, 202, 203, 256-265
Lipase 257, 270
liposome 38, 39, 102, 140, 141, 162, 164, 165, 172, 175-178, 182-183
lecithin 155, 156, 175, 182, 183
PC/DLPC/DLPG 177-180
lipoxygenase 56, 188-190, 210, 220-222, 264, 269, 271, 274, 276, 281, 282
n-9/12-/15- 193, 211, 217
pathway 262, 273
liquid chromatography-mass spectrometry (see LC-MS)
low-density lipoproteins
see LDL
lysine 170, 200, 201, 263, 264

Macromolecules 196, 199
macrophage 250, 260-269, 274
Maillard reaction, products 148, 197, 198, 210, 212, 215, 220, 230
malonaldehyde 32, 56, 57, 58, 70, 74, 85-87, 94, 254
TBA adduct 57, 86
mayonnaise 163
mechanism(s) 13, 24, 27, 35, 49, 147, 161, 204, 228, 286
antioxidation 129, 141
Russell 15, 44
meat 39, 94, 173, 187, 192, 193, 198, 198, 201, 209, 212
melanoidins 200
membrane, charge 171
cellular 209, 211
fat globule 202, 203, 205
metabolism 268, 278, 287
lipoprotein 257, 270
metal 29, 43, 82, 94, 99, 100, 104, 169-173, 175, 187, 192, 210, 251
catalys(ts)(is) 20, 59, 62, 102, 104, 110, 120, 135, 137, 141, 147, 161, 165, 169, 171, 172, 190, 195, 203, 204, 209, 210, 216, 227, 231, 255, 264
complexes 20, 43, 116, 135, 174, 187, 190, 198, 203, 217
contamination 115
decomposition 14, 69
ions / free 171, 190
inactivation /chelation 115, 135, 136, 139, 142, 147, 148, 155, 168, 173, 203, 204, 205, 213, 214, 219, 252, 283, 284
metallic 203, 207
methionine 170, 199
methods 19, 79, 82, 99, 100, 105, 107, 109, 111, 112, 137, 144, 220, 236, 265, 282
frying 234, 235, 238

GC 87, 89, 91, 93
   manometric 106, 109
   sensory 79, 108
   stability 99, 106
methylene blue 44, 52
micelles 109, 147, 162, 164, 165, 169, 180, 182, 184
   emulsifier 164, 165, 180, 184, 185
   fatty acids 164, 165
   inverted 164, 169, 170
   mixed 165, 180
microsome 209, 211, 251
milk 163, 191, 198, 202, 205, 208, 209
   fat 158, 158, 163, 202, 205
   products 199, 202
mitochondria 209, 216, 250, 251
moisture 125, 194, 195, 197, 238
monoacylglycerols 3, 4
monocytes 250, 260
monoglycerides 4, 163, 167, 168, 230, 238
multiphase systems 161
muscle 94, 192, 193, 211, 267
myoglobin (met-/oxy-) 172, 187, 191, 192, 209, 210, 214, 214, 217

Neutrophils 250, 252
nitroxide radicals 148
NMR
   ^{13}C- 23, 24, 27
   ^{1}H- 24
nonadienal 67, 68, 189, 211, 218
nonanal 63, 70
nonanoate, methyl 64, 66
2-/3-/6-nonenal 64, 69, 70, 207, 213, 229, 264
nucleophilic 7
nuclear magnetic resonance (see NMR)
nutritional 124, 126, 198, 231, 269, 283
   anti- 190

Oats 222
octanoate, methyl 64, 65, 71, 73
   method 95
9,15-octadecadienoate 32, 121
octane 63
octenal 65, 122, 189
1-octen-3-ol 92, 229
odor (s) 79, 108, 122, 207, 212, 223, 244
oils 1, 36, 76, 112, 123-125, 207, 217, 230
   bulk 82, 101, 138, 149, 161, 176
   corn 3, 75, 101, 124, 133, 145, 146, 148-151, 182
   cottonseed 3, 239, 240
   fish 12, 36, 82, 104, 105, 115, 119, 120, 126, 139, 152, 267, 274, 275, 278
   frying 229, 230, 237, 241
   heated 227, 230
   high-oleic 11, 123, 246, 278, 279
   hydrogenated 240
   olive 3, 53, 112, 124, 286
   palm 3, 140

peanut 3, 151, 152
polyunsaturated 82, 115, 249
rapeseed 3, 52, 76, 82, 101, 120, 123, 274
safflower 3, 75, 123, 170
soybean 3, 52, 75, 76, 79, 80, 82, 92, 93, 101, 104, 115, 120 123-125, 138, 140, 142, 143, 151, 152, 240, 274
sunflower 230, 234, 239, 240, 278
vegetable 3, 75, 80, 103, 104, 115, 125, 129, 152, 205, 234
oleate
   hydroxy/epoxy 54, 56
   hydroperoxides 3-25, 45, 63
   keto 55
   methyl 23-25, 55
   monomeric products 55
   photosensitized oxidation 75
oleic acid 1, 2, 12, 202, 267, 270, 272, 279, 286
oligomer(ic)(s) 70, 76, 104, 155, 231
   products 59, 61
oxidant-antioxidant balance 254, 255, 286
oxidation 103, 109, 115, 131, 143, 202, 203, 208, 210, 215, 222, 228, 286
   accelerated test 102, 104
   anti- 129
   cholesterol / sterol 214, 246, 267
   emulsions 161, 169
   enzymatic (non-) 187, 211, 218, 222
   free radical 13, 44, 263
   hidden 62, 119
   interfacial 161
   in vivo 91
   LDL 259-261, 277, 275, 280, 286
   light 104, 204
   methods 79, 107
   photosensitized 43, 44, 48, 50, 75, 104
   processing 117-119
   products 55, 62, 69, 95, 96, 103, 109, 134, 220, 277, 283
   protein 210
   rates 37, 100
   susceptibility to 202, 279
oxidative damage 11, 249
oxidative stress 100, 252, 255, 261, 283, 285
oxidizability 17, 19, 36, 123, 124, 175, 180
9-oxononanoate 67, 71, 73
oxygen 13, 49, 205, 227
   absorption 18, 105, 107, 109, 208
   active/reactive species 250, 264, 278
   scavenging 269
   attack 34, 38
   concentration/level 105, 131, 201, 208, 220, 227
   copper-ascorbic acid 204
   diffusion 197
   headspace 106
   complexes 20
   pressure 17, 102, 252
   scavenger 147, 148, 209

singlet 43-49, 250, 254
  quenchers 252, 253
  solubility 102
oxysterols 262, 265, 267

Packaging 115, 125, 194, 195, 215, 220
painty 80, 207
palmitic acid 1, 38, 39
partition 137, 150, 156, 169, 182, 184
pentadienyl radicals 26-28
pentane 64, 74, 75, 91, 92, 189, 223, 267, 276
pentanal 64, 276
pentenal 66, 122, 152, 189, 218
2-pentyl furan 65, 75, 229
peptides 94, 212
peroxidases 187, 188, 202, 252, 243, 264
peroxide bonds 61
peroxide destroyers 253
peroxide tone 272
peroxide value 81, 82, 83, 100, 101, 105, 108, 113, 119, 137, 219, 236
peroxyl, radicals 14, 24, 26, 28, 129
  hydro- 250, 253
PG 129, 138
phenolic
  antioxidants 19, 244, 256, 281, 286
  compounds 130, 154, 213, 215, 280-282
  diterpenes 148
pH 170, 171-173, 179, 198
phagocytes 250, 251
phagocytosis 252
phases 87, 88, 182
  lag 266
phosphatides 4, 116
phosphatidylcholine 5, 39, 168, 177, 179
  liposome 175-178
phosphatidylethanolamine 5, 168
phosphatidylinositol 5, 168
phosphoglycerides 4
phospholipase 5, 6, 211, 219, 264
phospholipids 4, 93, 94, 102, 112, 117, 137, 139, 148, 156, 165, 166, 168-170, 202, 204, 209, 256-258, 262, 264, 271
  autoxidation 39
  hydrolysis 219
  polar 179
phosphoric acid 5, 116, 135
  glycero 5
photooxidation 43, 117
photosensitized oxidation 43, 75, 104
  type I, II 44
  inhibition 48-50
photosensitizers 43, 205
pigments 43, 136, 209
polar (non-), molecules 166, 176, 177, 179, 231-233, 236, 238, 239, 244, 245
  paradox 176
  total polar materials 235, 240, 245
polarity 176, 177
polymers 196, 228, 235, 245, 246

polymerization 9, 105, 228
  inhibitors 244
polyunsaturated fatty acids
  (see PUFA) 180, 233, 249, 271, 272, 284, 286
  autoxidation 36
porphyrin 191, 250
potato chips 195, 240, 242
poultry 173, 209
processing 53, 95, 101, 117-119, 142, 203, 222
  fish 219
  soybean oil 142
prostacyclin 262, 272, 273, 277, 278
protection 169, 281, 286
prooxidant (s) 53, 101, 102, 116, 147, 162, 185, 202, 209, 210, 216
  activity 131, 143, 149, 204, 242, 269
propagation 14, 16, 142
propanal 65, 68, 71, 74, 75, 92, 122, 152, 189, 218, 267, 276, 277
propylene glycol (esters) 152, 167
propyl gallate (see PG) 133, 157, 180, 181, 182-184, 242
prostaglandin (s) 34, 262, 271-276
protein (s) 93, 94, 163, 167-171, 190-193, 195, 197, 198, 200-204, 212, 215, 227, 252, 268
  carbonyls 210, 265
  complexes 203, 231, 255
  denatured 187, 209, 219
  heme 43, 44, 172, 192, 216
  iron 172, 173, 210
  metal 187, 190, 203, 217
  oxidation 211, 265
  whey 167, 201
n-3 PUFA 12, 36, 84, 105, 126, 249, 267, 274-279, 285
n-6 PUFA 249, 267, 271, 285, 286

Q_{10} 110, 111
quality 79, 81, 99, 113, 194, 198, 216, 239, 244
  flavor 238, 239, 243
quenching singlet oxygen 49-51

Radical traps 132, 134
  methods 282
rancidity 11, 15, 99, 106, 107, 111, 195, 215, 221, 223, 240
  foods 100
  hydrolytic 202, 221
  non-enzymatic 222
  oxidative 202
Rancimat 105, 109, 155
rate(s) 18, 100
raw materials 100, 187, 222
rearrangements 9, 65, 198, 212
receptor 257
redox 13, 135, 147, 202, 238
reducing agents 94, 135, 139, 147, 169, 209, 213

properties 198
reduction 95, 135, 148, 254
refining, alkali 116, 117
regeneration/recycling 136
resonance 8, 27, 140
   electron spin 147, 243
   nuclear magnetic 23
   stabilization 8, 131
riboflavin 43, 44, 205, 250
risk 270, 287
rosemary extracts 148, 214, 222, 243, 282
rosmarinic acid 150, 182-184
rubbery state 196, 197

Salad dressings 163
Schiff base 94, 94, 200, 199, 211, 263, 264
scavenger 253
   oxygen 147, 148, 208
   receptors 260, 262
screening test 104
SDS 170, 171, 181, 184
   micelles 180, 184
sensitivity 80-82, 107
sensitizer 14, 43, 49, 252
   photo 43, 205
sensory 79, 80, 93, 99
   evaluation 79, 113, 220, 239
   methods 79, 108
shelf-life 100, 102, 105, 111
   foods 99, 233
   testing / prediction 109, 111, 112
silicone (methyl) 238-240, 244
singlet oxygen 43, 44, 46, 49, 51, 250, 254
   quenchers 252, 253
sodium dodecylsulfate (see SDS)
sodium nitrite 213, 214
solubility 102, 137, 182
soybean(s) 94, 124, 125, 168, 188, 190, 234
   lecithin 38, 39
   oil 52, 51, 74, 76, 78, 80, 82, 92, 93, 101, 104, 115, 120, 123-125, 138, 140, 142, 151, 152, 240, 274
Spans 163
sparing, oxidation 53, 204
spice (s), extract 148
squalene 7
stability 10, 11, 62, 99, 101, 106, 119, 185
   emulsion 167
   flavor 101, 106, 108
   food 194, 197
   oxidation products 103
   oxidative 62, 99, 101, 123, 143, 162, 198
   protocol 111
steam 227, 241
stearic acid 1, 2, 270
stereochemistry 23, 24, 26
   cis / trans 36
sterols 7, 244
   oxy 262, 265, 267
   storage 85, 100, 101, 118, 194, 201, 219, 223

conditions 100, 104, 112
   normal 99, 100
   temperature 105, 111, 112, 201, 208
stripped oils 112, 133
substrates 112, 137
   lipid 134, 137
sucrose 170
   esters 163
sulfhydryl, compounds 169, 171, 205, 209
superoxide 193, 253, 254, 262, 278, 283
   dismutase 158, 192, 213, 253
   radicals 191, 250-252
surface 137, 161, 162, 178, 179, 258
surfactants 162, 163, 167
synergism (mixtures) 116, 136, 138, 140, 148, 169, 177, 243, 244
synthetic
   antioxidants 115, 129, 130, 138, 139, 205
   emulsifiers 163, 167
   triacylglycerols 38

Tallowy 202, 207
taste, panels 78, 108
TBA 210, 220
   TBARS 32, 85, 208, 210, 265-267, 277
   malonaldehyde adduct 57, 86
   test/method 57, 58, 85-87, 109, 210, 215, 220, 265, 266
   value 85, 208
TBHQ 129-131, 138, 177, 220, 222, 240-242, 244
temperature 25, 28, 32, 82, 100, 102, 110, 137, 144, 195, 196, 217, 219
   coefficient 103
   elevated / high 62, 105, 132, 227
   glass transition 196, 196
   storage 111, 112, 201, 208
termination 15, 175
tertiary butyl hydroquinone (see TBHQ)
test system 144
tetramethyl acetal(s) 57, 73
tetroxide, intermediate 15, 44
thermal 13, 15, 61, 63, 106, 107, 235
   decomposition 66, 74, 228
thin-layer chromatography (see TLC)
thiobarbituric acid (see TBA, TBARS)
thiols 204, 210
threonine 170
threshold value 73, 76, 80, 101, 218
thrombosis 273-276, 286
thromboxane 273, 274, 278
TLC 23
TMS 23
tocopherols 6, 105, 112, 116, 136, 139, 143, 146, 187, 212, 214, 220, 222, 242
   α- 18, 19, 28, 32, 34, 46, 49, 51, 58, 75, 115, 133, 134, 139-148, 176-179, 181, 205, 212, 214, 252, 254, 268, 269
   β- 139, 144
   γ- 49, 53, 123, 139, 144, 146

δ- 51, 53, 139, 140, 144, 146, 170
analog 179
concentration 142, 143, 145
dimers 140
loss of 115, 143, 242
natural 138, 142, 212, 214
oxidation 144
radical 140
tocotrienols 6, 140
toxicity 138, 246
transferrin 210, 216, 252
trapping 89, 91
radicals 132, 134, 281
triacylglycerols (triglycerides) 3, 61, 74, 112, 137, 166, 182, 202, 219
autoxidation 37, 95
blood 270
emulsion 180
in LDL 257, 258
linoleate/linolenate 120
lipoxygenase 188
polymeric products 61, 246
polar 244
synthetic 37, 38
trilinolein 19, 37
mono-hydroperoxides 37, 74
trilinoleylglycerol 61, 73
trimethylsilyl ethers (see TMS)
triolein/ trioleylglycerol 3
tripalmitin 3
tripalmitoylglycerol 3
triplet oxygen 49
triplet state 43
Trolox 145, 147, 176, 179-184
turnover rate 233, 241
Tweens 163, 182-184

Ultraviolet light 43
absorption 108
deactivator 43, 135
detector 106
uric acid 214, 254

Vegetables 129, 198, 256, 281, 286
very low-density lipoproteins (see VLDL)
viscosity 39, 171, 234, 241
vinyl compounds 80, 81, 189
vitamins 1
vitamin A / D 6
vitamin C 177, 178, 252, 268, 269, 281, 283
vitamin E 144, 177, 178, 212, 214, 254, 255, 268, 269, 271, 279, 281, 286
VLDL 256-258, 270
volatile(s) 63, 72, 74, 108, 113, 238-240
acids 70, 105, 109
analysis 106, 108
compounds 61, 88, 93, 137, 189, 218, 217, 228, 229
flavor 211
decomposition products 21, 104, 189, 227, 241
linoleate/linolenate 101
monomers 228
products 62, 69, 106, 220
total 277

Warmed-over flavor 210, 214
wheat 221-223
whey proteins 167, 201

Xanthine oxidase 202, 251